UNDERSTANDING THE SKY

by DENNIS PAGEN

A SPORT PILOT'S GUIDE TO FLYING CONDITIONS

ILLUSTRATIONS BY THE AUTHOR

UNDERSTANDING
THE SKY

A SPORT PILOT'S GUIDE TO FLYING CONDITIONS

First Printing: January 1992
Printed in the United States of America
6 7 8 9 10 11 12 13 14

Published by:
Dennis Pagen
Sport Aviation Publications
P.O. Box 43
Spring Mills, PA 16875 - USA
Web page: www.lazerlink.com/~pagenbks
E-mail: pagenbks@lazerlink.com

Books by the Author:

Hang Gliding Training Manual
Performance Flying
Towing Aloft
Powered Ultralight Training Course
Powered Ultralight Flying
The Art of Paragliding

ISBN 0-936310-03

ERRATA

Page 8	-	Figure 8: below EQUINOX read numbers 6, 12, 18
Page 13	-	Line 21: read 3.6° per 1000 feet
Page 19	-	Line 4: read 3.5° per 1000 feet
Page 74	-	Figure 62: reverse the words LOW and HIGH in right half of figure
Page 94	-	Line 16: read precede not proceed
Page 95	-	Line 3: read preceding not proceeding
Page 101	-	Figure 84: read 45° not 15° in lower left corner
Page 108	-	Figure 81: read Figure 91

ACKNOWLEDGEMENT

I wish to thank Kathy Lee of Taos, New Mexico for her generous offering of photographs. She has truly captured much of nature's beauty with her discriminating eye.

I also must thank my wife Claire for her help in all phases of this book. Her editing and layout skills are second only to her piloting skills.

Dennis Pagen
January, 1992

Cover Photo: Claire Vassort-Pagen
Photos in text: Kathy Lee
Dennis Pagen
Claire Vassort-Pagen

CONTENTS

PREFACE vii

CHAPTER I—THE AIR AROUND US 1
The Big Picture...The Structure of the Air...Properties of the Air... Pressure in the Atmosphere...Temperature in the Air...Solar Heating...Cooling Cycles...Daily Changes...Seasonal Changes... Water Vapor...Humidity...Water's Amazing Properties...Latent Heat...Summary

CHAPTER II—THE LIVING ATMOSPHERE 13
The Lapse Rate...Stability and Instability...Indications of Stability... The Stability of Layers...The Moist Lapse Rate...The Standard Atmosphere...Density Altitude...The Wind We Feel...Coriolis Effect...Summary

CHAPTER III—THE MESSAGE OF THE CLOUDS 25
Cloud Causes...Lifting the Air...The Dew Point and Cloud Height... Cloud Formation...Cloud Bases/Cloud Tops...Cloud Life...Old Clouds ...Rain...Visibility...Cloud Types...Altitude Classification... Seasonal and Location Variations...Lesser Known Clouds... What They Tell Us...Flying in Clouds...Summary

CHAPTER IV—THE BIG PICTURE—GENERAL METEOROLOGY 53
Movement of the Atmosphere...Air Masses...The Meaning of Fronts...Front Characteristics—Cold...Front Characteristics—- Warm... Frontal Action...Variations in Fronts...Occluded Fronts...Seasonal Changes...High Pressure Systems...Low Pressure Systems...Local Lows...Highs and Lows Together...Isobars...Isobars and the Wind... Circulation Aloft...A Trough Aloft...The Jet Stream...Steering Winds and Pressure Patterns...Tropical Weather...Island Weather... Summary.

CHAPTER V—WIND PATTERNS 87
Measuring the Winds...Watching the Wind...Prevailing Winds...Wind Flow Near the Surface...Winds Around Fronts and Pressure Systems...Moving Weather...Areas Beyond the Fronts... Backing and Veering Winds...The Wind Gradient...Winds Aloft...Jet Streams... Diurnal Variation of the Wind...Layering of the Air... Low-Level Jets...Special Winds...North American Winds...Old World Winds ...Summary.

CHAPTER VI—TURBULENCE—UNSTEADY FLOW 113
The Meaning of Turbulence...The Nature of Turbulence...The Causes of Turbulence...Vortex Turbulence...Rotors...Detecting Turbulence...Conditions and Turbulence Cycles...Surface Conditions...Wind Shadow...Flying in Turbulence...Summary.

CHAPTER VII—LOCAL WINDS 137

Heating and Circulation...The Sea Breeze...Sea Breeze and General Wind...Sea Breeze Prediction...Sea Breeze Strength, Reach and Seasons...Sea Breeze Effects...Sea Breeze Fronts...Complicated Coastlines...Island Effects...Land Breezes...Heat Fronts...From Forest to Field...Upslope Winds...Downslope Winds...Gravity Winds...Magic Air and Wonder Winds...Up Valley and Down Valley Winds...Complicated Systems...Summary.

CHAPTER VIII—SOARING CONDITIONS—LIFTING AIR 161

To Soar...Ridge Lift...Complex Shapes...Canyons and Gaps...Ridge Soaring Problems...Variable Lift...Wave Lift...Wave Properties...Wave Clouds...Wave Dangers...Flying in Waves... Finding Waves...Other Wave Sources...Convergence Lift...Flying in Convergence...Frontal Lift...Lift Indicators...Summary.

CHAPTER IX—INSTABILITY AND THERMALS 189

The Birth of a Thermal...Thermal Triggering...Lee Side Thermals...Thermal Sources...House Thermals...Terrain Sources...Wet Ground Sources...Thermal Cycles...Evening Thermals...Thermal Rising...Real Lapse Rates...The Lapse Rate Aloft...Thermals and Lapse Rate Variations...Dust Devils...Ideal Thermal Conditions...Lift in a Thermal...Thermal Sink...Summary.

CHAPTER X—THERMAL LORE 213

Thermal Sizes and Strengths...Thermal Heights...Thermal Duration and Daily Variation...Thermals in Wind...Thermal Tracks and Cycles ...Thermal Types...Thermal Spacing...Thermal Patterns...Thermal Streets...Cloud Streets...Cloud Street Behavior...Thermals and Cloud Character...Near Cloud Features...Wind Effects on Clouds...Locating Thermals...Blue Holes...Summary.

CHAPTER XI—THUNDERSTORMS 237

Where and When...The Cause of Storms...A Thunderstorm Life Cycle...Thunderstorm Variations...Thunderstorm Dangers ...Downdrafts and Gust Fronts...Judging the Storm...Escaping a Thunderstorm...Tornados...Summary.

CHAPTER XII—WATCHING THE WEATHER 253

Weather Cycles...Weather and the Barometer...Weather Wisdom...Weather Information Sources...Weather Maps...Soaring Forecasts...Reading an Area...Site Examples...Summary.

APPENDIX I—DENSITY CHANGES IN THE ATMOSPHERE 269
APPENDIX II—CORIOLIS EFFECT 271
APPENDIX III—DRYING WINDS 272
APPENDIX IV—THERMAL BOUYANCY 273
APPENDIX V—TRIGGER TEMPERATURE 274
GLOSSARY 276
INDEX 279

PREFACE

In the closing decades of the twentieth century a fortunate coincidence of timing and technology has allowed our species—designed for life on the surface of planet earth—to enter the atmosphere and cavort among the clouds. Flying for fun has come into its own as a reasonable and legitimate pastime. But inhabiting the realm of the sky requires a certain amount of understanding. The air is an ever-changing environment and we must know its ways and wiles in order to fly safely and become excellent pilots.

This book is designed to present a clear picture of how the atmosphere works. Naturally some simplifications must be made, for the subject of weather is a complex one requiring many years of study to master. Consequently we have distilled the important lore and knowledge necessary for pilots who fly for fun. The best way to use this book is to read it through, experience flight then reread the pertinent portions to gain a deeper understanding.

We have tailored the material to suit the needs of balloonists, RC modelers, paraglider, hang glider, sailplane and ultralight pilots alike. Hopefully each reader will discover new insights and ideas within these pages to enhance the enjoyment of flight.

Besides embracing all air sports, the material herein is written with an international viewpoint, for many pilots today travel in pursuit of their aerial endeavors. We include the perspective of both the northern and southern hemispheres, where appropriate, and give some attention to regional and continental specifics. We also use both English and metric equivalents in the text as well as the charts and figures.

While we begin the chapters on the air's properties and general weather, we would like to point out that the emphasis in this book is on smaller-scale conditions known as local effects or micrometeorology. The reason for this emphasis is that recreational flying normally takes place within a relatively small volume of airspace where local effects play a major role. Most weather books written for general aviation do not address the small-scale conditions in enough detail to satisfy recreational pilots. This book is intended to fill this void.

The background material for the information in the following chapters comes from many sources. Certainly textbooks have been very useful, but most important is the experience of almost two decades of flying and the sharing of ideas with other pilots from all forms of aviation. It is my wish to pass on some of the knowledge I have gleaned from these experiences so that we can all better savor our time in the sky.

CHAPTER I

The Air Around Us

We grew up on a planet that is surrounded by a life-giving mixture of gases. We call this mixture air and we refer to the entire gas cloud around the earth as the atmosphere.

Most of us take this air and atmosphere for granted as we pass it through our lungs to borrow some oxygen, or slip through it while on the move. For the most part, the air is just simply there. But give us a set of wings and a whole new world opens up. New challenges, new vistas and new experiences alter our viewpoints forever. We become pilots with the realm of the sky to explore.

We quickly become aware of the constant changes that take place in the atmosphere and the need to understand what these changes mean. With understanding we become comfortable in our new environment. With understanding we leave our fears behind and free ourselves from the limitations of an earthbound existence.

In this chapter we begin to study the nature of the sky so we can later predict its behavior as we enter its domain and cast our fate to the winds.

THE BIG PICTURE

The atmosphere is held to the earth by gravity. Although the total thickness of the atmosphere exceeds 500 miles (800 km), most of the air is packed near the earth's surface since air is compressible and gravity pulls each molecule downward. In fact, fully half of the atmosphere's total weight of over 5.6 quadrillion (5,600,000,000,000,000) tons is below 18,000 feet (5,500 m)!

The atmosphere can be divided into different levels like the layers of an onion according to certain characteristics. We are mainly interested in the lowest layer which is known as the troposphere (tropo means change). It is here where the changes take place that we identify as weather. It is here that we live and breath and fly.

The troposphere extends from the surface to 5 or 6 miles (7 to 9 km) at the poles and 10 to 12 miles (17 to 20 km) at the equator. The reason for this difference is centrifugal effects due to the earth's spin (see figure 1). The extent of the atmosphere is greatly exaggerated in the figure for clarity. To put matters in perspective, the entire atmosphere compared to the earth would only be about the same relative thickness as the peel of an orange while the troposphere's thickness would be equivalent to the skin of an apple.

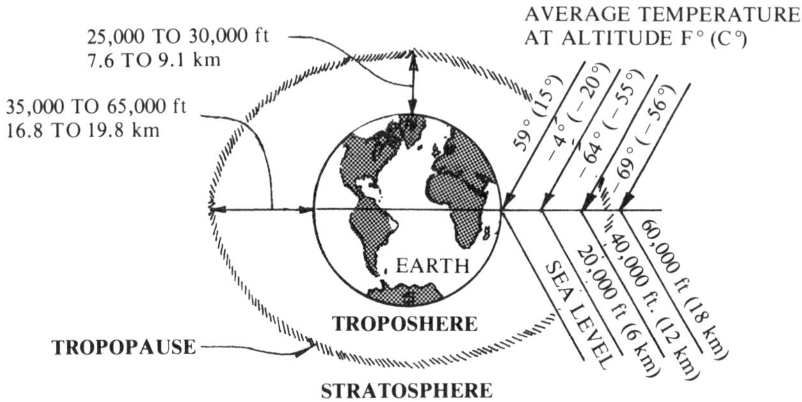

Figure 1 - The Lower Atmosphere

On top of the troposphere lies the stratosphere and the transition between the two is known as the tropopause. The way we differentiate these two lower layers is that the temperature drops steadily with height in the troposphere but it remains nearly constant as we climb into the stratosphere. Thus the stratosphere is stable and clear but the troposphere exhibits clouds and a wide variety of conditions. The troposphere is our sphere of interest in this book.

THE STRUCTURE OF THE AIR

We know that the air is a mixture of gases. Most of it is nitrogen (78%) and oxygen (21%) with the remaining 1% being mainly argon with a little carbon dioxide and some pollutants thrown in.

Water vapor is also a highly variable constituent of air. It can be from zero (dry air) to 4 or 5 percent by weight (saturated air). As we shall see later, water vapor is an extremely important part of the weather process for without it there would be no clouds or rain. Almost all the water vapor in the atmosphere is concentrated in the troposphere for it enters the atmosphere through evaporation from ground sources and is carried aloft by vertical air currents which are limited to the troposphere. Ninety percent of all water vapor remains below 18,000 feet.

The pollutants mentioned above, including smoke, dust, salt particles and industrial exhaust, are important for they serve as condensation nuclei which promotes cloud formation. Clouds are of great interest to us creatures of the sky for they help point out lift and generally give us clues to the atmospheric behavior as we shall see in chapter III. Clouds and pollutants in general can also present visibility problems which are pertinent to our flying.

PROPERTIES OF THE AIR

The air is fairly wispy stuff, but just how insubstantial it is depends entirely on its density. As we noted earlier, the air can be compressed so its density depends on its composition and how much compression takes place. It is this density that interests us most for it directly affects our flying.

The three features that determines the air's density are its *temperature, pressure* and *water vapor* content. The two main factors that control these features are *gravity* and the sun's *heating*. Before we look at the importance of each of these items, let's review what we know about how a gas (air) works.

The molecules in a gas are bouncing around like hyperactive kids on a chocolate diet. All this scurrying about causes them to knock into their neighbors and ricochet off in random fashion. If the molecules encounter a solid they leave some energy behind as they hit the solid. In fact, this exchange of energy is what we feel as heat. The more excited the molecules are in the gas, the faster they are moving and the more energy they impart to any solid they contact so the warmer the gas feels. What we know as temperature is simply the state of excitement of the gas molecules.

It isn't too hard to imagine that if we add heat energy to a gas we raise the temperature by causing its molecules to move around more vigorously which in turn makes it want to expand, for each madly careening molecule is knocking its neighbors farther apart with each contact. Also we can see that if we allow a portion of gas to expand the molecules will spread out so it becomes less dense and at the same time cooler since there are fewer molecules in a given volume to knock into one another or a nearby solid. Conversely, if we compress a gas it becomes denser and its temperature rises as the molecules become more jittery (see figure 2). These properties should be well understood for they are of great importance to soaring pilots.

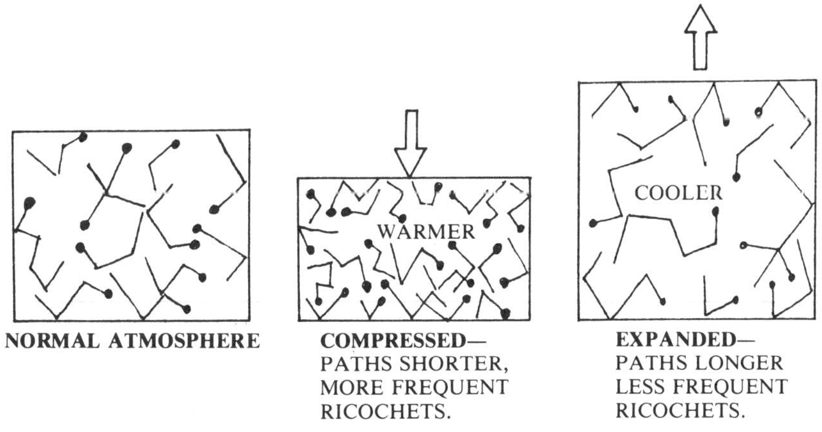

Figure 2 - The Properties of Air

PRESSURE IN THE ATMOSPHERE

We all do our daily chores under considerable pressure—from the atmosphere. In fact, at sea level we experience about 14.7 pounds per square inch (1.03 kg/cm2) on our bodies. That's almost 200 tons (!) on an average size adult. Of course, the air pushes on us equally from all directions, and we are basically water balloons with a rigid internal framework, so we don't notice atmospheric pressure unless it changes suddenly.

We can think of presssure in the atmosphere as simply a measure of the weight of air above us. This weight is caused by gravity pulling down on the air's mass as mentioned previously. At sea level the air weighs .076 pounds per cubic foot (1.22 kg/m3) so a medium sized bedroom (20x10 ft floor plan) contains over 120 lbs. of air. When we consider how high the atmosphere extends, it is no wonder there is so much pressure here at the bottom of the ocean of air.

It stands to reason that the lower our altitude, the higher the pressure since more air is pressing down above us. Likewise, the higher we go the lower the air's pressure. We can also see that higher pressure results in more dense air since the air's molecules will be compressed together by the greater weight they must support.

We measure the air's pressure with a barometer which is simply a cavity with some of the air removed so a partial vacuum exists. As the outside pressure changes (the air's weight changes) the walls of the cavity move in or out in response. A suitable linkage turns a needle to register the correct pressure (see figure 3). Another type of barometer uses a tube filled with mercury suspended by a vaccum at the top of the tube. The mercury moves up and down the tube to register pressure changes. Weather reports for the public often report pressure in inches of mercury in the English speaking countries. On the other hand, the rest of the world and weather maps use millibars to report pressure (1 millibar equals 1000 dynes per square centimeter).

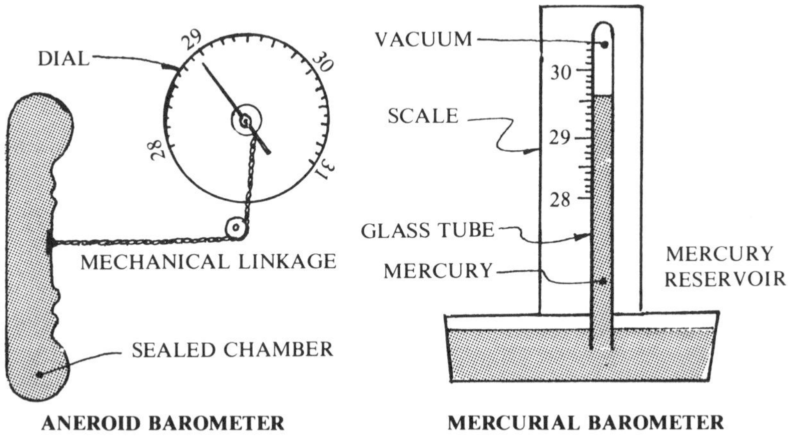

Figure 3 - Barometers

The altimeters we use as pilots are nothing more than sensitive barometers. They sense the pressure drop as we go up and the pressure increase as we go down. Some altimeters used by sport pilots can detect the difference in pressure of as little as one foot of altitude change—that's remarkably only .03 millibar or .001 inch of mercury at sea level.

Here is a summary of some important points:

When air is *lifted* it feels *less pressure* because there is less air pushing down above it, so it *expands* and *cools* and becomes *less dense*. Conversely, when air *sinks* it experiences *more pressure,* which *compresses* it, *heats* it and makes it *more dense* (see figure 4).

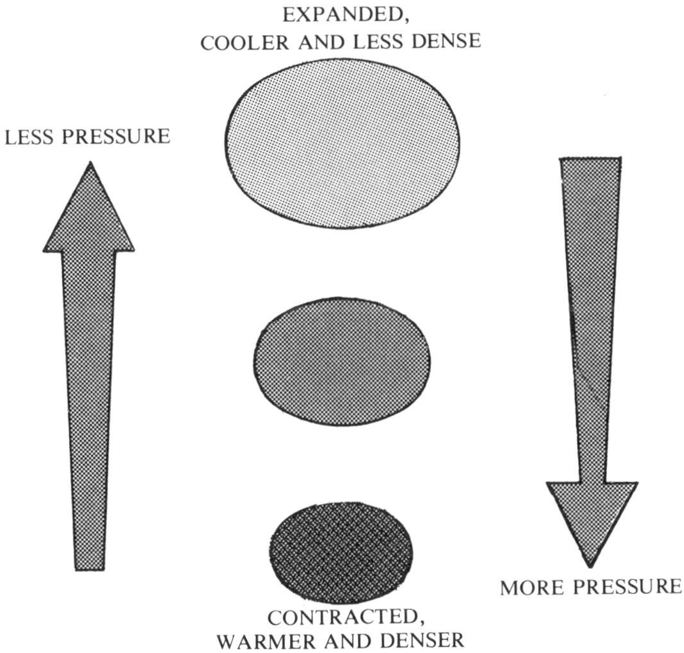

Figure 4 - Expansion and Compression of Air

TEMPERATURE IN THE AIR

We are not used to thinking that cooler air is less dense and warmer air is more dense as we indicated in the box above. However, it is the compression or expansion of the air that causes the temperature change. When the air changes temperature through compression or expansion alone—without the addition or subtraction of outside heat—it is known as an *adiabatic* process. This is the case in general when a thermal rises or convergence, ridge and frontal lift occur. In later chapters we will explore the cause and use of such lift.

Near the earth's surface the air is heated indirectly by the sun. This is a non-adiabatic process since the heat is from an outside source. Such solar heating is the main generator of motion in the atmosphere because air warmed from the outside expands and becomes less dense while air cooled by the surface becomes more dense. In general, air flows from the cool areas to the warm areas.

The sun's radiation does not heat the air from above, but passes through the air to heat the ground which in turn heats the air from below.

We measure this heat with a thermometer which reads in either Celsius (C) or Fahrenheit (F). Water freezes at 0° Celsius or 32° Fahrenheit and it boils at 100° Celsius or 212° Fahrenheit. The conversion formula is: $9/5\,C + 32 = F$.

To avoid the direct effects of reflection from the ground and other objects, the standard temperature reading is taken from a thermometer located 1.25 to 2.0 meters (3 3/4 to 6 ft) above a short grass surface. The thermometer should be shielded with a well-ventilated white box and located in the shade. Only by these means can we acquire a true air temperature reading.

SOLAR HEATING

Most of the sun's radiation passes through the air to the ground. It heats the air directly only 0.5 to 1 °F per day, depending on the amount of water vapor and pollutants present. Much of the sun's radiation is absorbed or reflected back into space by clouds. The amount of reflection naturally depends on the amount of cloud present. Only about 43% of the sun's insolation actually reaches the ground as shown in figure 5.

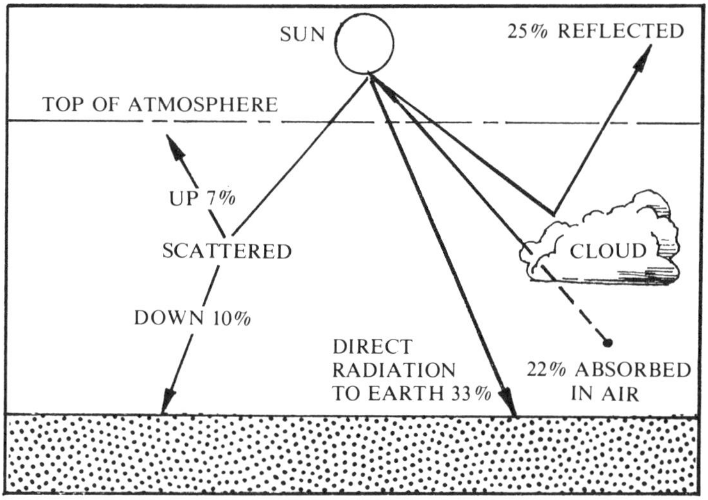

Figure 5 - Solar Heating

The fate of all this sunshine depends on what it meets at ground level. South facing slopes absorb more heat than level ground or northerly slopes. Concave shaped areas take on more heat than flat or convex areas. Trees and grass reflect light in the green wavelengths while sand reflects about 20% of the incoming radiation. Snow and ice reflect from 40% to 90% while dark surfaces such as parking lots or plowed fields reflect only 10 to 15% of the incident radiation. Water reflects the sun's rays according to their angle of arrival—about 2% when the sun is straight up and over 35% when the sun is just above the horizon.

All the radiation that is absorbed by the ground is spent in the process of making heat. Some of it directly heats the air next to the ground by conduction. Some of it heats the lower atmosphere through convection whereby currents or bubbles of warm air rise and spread outward. Some of it evaporates water which later gives back the heat to the atmospere as the water vapor condenses to form clouds.

The nature of the surface of the earth affects how the heat is absorbed or imparted to the air. For example, sand heats up in a shallow layer very readily while water allows the sun's rays to penetrate deeply so the surface temperature doesn't rise significantly. Generally, the hotter the earth's surface, the warmer the air will become above it.

It should now be clear that different types of surfaces heat at greatly different rates given the same incoming radiation. We shall study such properties in detail in Chapter IX for they are extremely important to thermal generation. For now let us note that the daily dose of sunshine keeps our atmosphere warmed from below and this is the main source of energy for our weather and soaring conditions.

COOLING CYCLES

Just as the air is heated from below by the sun heating the ground, so too is the air cooled. When the sun drops from the sky, heat from the ground radiates off into space in the form of infrared radiation. This radiation passes readily through the dry air with little absorption. Consequently the ground cools steadily through the night and in turn cools the air above it.

If a wind is blowing at night, the mixing of the air spreads the loss of the heat upward so it doesn't become as cold near the ground. If clouds or humidity are present they scatter the escaping radiation, sending some of it back down which slows the cooling process. This is the reason it takes a clear, still night to produce dew or frost. This is also the main reason that desert areas get so cold at night. The nighttime air and earth heat exchange is shown in figure 6.

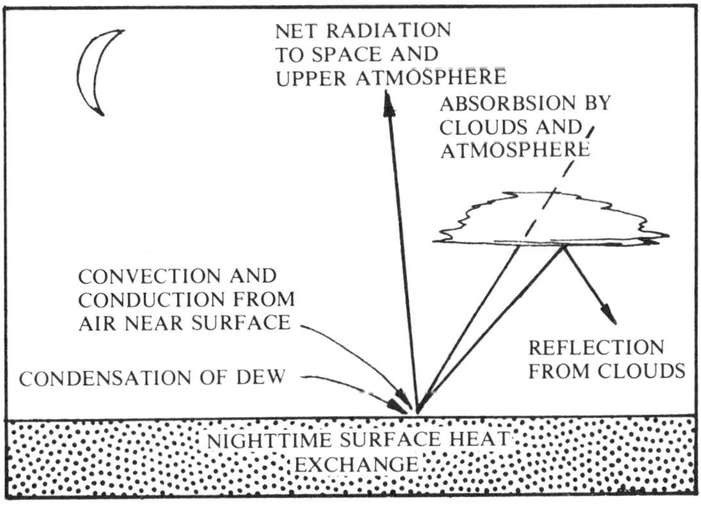

Figure 6 - Heat Radiation

DAILY CHANGES

The daily or diurnal variation of solar heating is an important concept to pilots whether they are looking for soaring conditions or smooth air. To see how this works we need to note that the sun's heating effects begin as soon as it looks upon the surface in the morning and increases to a maximum at noon (local sun time) when it is directly overhead, then diminishes to zero as the sun sets.

As long as more radiation is incoming than outgoing, the surface will heat. Now the outgoing radiation varies directly with the temperature of the surface, so the sun's heating reaches a maximum before the outgoing radiation does and thus the maximum surface heating occurs around 3:00 pm as shown in figure 7. This is also the usual time of maximum thermal production.

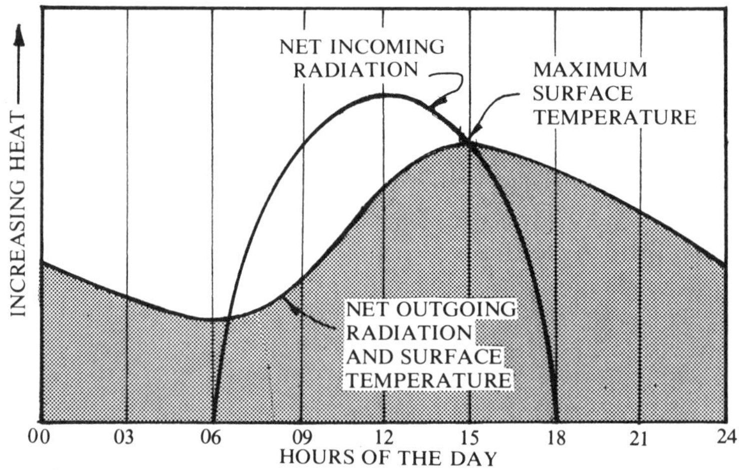

Figure 7 - Surface Heating Cycles

SEASONAL CHANGES

In figure 8 we see the seasonal differences in solar heating. The peak heating during the day is still at noon (local sun time) but it is much less during the winter solstice (when the sun is the furthest away) and at a max-

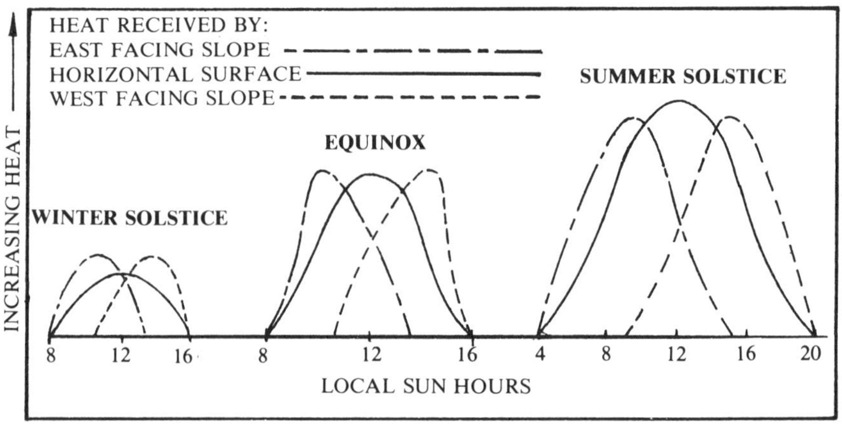

Figure 8 - Seasonal Heating

imum during the summer solstice (when the sun reaches its maximum height in the sky). The time of the equinox is when the sun is passing above the equator. Naturally this is when the heating is maximum at the equator. Note that during all these different heating cycles the maximum ground temperature and thermal production lags the maximum sunshine just as it does in figure 7.

A very important matter to see from these graphs is the difference in heating on various slopes. For example at the time of the equinox the east-facing slope receives the same heat at 8:00 AM as the horizontal surface at noon and the west-facing slope at 4:00 PM.

The cause of the seasonal change in solar insolation is twofold: the tilt of the earth's axis of rotation with respect to the plane of its orbit around the sun and the elliptical shape of this orbit. These features are illustrated in figure 9. Here we see that when the earth is tilted away from the sun in the northern hemisphere, the sun shines less directly on this hemisphere and it shines for a shorter time each day. At the same time it is summer south of the equator and the sun's rays are more direct with longer days.

At the other side of the orbit summer visits the north and winter assails the south. The interesting part of this discussion is that when the northern hemisphere is tilted away from the sun the earth is actually closest to the sun in its orbit as shown in the figure. When the north is tilted toward the sun the earth is furthest away in orbit. This results in making the winters milder and the summers less torrid. This arrangement wasn't always so as past ice ages testify.

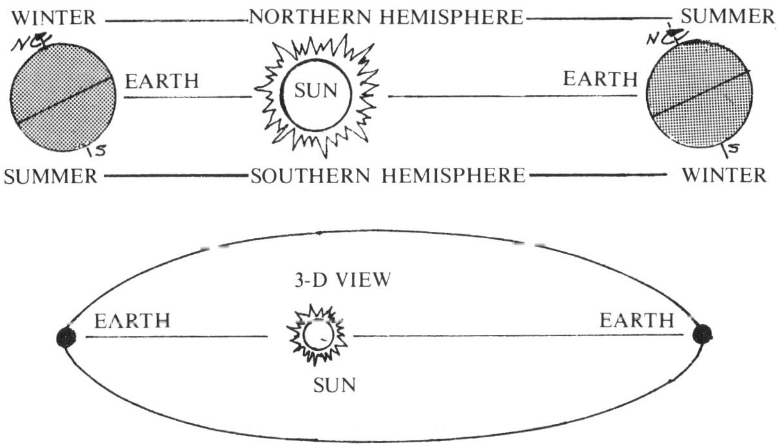

Figure 9 - Orbital Heating Changes

The opposite situation occurs in the southern hemisphere: the sun is closest in the summer and furthest in the winter. This would create more severe weather for southern pilots except there is much more ocean compared to land mass in the southern hemisphere which tends to moderate the temperature. Furthermore, not many people live below the 38th parallel in the southern hemisphere so the most severe winter weather is avoided.

These seasonal changes are important to pilots for the well-known general conditions they bring:

Winter—Cold, dense air, strong winds at times with stable air.
Spring—Changing conditions with cold fronts bringing unstable air and great thermal soaring.
Summer—Hot and humid with poor soaring in the wetter areas but good thermal production due to intense heating in desert areas.
Fall—A return of the cold fronts and unstable air with thermals in northern areas.

WATER VAPOR

Water continuously and universally affects the weather because of its widespread presence both as water vapor and cloud. An estimate of the total amount of water vapor drifting across our land is more than six times the amount of water carried by all our rivers! Even the smallest shower involves thousands of tons of water and one inch of rain falling over an area the size of Oregon state is equivalent to about 8 million tons. All this water vapor and rain comes from evaporation from open bodies of water and transpiration from vegetation.

Water vapor is the gaseous form of water and clouds consist of tiny water droplets that have condensed from water vapor. Water vapor forms clouds when the vapor is cooled to the point of condensation. This point is called the *dew point* and is given as a temperature. The dew point for a given parcel of air depends on its relative humidity.

HUMIDITY

Absolute humidity is a measure of the amount of water vapor in a given volume of air. This is frequently given in pounds per 1,000 cubic feet or grams per cubic meter. Absolute humidity varies from 1 part in 10,000 to 1 part in 40 according to the air's evaporation and temperature history.

Relative humidity is a measure of the percentage of water vapor present compared to how much the air can hold at its present temperature. Relative humidity is given as a percentage and ranges from near zero for warm, dry air to 100% for saturated air.

We must understand that warm air can hold more water vapor than cold air. Consequently the warm air will have a lower *relative* humidity than the cold air even though their *absolute* humidity (actual vapor content) is the same. For this reason we can increase the relative humidity by cooling a parcel of air. If the air is cooled enough its relative humidity reaches 100% or saturation and cloud forms. This saturation temperature is the dew point identified earlier. We will look deeper into this cloud-producing process in Chapter III. For now we note that the most common way that air is cooled in the atmosphere is by lifting which causes expansion and cooling.

The cold air of winter is always more nearly saturated than summer air because it can hold less water vapor. This fact is bad news to a pilot for the result is more clouds and precipitation in winter in general and also lower cloud bases because less lifting is needed to cool the air to saturation. When we heat this cold air and bring it into our homes in winter we decrease its relative humidity and our bodies lose moisture to the air causing us to think of winter air as dry. Relative humidity, not absolute humidity is in charge here.

WATER'S AMAZING PROPERTIES

Water in its various forms—solid, liquid and gas—has some unique properties that give it a special place in our understanding of weather (see figure 10). To begin, water has a high heat capacity. This means it is very happy to accept and store heat. Water absorbs all the sun's radiation it can get without increasing much in temperature. Consequently it tends to be cooler than land areas in the day but warmer than land at night when the lands quickly releases its stored heat. At night the slow release of heat from water can warm the air at the surface to cause instability and convection. The stored heat of water can likewise warm cold winter air moving across it to create "lake thermals," a topic we explore in Chapter IX.

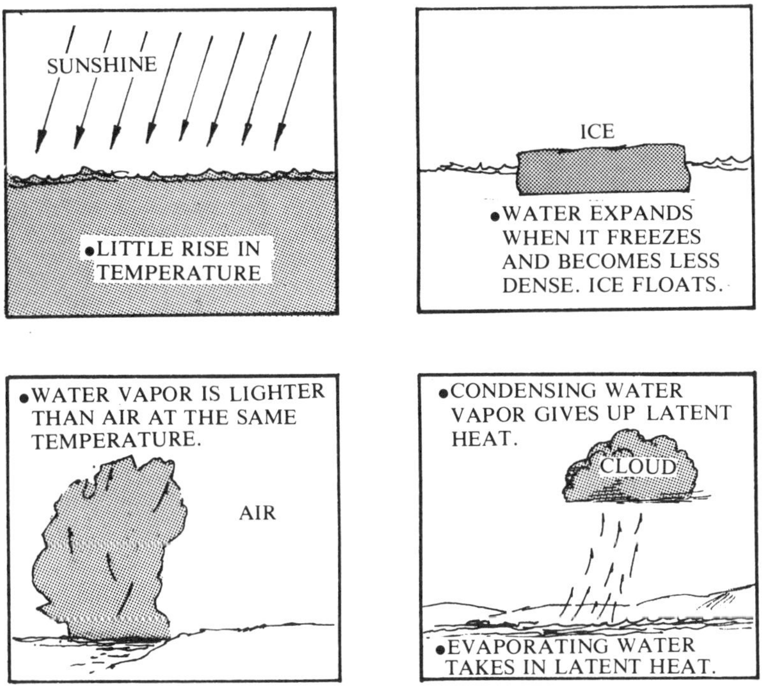

Figure 10 - Properties of Water

The temperature modifying effects of water result in warmer winter air and cooler summer air in its immediate locale. This gives England and France their mild climates despite their high latitudes and allows northern areas such as New York state, Ontario, and British Columbia to support orchards and vineyards. But the next property of water is even more important in terms of modifying our climate.

Water has the unique behavior of expanding when it becomes a solid (ice) so that it is actually lighter as a solid than a liquid. Ice floats. As a result, only a relatively thin layer of ice rests on the top of open bodies of water—a layer that can easily be melted with the return of warm weather.

If ice didn't float it would gradually accummulate on the lake bottoms and build up until the lakes were frozen solid. They would barely thaw in the course of a summer and worldwide temperatures would be considerably cooler, at least in the temperate areas.

The next property of water is its relative lightness as a gas (water vapor). Water vapor is only about 5/8 as heavy as the rest of the air due to its lighter molecules (two hydrogen atoms and one oxygen atom compared to two united nitrogens or two oxygens). As a result, humid air rises in the presence of dry air. This property accounts for the continued progress of thermals and thunderstorms in many instances.

LATENT HEAT

The final property of water we'll investigate is its latent heat. Latent means hidden and this heat is acquired by water vapor during the evaporation process and is "hidden" or stored away to be released later to the surrounding air when the vapor condenses back into water.

The process of releasing heat upon condensation or absorbing heat during evaporation is very important to the behavior of clouds, thermal formation and downdrafts (see Chapter XI). Since the source of the latent heat is usually the air into which the water vapor evaporates, the air above water tends to be cooled and thus rendered more stable (unless the water is much warmer than the air as noted previously). The subject of stability will be explored in more detail in the next chapter.

SUMMARY

In this chapter we have gained an understanding of the makeup and mechanics of the atmosphere. We separate each aspect of the air so we can investigate it, but really all facets of our study are intricately affected by one another. The composition of the air along with its temperature, pressure and humidity all interact and are modified by the sun's input to our little planet with further alterations by that planet's gravity.

As we gain understanding we begin to put together the big picture that allows us to make predictions and judgements as to what we may encounter in the sky. But before we can look at general weather conditions we must learn about a few more forces and effects that take place in our atmosphere. We do this in the next chapter.

CHAPTER II

The Living Atmosphere

Any breathing being on earth knows that the atmosphere is not simply a big blob that squats over us like a brooding hen. It is a dynamic ever-changing mass, more-or-less in constant motion. The air has its ups and downs and moods if you will.

In this chapter we will look at some of the traits of the atmosphere that modify its behavior. A few of these important traits are stability, pressure imbalances and Coriolis effect. These three factors are the main causes of air currents in both the vertical and horizontal dimension.

THE LAPSE RATE

Stable and unstable air is a concept we must explore in great depth to understand how convective lift (thermals) is created. But first we must picture the temperature profile or *lapse rate* of the air.

As mentioned earlier, the air is heated from below by the ground. Also, the atmosphere becomes less dense with altitude. These two factors combine to create the normal situation of warmer air at the surface and gradually cooling air as altitude is gained.

Look at figure 11. The curve drawn at A is the ideal temperature profile or *lapse rate* of a "normal" atmosphere. The atmosphere is rarely normal, but this lapse rate is the average found around the earth. This average lapse rate is known as the *standard lapse rate* (SLR) and exhibits a drop of 3.6°F per 100 feet or 2.0°C/1000 ft or 2.0°C/300 m.

Now look at the curve B. This is a more realistic situation at night. Here we see the air is much cooler near the ground due to contact with the cold ground. This feature is called a *ground inversion* and is the typical state of affairs at night. This inversion may extend upwards to 1000 feet (300 m) or more depending on the amount of wind present to create mixing. The word *inversion* refers to the fact that the air's temperature actually increases or at least doesn't cool as much as normal for a given gain in altitude in the inversion layer. An inversion layer contains stable air as we

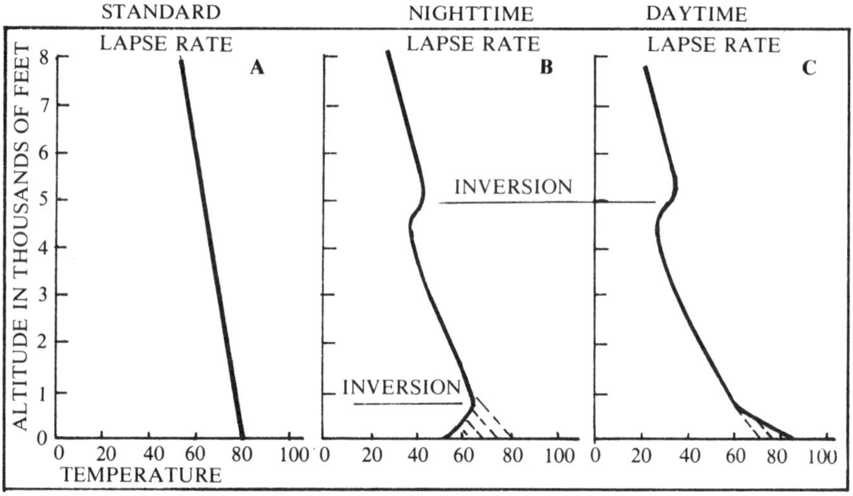

Figure 11 - Lapse Rate Curves

shall see.

If we look higher up in the graph we see another inversion layer at about 5,000 ft. Here the air gets warmer as we climb then drops off again. This is a common feature of the atmosphere and will be explained below.

The daytime situation is very similar to the curve at C. Here we see the air near the ground heated well beyond the normal amount. This intense ground heating spreads its warmth upward increasingly as the day progresses through convection currents. The dashed lines in curve B and C show the gradual change of the lapse rate from night through morning to the maximum mid-day heating. As evening falls the reverse process takes place.

Such a lapse rate as shown in the lower part of curve C is known as unstable for reasons we shall see.

STABILITY AND INSTABILITY

Stable air is air that wants to stay where it is in the vertical dimension. Lets see how this works. Imagine a bubble of air rising in the atmosphere as pictured in figure 12. As it rises it expands due to reduced pressure. This pressure drop is fairly linear in the lower 10,000 ft (3,000 m) and the uniform expansion of the air bubble cools it at a rate of 5.5 °F per 1,000 ft or 3 °C per 1000 ft or 1 °C per 100 m. The same cooling will occur in a helium or hot air balloon as they rise without added heat.

The rate of cooling a rising parcel of air undergoes—5.5 °F /1,000 ft—is known as the Dry Adiabatic Lapse Rate or DALR. It is dry not because there is no water vapor in the air parcel, but because this vapor is not condensing or changing to visible cloud. It is adiabatic because no heat is gained from or lost to the surrounding air. In the real-life situation some mixing with the surrounding air may occur, but this is usually of limited extent.

Now we know that warmer air at a given level is less dense than cooler air at that level because they both experience the same pressure but the warmer air is more energetic so the molecules are spread further apart. Thus warmer air wants to rise when it is surrounded by cooler air because it is lighter and cooler air wants to sink when it is surrounded by warmer air because it is heavier. This is the same principle that causes less dense wood to rise in water and more dense rock to sink.

If our happy bubble of air was rising in an atmosphere whose lapse rate cooled off less than 5.5°/1,000 ft, then the bubble would be cooling faster than the surrounding air as it climbed and eventually it would reach a height where it was the same temperature as its surroundings as shown in the figure. In fact if it was forced to go higher than this equilibrium point it would get the urge to drop back down to the equilibrium level because it would be cooler than the surrounding air. This is the meaning of stability.

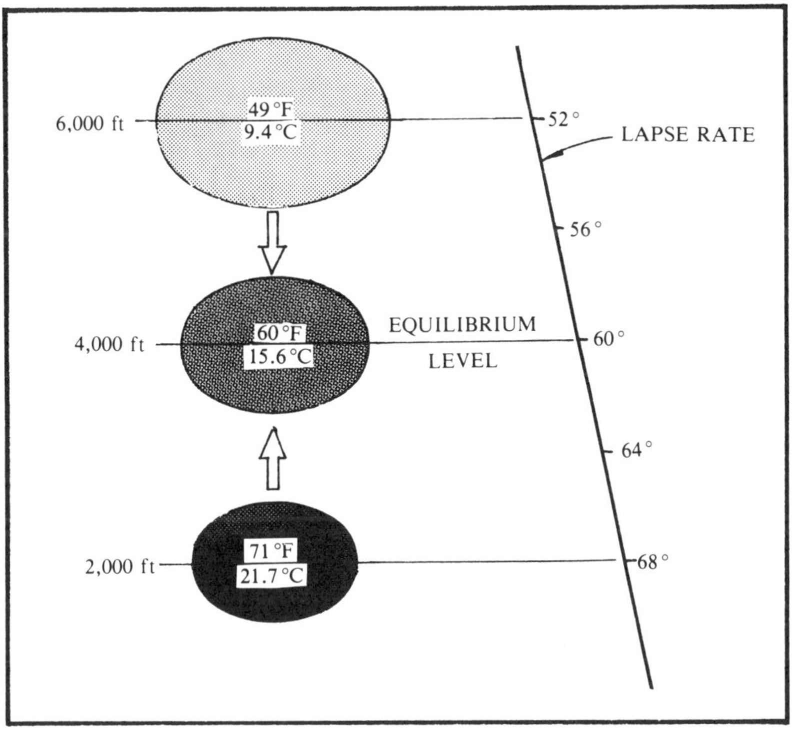

Figure 12 - The Meaning of Stability

Unstable air is just the opposite. If the lapse rate of the air cools more than 5.5 °F/1000 ft, a parcel of air forced upward will not cool as much as the surrounding air so it will continue to rise (see figure 13) . Instability of the air means that it is out of balance, for the air in the lower layers is too warm for it to remain tranquil in the vertical dimension (note that horizontal wind blows in stable and unstable conditions). Unstable air wants to turn turtle to distribute the heat upward.

We can now form the concise definitions:

> *Stable air*—occurs when the lapse rate is *less* than the DALR (5.5 °F/1000 ft).
> *Unstable Air*—occurs when the lapse rate is *greater* than the DALR.

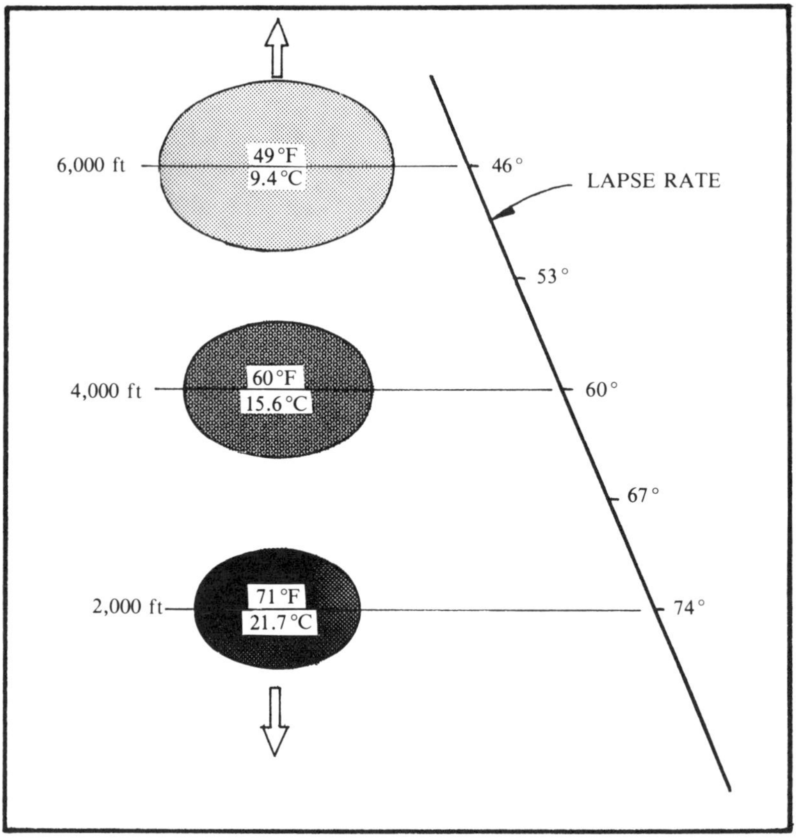

Figure 13 - The Meaning of Instability

It is important to note that in stable conditions a parcel of air moved downward will want to move back up to equilibrium while a parcel moved downward in unstable conditions wants to continue downward. The latter point accounts for much of the widespread sink found on unstable days. Also the nature of stability and instability is what causes some days to exhibit more bouyancy than others in ridge lift. Of course, unstable conditions lead to thermals (bubbles of convective lift) which are featured in later chapters.

Now look back at Figure 11. The solid curve A which shows the SLR can be seen to be stable because the temperature drops less than the DALR which is shown by the dashed line. If the lapse rate is greater than the DALR it is known as *Superadiabatic*. Such a lapse rate is shown in the lower portion of curve C. A superadiabatic lapse rate generally only occurs over hot deserts or close to the ground on sunny days in less torrid climes.

INDICATIONS OF STABILITY

Pilots of all sorts should be able to detect the general stability of the air before they commit body and soul to its mercies. Perhaps you are a soaring pilot and wish to hunt thermal lift, or possibly you want to motor around in glassy air. In the first case you need unstable conditions and in the second you must look for stable and slow moving air.

In general, a clear night followed by a clear morning will bring unstable conditions for the clear night allows a thick layer of cold air to form which is unstable with respect to the air warmed at ground level in the morning. However, a very cold night delays the onset of deep convection because of the low level inversion as shown at the bottom of curve B in figure 11. On the other hand, overcast days and periods of days where the air has continuously warmed tend to be more stable.

Cloud types (see Chapter III) are always indicators of stability. Cumulus or tumbled clouds are caused by vertical currents and always imply instability. Stratus or layer clouds are usually signs of stability. Likewise, smoke that rises to a certain level then spreads out signifies stable conditions while high rising smoke means unstable conditions.

Dust devils, gusty winds (away from turbulence inducing structures) and good visibility are also signs of unstable air while steady winds, fog layers and poor visibility due to haze and smoke denote stable air. These effects are summarized in figure 14.

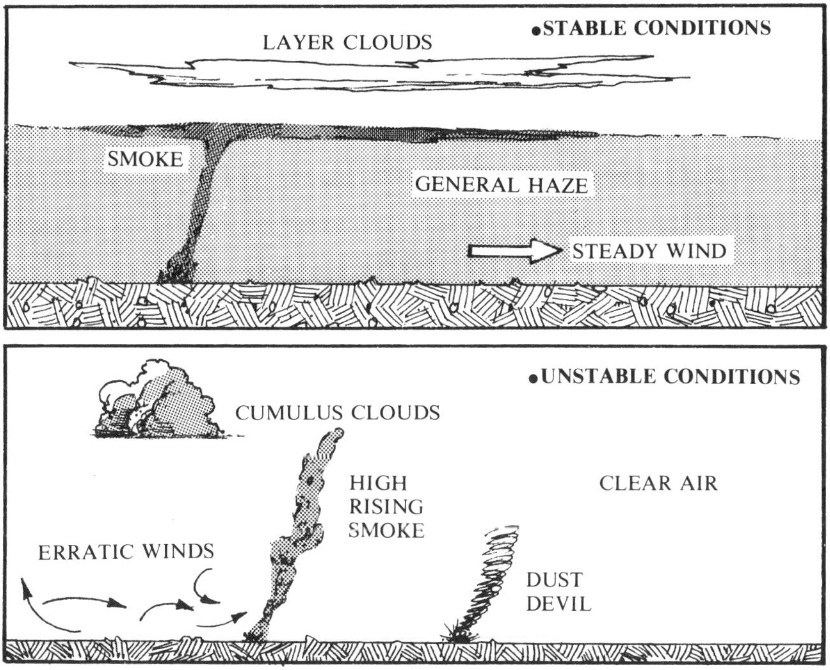

Figure 14 - Stability Indications

THE STABILITY OF LAYERS

Here we turn our attention to the ways which the stability of the air mass or that of certain layers changes. First we note there is a daily variation from the sun's heating. Also, whole new air masses can enter the area as when cold and warm fronts bully their way through. These new air masses typically have different temperature profiles and thus different stability (we cover fronts in Chapter IV).

Along coastal areas marine air usually invades the land in the warm season. This air is cool and pushes under the warmer land mass. The result is cool, stable air near the ground topped by warmer unstable air. This is known as the marine inversion. It is "inverted" because the cool air is below the warmer air. Typically stratus clouds are formed at the top of this marine layer if it is thick, or fog if it is thin.

In mountainous terrain, warm air moving into the area may flow across the tops of the valleys rather than descend the slopes. This leaves pools of cool air below the upper warm layer which again results in an inversion layer at mountain top height. Lift is surpressed above the bottom of the inversion layer.

One of the most common and important ways that the stability of air masses changes or inversion layers are formed is by lifting or sinking of the entire air mass. This is such an important point that we note it specially:

- When an airmass is *lifted* it becomes *less stable.*
- When an airmass *sinks* it becomes *more stable.*

To see why this principle is so, look at figure 15. Here we see a rising layer of air. As it is lifted it stretches vertically and the top expands faster than the bottom, thus cooling more. In the example shown, the layer starts at 5,000 ft with a lapse rate identical to the entire air mass (3.5 °F/1,000 feet). As it rises the layer cools at the dry adiabatic rate (DALR) of 5.5 °F/1,000 ft.

After some time the bottom of the layer reaches 15,000 ft but the top is nearly at 18,000 ft due to vertical stretching. Thus the bottom of the layer has cooled to 22 °F which is 77 °F – 5.5 ° × 10 (thousand feet). The top has cooled

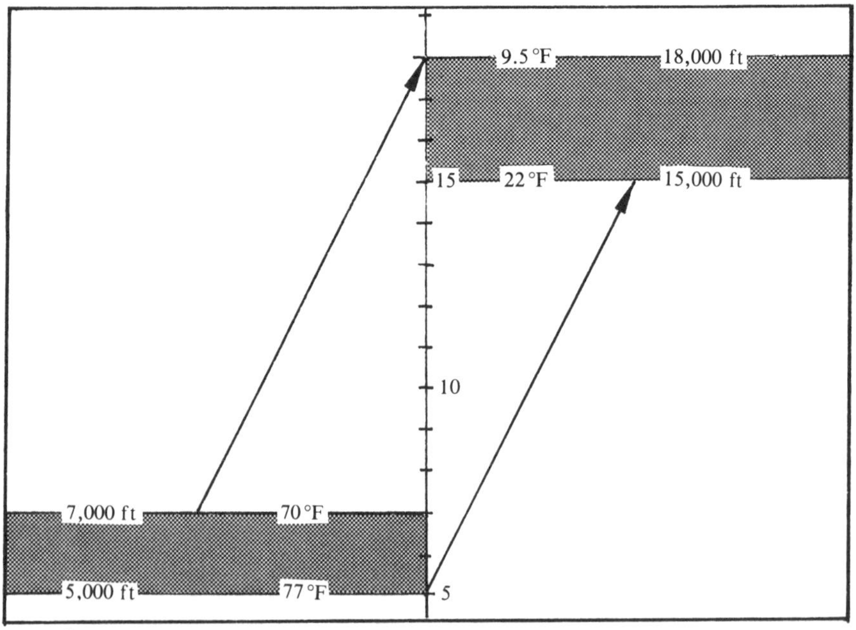

Figure 15 - Layer Stability Changes

to 9.5 °F which is 70 °F – 5.5 ° × 11 (thousand feet). The difference in the top and bottom layers is now 22 ° – 9.5 ° or 12.5 °F. Since 3,000 feet separates the top and bottom, the lapse rate is now 12.5 ÷ 3 or 4.2 °F per 1000 feet. This is considerably less stable than the 3.5 °F/100 ft we started with.

In a similar manner a sinking layer will become more stable. The result of a descending layer is often an inversion if the descent lasts long enough. The upper level inversions shown in figure 11 are caused precisely by this mechanism. Very often such an inversion puts a cap on thermal heights and is especially found in high pressure dominated weather (see Chapter IV).

The results of an ascending layer at different times can be widespread lift bands, fat, gentle thermals, improved soaring conditions, alto cumulus clouds and mackerel sky (clouds that look like the scales of a fish). Ascending layers are caused by the lift created by moving fronts, surface warming and low pressure systems. Descending layers are notably associated with high pressure systems and surface cooling.

THE MOIST LAPSE RATE

In the previous chapter we discovered that rising air that contains water vapor expands and cools so that its relative humidity increases. If this process continues, the relative humidity reaches 100%, saturation is said to occur and the air's temperature has reached the dew point. If this air is lifted further, condensation begins that causes the release of latent heat. The release of latent heat warms the air so it no longer cools at the DALR as it continues to rise.

The lapse rate that occurs when condensation takes place is called the moist adiabatic lapse rate (MALR) This lapse rate is between 2 ° and 5 °F per 1,000 feet (1.1 ° to 2.8 °C/300 m) depending on the original temperature of the rising air, and averages about 3 °F per 1,000 feet (0.5 °C/100 m).

The average MALR along with the DALR and the standard lapse rate (SLR) is shown in figure 16. When the temperature profile of the air lies

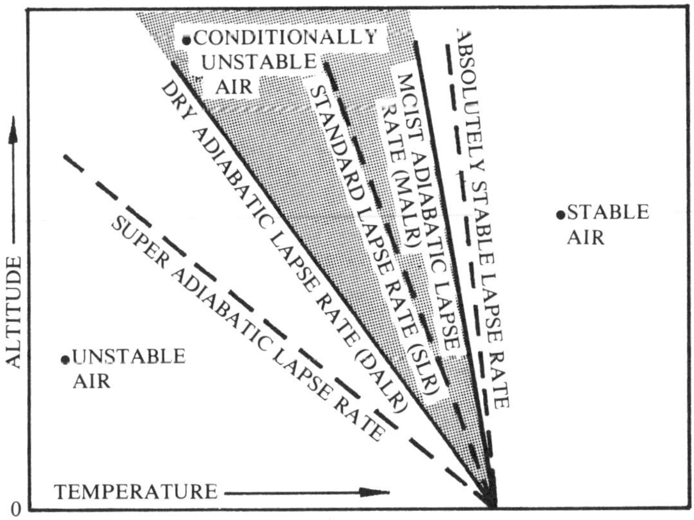

Figure 16 - Relationship of Important Lapse Rates

between the DALR and the MALR it is said to be "conditionally unstable." This means that it will be unstable if the air is saturated and further produces condensation. This is the case in clouds that form in stable air and grow vertically.

Also in the figure we have labeled the area to the right of the MALR as *absolutely stable* for a parcel of air rising in an air mass with a lapse rate in this region will always want to return to its point of origin, even if condensation occurs. The area to the left of the DALR is where *absolutely unstable* conditions occur with the spontaneous generation of thermals. A lapse rate in this region is termed *superadiabatic* as mentioned earlier. Such a super lapse rate condition rarely lasts long in nature except very close to the ground on sunny days, for thermal currents distribute heat upwards and thus modifies the lapse rate.

The whole process of water vapor rising and exchanging heat with the atmosphere is very important to the weather process. For each ton of water that condenses, almost 2 million BTUs of latent heat is released to the atmosphere. This energy is the main thing that powers thunderstorms, tornados, hurricanes and other strong wind sources. We can think of water vapor as a transporter of heat in our atmosphere that causes heat imbalances that "weather" works to straighten out. Water is the great modifier.

THE STANDARD ATMOSPHERE

Over the expanse of time and space scientists have measured and probed the air enough to declare a *standard atmosphere*. This is a great aid to pilots for altimeters can be calibrated to a standard. At a given airport with a known altitude above sea level, a given standard temperature and pressure exist. By reading the actual temperature and pressure from a thermometer and barometer, adjustments can be input to the pilot's altimeter to read true altitude above an airport.

A chart of the standard atmosphere is given in Appendix I. Note that the change in temperature with altitude is exactly the SLR. We can also note that there is a 3% drop in density per 1,000 ft (300 m) which leads to a 1.5% per 1,000 feet increase in all flying speeds. However, we know that the SLR is an average condition and in truth in soaring conditions the lapse rate is more apt to be closer to the DALR. Thus the relationship is:

> **Density, Altitude and Airspeed**
> There is a 4% drop in density per 1,000 ft (300 m) which leads to a 2% increase in all flying speeds per 1,000 ft (300 m) of altitude.

DENSITY ALTITUDE

Of course the standard atmosphere is only an illusion. The real sky is not so tidy. If an area is warmed or cooled an amount different than the standard atmosphere our altimeters will read high or low respectively. If the area becomes more or less humid or high or low pressure systems move into the area then the local pressure will again be altered and affect our altimeters. These aberrations can be corrected for by readjusting our altimeters before takeoff, but caution must be awarded to the possibility of significant changes during long duration or distance flights. Here is our rule of thumb:

> **Density Changes**
> An altitude change of 300 ft is equivalent to 1% change in density which equals a pressure change of .3 inches or 10 millibars, a temperature change of 5°F (2.8°C) or the addition of water vapor at .8 inches (27 mb) of pressure.

Thus, for every millibar you cross when flying a given route, your altimeter reading changes 30 ft—higher if you are going towards a low pressure system and lower if you are going towards a high. An altimeter is essentially a barometer calibrated to read altitude. Most altimeters are compensated to eliminate temperature effects, so the temperature changes are not a problem. Pressure changes are not a problem either, as long as we can see the surface and do not rely blindly on our altimeter.

What is a problem with density altitude is the effect it has on our takeoff and landing performance. When the air is hot and humid and low pressure is in the area, takeoff and landing speeds are increased. Higher altitudes especially affect all these critical speeds. Appendix I covers density altitude and these considerations in more detail.

THE WIND WE FEEL

One aspect of weather that affects our daily lives and especially our flying is the wind. The air is rarely perfectly still, but it usually takes motion of a few miles per hour before we detect it readily. Wind can carry various characteristics such as humidity and temperature for great distances and so has an important role to play in weather. Because it is also an integral part of soaring conditions, we devote two chapters to its study. For now let us simply identify its cause and nomenclature.

Wind is produced simply by an imbalance of pressure, usually in the horizontal dimensions. This imbalance itself is caused by temperature differences in adjacent areas or circulation aloft that piles air up in some spots. Ultimately it is the uneven heating of the sun that causes the temperature differences and the circulation that creates the pressure differences on both the small and large scale. So again we have the sun to thank for our soaring weather.

Wind is usually identified by the direction it comes *from*. For example, a north wind comes from the north, a southwest wind comes from the southwest and so on (see figure 17). Likewise a mountain wind flows from the mountains, a valley wind flows upslope from the valley, a seabreeze flows from the sea and a land breeze flows from the land.

In aviation terminology it is standard practice to give the wind direction in degrees and the velocity in knots. Thus a north wind is 360° (the same as zero degrees), and east wind is 90°, a south wind 180° a west wind 270° and a southwest wind is 225° as shown in the figure. A knot is based on the nautical mile and is equal to 1.15 mph or 1.85 km/h. Note that a compass does not point exactly at the north pole since the earth's magnetic field is not aligned with the earth's axis of spin. The difference in magnetic north and true north is called variation. Surface winds are given according to magnetic compass readings while upper winds are given according to true directions.

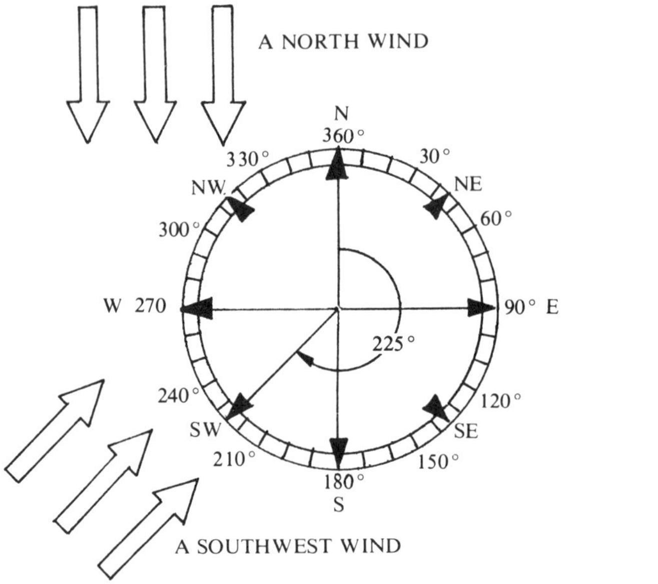

Figure 17 - Wind Directions

CORIOLIS EFFECT

The last matter we'll investigate is the Coriolis effect. This effect is very important to the understanding of weather on the large and intermediate scale. Coriolis effect results in a tendency for all objects moving in the northern hemisphere to turn to the right and all moving objects south of the equator to tend to the left. Coriolis effect is strongest at the poles and is reduced to zero at the equator.

The cause of Coriolis effect is simply the turning of the earth below the moving object. It is not a real force, but the earth's motion interacts with the force of gravity to produce an illusion of a right turn in all freely moving bodies. The air's flow and ocean currents on the large scale are affected by the Coriolis effect. Large artillary pieces whose shells have a long transit time require a significant correction to hit their target due to Coriolis effect.

To better visualize Coriolis effect, look at figure 18. Here we see an object projected from the center of a rotating disk. As it moves to the outside it tends to keep its same direction of travel to an outside observer as in 18a. However, to an observer on the disk, the moving object appears to describe a curved path to the right as in 18b because the observer is moving away from the object.

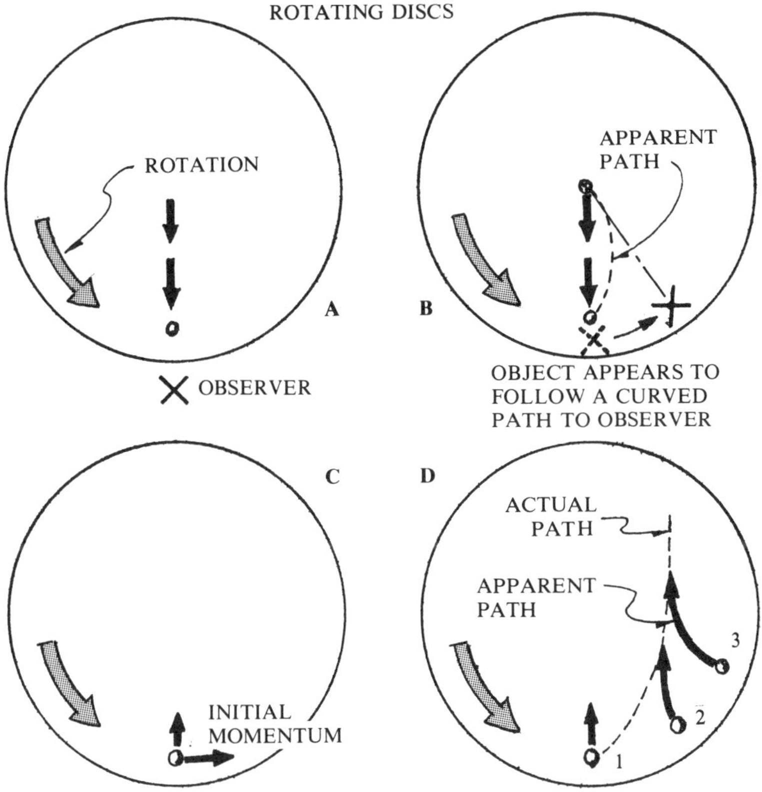

Figure 18 - Coriolis Effect

In the case where the object is moving towards the center of the disk as in figure 18c, it has angular momentum which causes it to circle around the center at the same time it maintains its orientation in space. This combination of motions results in a path shown in 18d. This is again curved to the right from the viewpoint of an observer on the disk.

It's not hard to see that the disk can be the sphere of our earth as viewed from space above the north pole. There are three-dimensional effects in the real world which account for the diminishing of the Coriolis effect from the pole to the equator, but the principles remain the same.

Coriolis effect accounts for the flow around pressure systems (see Chapter IV) as well as many other actions of the wind which we shall discover. We repeat the important ideas here:

> **Coriolis Effect**
> Causes the wind to turn to the *right* in the *northern* hemisphere and *left* in the *southern* hemisphere.

SUMMARY

Now we have reached an understanding of the basic rules that govern the atmosphere on both the small and large scale. The effort to gain this understanding is necessary for it is the combination of these physical rules that create the weather effects we pilots are eager to avoid or exploit as the case may be. We will be continually referring to the stability of the air as well as pressure or heating differences and Coriolis effect as we pursue our adventures in the sky.

In the following chapters we take a practical look at actual weather conditions, then we progress to learn about smaller scale matters that directly affect the quality of the air we traverse.

Cumulus clouds building in unstable conditions.

CHAPTER III

The Message of the Clouds

Clouds have figured in the imagination of mankind from the beginning of history. Their ever-changing shape and habit of floating on high seem to invite images of freedom and flight. In fact, birds and clouds are the models that we have always looked to in our dreams of floating free of the earth.

But clouds themselves are not entirely free for they too must obey laws of gravity, inertia and heat exchange. If we gain a little knowledge of how they conform to these laws we can learn to read the message of the clouds. It is this message that is important to pilots for it carries hints of safety, future conditions, poor soaring or great lift.

CLOUD CAUSES

Clouds are made up of countless microscopic water droplets of various size ranging from 1/2,500 inch (.001 cm) near saturation and increasing to a maximum of about 1/100 inch (.025 cm) as condensation continues. As you recall, saturation is when the air reaches 100% relative humidity which varies with the air's temperature. Air containing a given amount of water vapor can reach the saturation point if it is cooled. The main way in which clouds are formed in the atmosphere is by cooling of air containing moisture. The way this cooling occurs is by lifting of the air. Thus we can make the important observation:

> **Cloud Formation**
> Except for fog which is formed in air cooled from contact with the ground, *all clouds are formed by air that is lifting or has lifted.*

It would seem then that clouds are most welcome by a soaring pilot for they are signs of lift, but this is not entirely true. Certain types of clouds are created by air that rises too slowly to sustain flight while widespread clouds block the sun and prevent further lift from developing. We must view clouds from both sides—good and bad. A bit later we will identify the friends and enemies of pilots.

LIFTING THE AIR

There are three main causes of lifting in the air. They are frontal movement, rising terrain and surface heating. Let's look at each one to understand how it works (see figure 19).

Frontal Movement—When large masses of air move an appreciable distance they generally encounter air of a different temperature. As the air masses continue to flow, the warmer, lighter air will ride up over the cooler, denser air. This warmer, lifted air will form clouds if it reaches the dew point.

The upward velocity of air rising due to frontal movement alone (neglecting heating effects) is typically from 30 to 300 feet per minute (10 to 100 meters per minute). This lifting is relatively slow and occurs fairly uniformly over a wide area and thus creates layer clouds.

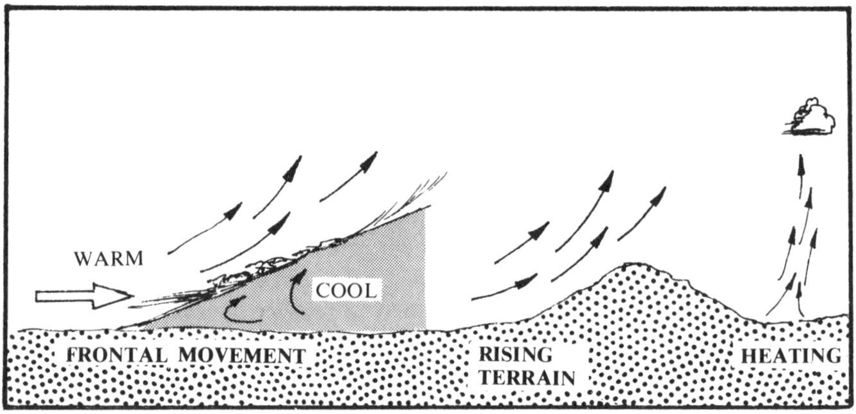

Figure 19 - Sources of Lift

Rising Terrain—When air flows over features of the earth that are themselves rising, the air cannot avoid being lifted. Such lifting occurs commonly in mountainous areas. For example, clouds often appear in the lifted air above the coastal ranges along the west coast of the US, over the Alps in Europe and above the Allegheny Plateau which gradually rises from the Mississippi valley to the eastern United States.

Heating Effects—We have already outlined how surface heating produces instability which results in upward convection currents. We also include here the rising air in the area of a low pressure system although this air is lifting due to the combined effects of heating, frontal activity and convergence (coming together of air). Isolated convection currents produce relatively small (except in thunderstorms) puffy clouds while low pressure systems produce widespread layer-type clouds.

THE DEW POINT AND CLOUD HEIGHT

We have learned before that when air cools to the saturation point it condenses and forms cloud. The point of saturation is called the *dew point*. Because the greater the relative humidity the higher the dew point temperature, the dew point is in essence one measure of the air's humidity.

The dew point can also be used to tell the height of the bottom or base of a certain cloud. The puffy cotton boll clouds that are created by thermals (see below) are born from moisture that orginates in air at ground level. We have already found out that air rising in such a manner cools at 5.5 °F per 1,000 feet (1 °C per 100 m). However, the dew point lowers only about 1 ° F per 1,000 ft. (.55 °C per 300 m). So the temperature of the lifting air and the dew point of that air approach each other at 4.5 ° F per 1,000 ft (.8 ° C per 100 m). When the air's temperature and the dew point become identical, cloud is formed.

We can see how to use the above fact by example as shown in figure 20. Here the air's temperature at the surface is 82 °F; the dew point is 59.5 °F. If we subtract 59.5 from 82 we get 22.5, the difference between the temperature and the dew point. Dividing 22.5 by 4.5 gives us 5 so we can expect cloud base to be at 5,000 feet above the surface. In the figure we can see how the lifted air's temperature and the dew point approach each other to coincide at 5,000 feet.

In practice you can get the dew point from various weather services outlined in Chapter XII. Weathermen use a dry and wet bulb thermometer to get the surface temperature and the saturation (wet-bulb) temperature then consult detailed charts which vary for each altitude to find the dew point.

Why are we so interested in the height of clouds? Because the higher the clouds are, generally the higher the usable thermal lift extends which improves the cross-country soaring prospects. Pilots flying motorized aircraft want to know cloud heights because flying above the clouds on a thermally day provides much smoother conditions.

Our eyes are not able to judge the distance to objects very effectively. Our distance judgement comes mainly from comparing the relative sizes of objects. Clouds with their multitude of fanciful shapes and sizes give the eye few clues as to their height or distance. With practice in a given area pilots can learn to make educated guesses as to the cloud height by com-

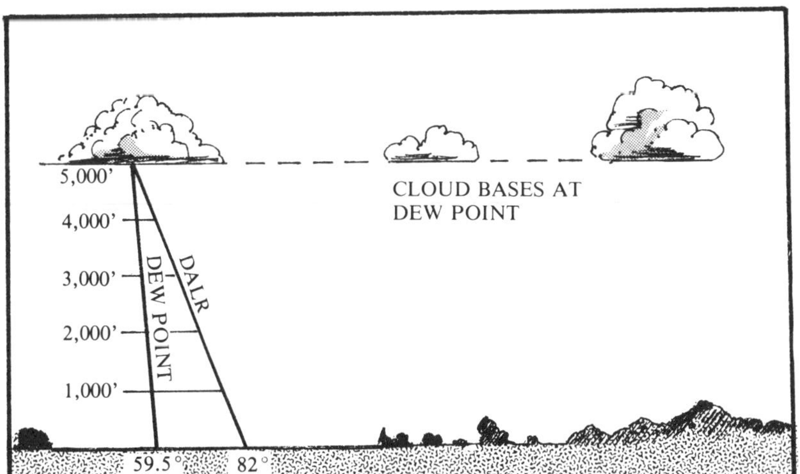

Figure 20 - Finding Cloud Base Height

paring the size and spacing of the clouds as they recede into the distance, but the dew point method remains to be the most reliable way to determine cloud heights.

CLOUD FORMATION

Once lifted air reaches the dew point and is 100% saturated it is ready to form cloud. But the funny thing is, it needs something to form on. In fact without assistance the air can become supersaturated and acquire a humidity above 100%. This assistance comes in the form of microscopic particles in the air.

These small particles are called condensation nuclei because they allow water vapor to condense, or sublimation nuclei when they allow water vapor to change directly to ice crystals (sublimation). We have all seen the condensation that forms on a cold glass or from our breath in winter. We have also witnessed the effects of sublimation as frost forms on a window pane, so these processes should be quite familiar.

Condensation nuclei, on which liquid droplets form are combustion products, sulfuric acid droplets and salt particles. The first two are pollution by-products and the latter is produced by waves pounding the shores of the world's seas. Sublimation nuclei on which ice crystals form are themselves crystalline in nature like dust or volcanic ash. These sublimation particles are relatively large so they are sometimes rare at high altitudes where sub-freezing temperatures occur. This explains why upper level clouds are abundant for a long period of time after a volcano spews its ash into the upper atmosphere.

Cumulus cloud bases at the dew point level.

CLOUD BASES/CLOUD TOPS

When an air mass overlies an area its temperature profile is generally the same over a wide area. Also the humidity of this air mass will be fairly uniform. Thus, the dewpoint will occur at about the same altitude throughout the airmass so that clouds formed in air rising from the ground (isolated, puffy clouds) will all have roughly the same bases—at the dew point (see figure 20).

Exceptions to the above rule occur when the humidity of the air near the ground varies greatly such as near lakes or swamps. Also, air originating over plateaus or mountains may exhibit higher bases when it rises than that originating over nearby low lands, for it will start higher at approximately the same temperature. This only applies to large landforms for the air is fairly well mixed from valleys to peaks of small hills and mountains.

Sometimes small puffy clouds can be seen to float below a higher layer. This is usually caused by rising air of greater humidity than the surrounding air. In rainy weather such lower clouds are caused by evaporating rain cooling the air it is falling through to the saturation point. This action can sometimes be seen at ground level in everyone's favorite weather: cold drizzle.

The bases of clouds that form in rising air not originating at the ground will still be at the same level as long as the local air mass was lifted uniformly. Of course, it is quite common to see clouds of various types in different layers, indicating that they arose in different horizontal air masses or through different lifting processes.

While bases of clouds tend to be uniform, the tops vary greatly in altitude. This is because nothing definite determines how high the lifting process in the cloud can extend. Bullets of lift may penetrate some clouds and carry the moist air much higher than its neighbors. Even extensive layer type clouds will often have greatly varied tops, especially if the air is unstable as shown in figure 21.

Figure 21 - Cloud Tops

CLOUD LIFE

The droplets that form clouds average about 1/2,500 inch (1/1,000 cm) at saturation and are abundant enough to form a visible mass. As condensation continues they grow to about 1/100 inch (1/40 cm) maximum which is about the size of light drizzle drops. Even at the larger sizes these droplets are so light that they are essentially suspended in space so the cloud lingers indefinitely.

However, there are several factors that determine a cloud's life span. To begin, a cloud formed from isolated shots of rising air (thermals) tend to mix vigorously with the surrounding air and thus dry out. When the updraft is rising initially it only mixes along its boundaries, but once the water vapor condenses to form cloud it releases latent heat of vaporization that energizes the system and turns the whole rising mass inside out which mixes it with the surrounding air much more thoroughly as shown in figure 22.

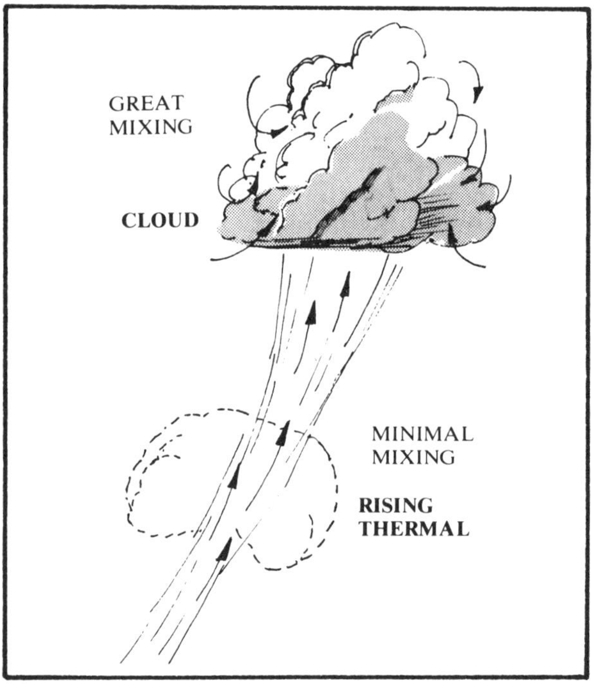

Figure 22 - Mixing of Clouds With Their Surroundings

A single isolated puffy (cumulus) cloud typically lasts only ½ hour from the initial angel's hair wisps until it dissolves in an amorphous mass. Even though the air may be filled with such clouds, they are being born, living a short life on high and dying in a continuous process.

Certainly there are times when the clouds do not dissipate quickly as any pilot or picnicker can tell. This occurs when the surrounding air at the level of cloud formation is itself moist so the clouds do not dry out even though they mix. In fact, in many instances the continuous transport of moisture aloft by thermal currents will result in a gradual or rapid spreading of the clouds to cover most or all of the sky. This condition is known as *overdevelopment* or OD in the abbreviated lingo of pilots. OD conditions are generally frowned upon by soaring pilots for such spreading clouds block the sun and tend to shut off lift.

Continuous thermal currents feeding a cloud can prolong a cloud's life beyond its allotted 30 minutes. In fact thunderstorms (see Chatper XI) are simply overgrown thermal generated clouds. They can last for many hours.

Layer type clouds also tend to last for hours or days for there is no surrounding dry air to mix with and in stable conditions there is nothing to induce such mixing. Layer clouds generally dissipate when the lifting force (front or pressure system) moves on or wanes.

OLD CLOUDS

Old clouds don't die, they just fade away. This disappearing act has special significance for soaring pilots for one trick these pilots use is to fly under active clouds and ride the thermal updrafts to new heights, new horizons. It is important for this reason to be able to distinguish newer clouds from older clouds.

A fading cloud is a drying cloud. In this drying process the smaller particles disappear first. This can change the appearance of the cloud, for different size particles reflect light differently. Generally an older cloud will take on a more duller or yellowish hue compared to a new cloud. This is a subtle difference but one that can be detected with practice. Also older clouds tend to be softer at the edges than the younger clearly defined clouds.

RAIN

Rain can certainly spoil our fun, especially if we are aviators, but it is most necessary to the regeneration of life as we know it. Rain of course comes from clouds. Most clouds, however, do not produce rain. The reason for this is that the release of latent heat during the condensation process warms the water droplets and stops their uptake of more water molecules from the water vapor in the air. Thus an equilibrium condition is reached and the cloud goes through its life cycle without producing rain.

In order for precipitation to occur either the lifting process has to continue or more humid air has to be fed into the cloud or both. When these conditions arise ice crystals will grow by robbing vapor molecules from the water droplets, or water droplets themselves will grow by bumping into their neighbors as they scurry about in the confusion of cloud. Ultimately these crystals or drops fall as snow or rain as they grow so large that gravity pulls them down (see figure 23).

Rain drops vary in size from about 1/50 inch (1/20 cm) to about 1/5 inch (1/2 cm) in diameter. A drop that grows much larger than this will usually break apart as it falls due to the friction of the air. Rain can deplete a small cloud of moisture quickly, for it takes nearly 30 million cloud droplets averaging 1/2,500 inch in diameter to make one rain drop of 1/8 inch in diameter. This means that one rain drop can deplete two cubic feet (.057 cubic meters) of cloud droplets. Those thousands of drops raining on your parade can equally ruin a cloud that is not being regenerated.

When rain falls through the sky it tends to evaporate on the way down which cools the surrounding air. This fact along with the sheer drag of tons of

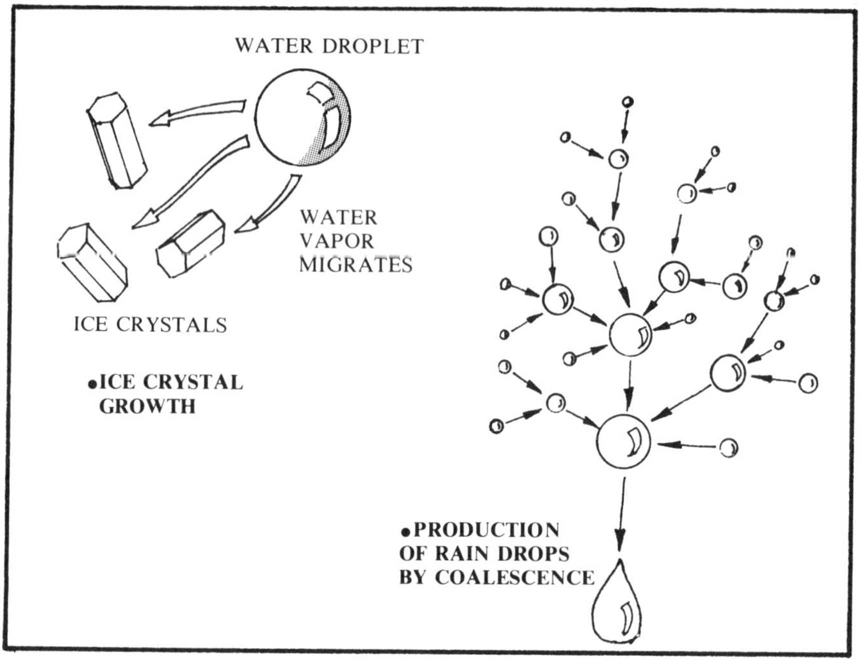

Figure 23 - Rain Formation

falling drops can create downdrafts that help kill thermal updraft production and the continued growth of a cloud. Also the spreading of rain on the ground can cool the surface, thereby stopping or subduing thermals. We cover these topics more thoroughly in the chapters on thermals and thunderstorms.

Rain can often be seen falling in vertical streaks known as *virga* (see figure 24). This usually happens only when isolated clouds are raining so that some sunshine is available to illuminate the falling rain. The rain falls

Figure 24 - Virga

Virga below a thunderstorm.

in streaks because a falling drop tends to entrain others by its disturbance of the air as shown in figure 25. Just as a race car experiences less drag when it drafts close behind another car, so too do following raindrops fall faster. They catch up with the leaders, coalesce, break apart and fall together in a mass we recognize as a vertical streak. Often these streaks dry out before they reach the ground. This does not mean that the falling air or downdraft has stopped. In fact, some of the most severe downdrafts can exist beneath drying rain for all the evaporation cools the air greatly and creates a dense cascade of localized falling air.

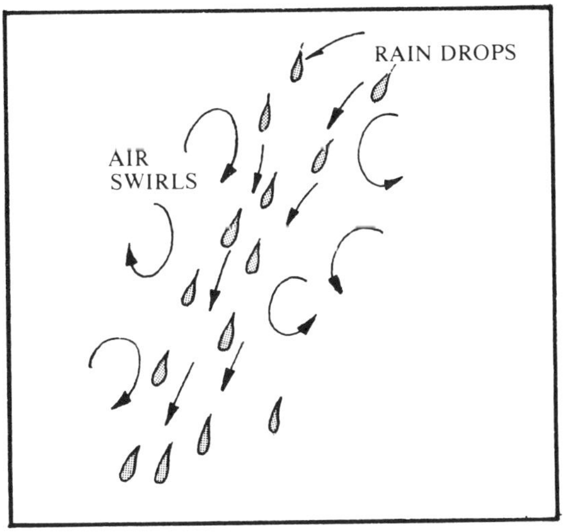

Figure 25 - Streaking of Rain

VISIBILITY

The ability to see where we are going is of utmost importance to pilots. The weather in general and clouds specifically can greatly alter the visibility or *viz* in pilot's parlance.

Fog can certainly reduce visibility to nil because the multitude of droplets scatters light so effectively. In a similar manner pollution products can scatter light or absorb certain wavelengths so that the air color changes from predominently blue to the browns, reds and yellows of the longer wavelengths. Pollution may give us crimson suns and kaleidescope sunsets, but it certainly ruins the vistas when we are flying and who knows how much it shortens our flying career?

Haze or moisture is a common foe of visibility. Water vapor is invisible, but on those hot and humid days of summer in areas that receive ample rainfall the air can be filled with droplets that have formed on condensation nuclei long before saturation humidity is reached. Such haze can be anything from a pale blue fade to the thickness of London fog. When the moisture haze is added to pollution we have the familiar condition known as smog (smoke and fog).

An inversion layer that limits the upward movement of the air can greatly reduce the viz by preventing vertical dispersion of the haze or smog. Stable conditions and light winds in general worsen a poor visibility situation because the production of water vapor, dust and pollution is more or less a continuous process that doesn't get carried away unless vertical currents or horizontal wind is active. Figure 26 illustrates the effects of an inversion layer or stability on visibility.

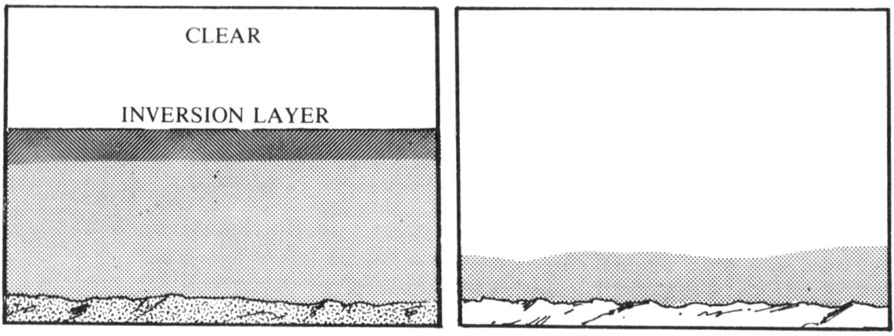

•INVERSION LAYER— POLLUTION OR HAZE IS CONCENTRATED NEAR INVERSION.

•STABLE CONDITIONS— POLLUTION OR HAZE IS CONCENTRATED NEAR GROUND.

Figure 26 - Visibility and Stability

CLOUD TYPES

Anyone with normal vision and slightly more sense than a box of rocks is aware that clouds come in many shapes, types and sizes. To the casual observer classifying clouds may seem like a hopeless muddle. However, the matter is really so simple when organized properly that you may become disillusioned with your grade school science teacher.

There are only two main types of clouds. These are *stratus* and *cumulus*. Stratus clouds are flat, layered clouds (think of flat or stratified) caused by the slow rise of widespread areas of air. These clouds cover broad areas of the sky and make the day gray. They often are found in stable conditions and are

Figure 27 - The Two General Cloud Types

normally caused by frontal lifting or the slowly rising air around large low pressure systems. Some low-level stratus clouds can be formed when low level turbulence mixes the air and raises it above the condensation level.

Cumulus clouds are piled or tumbled (think of *accumulated*) and look like giant cotton puffs or cauliflowers floating on high. These clouds are often found in good weather and when they cover 1/4 or less of the sky they are known as fair weather cumulus. Cumulus clouds are created by individual updrafts or convection currents carrying moist air aloft.

Here we emphasize the two main types of clouds as shown in figure 27.

Main Cloud Types

STRATUS—Layered widespread clouds with a fairly uniform base.
They often appear gray since they block the sun extensively.

CUMULUS—Separate clouds that have piled or rounded tops at various levels. These clouds can be very small or of great extent when they develop into thunderstorms.

Spreading stratus clouds with a few cumuliforms on the right.

Cumulus clouds. Note the presence of towering clouds and dissipating cloud in the top foreground.

ALTITUDE CLASSIFICATION

To provide more information we further distinguish clouds by their general altitude. *Cirrus* (meaning curl in Latin) are the highest clouds and consist of wisps or streaks of ice crystals at altitudes from 18,000 to 40,000 feet (6 to 13 km) in the temperate climates. Figure 28 shows classic "mare's tails" cirrus forms and indicates how ice particles falling from high wind layers into a lower velocity wind layer produces the wispy shape. Understanding this allows us to recognize the wind direction at the cloud's altitude.

We use the prefix *cirro-* to refer to stratus and cumulus clouds in the upper atmosphere as well. We use the prefix *alto-* (Latin meaning high like the singing voice) to refer to medium high clouds. No prefix is used when speaking of clouds below 7,000 feet (2 km). The chart below gives the general classification of clouds according to type and altitude in temperate climates:

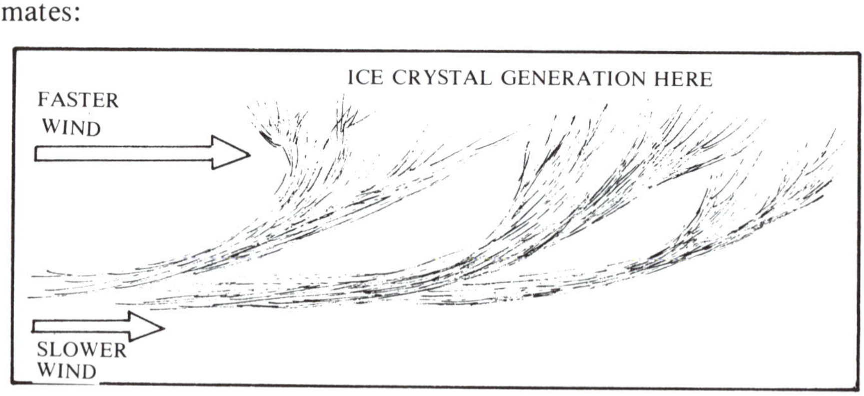

Figure 28 - Cirrus Clouds

Cloud Heights

Height	Level	Cloud types
40,000 ft 13 km	High Clouds	Cirrus Cirrocumulus Cirrostratus
18,000 ft 6 km	Medium High Clouds	Altocumulus Altostratus Nimbostratus Nimbocumulus
7,000 ft 2 km	Low Clouds	Cumulus Stratocumulus Stratus

The prefix *nimbo-* means a cloud from which rain is falling. These clouds may look like the others in their class except they are darker. Note that nimbocumulus clouds are also commonly known as cumulonimbus clouds. We have placed them in the medium high category but in reality they can be much lower and when it is a thunderstorm we are talking about it can have a base as low as 3,000 feet (1000 m) and tops up to 75,000 feet (25 km).

Stratocumulus clouds are often formed when cumulus clouds created by thermal currents reach an inversion layer which they cannot penetrate so they spread out into stratus layers. This is the overdevelopment situation and these clouds often have a somewhat lumpy bottom even though they are layer types. Figure 29 shows the different cloud types and their

Altostratus layers with altocumulus clouds along the edges. A layer of cirrostratus is seen in the upper right of the photo.

altitudes. The chart on the following page lists their characteristics as well as their abbreviation and the symbol by which they often appear on weather charts and reports.

Cloud Types and Characteristics

CLOUD NAME	ABBR. AND SYMBOL	HOW FORMED	HEIGHT (TEMPERATE ZONES)	APPEARANCE	RAIN
CIRRUS	Ci	Warm air lifting over colder air (warm front).	Usually above 25,000 ft-8 km	Thin wisps, delicate patches, Narrow bands, mare's tails	None
CIRRO-CUMULUS	Cc Ci-Cu	High lift above a warm front or wave-like action between layers.	20,000 to 25,000 ft 6 to 8 km	Wave-like or patchy Mackerel sky,thin sheet or layer cloud broken into small clumps or ripples.	None
CIRRO-STRATUS	Cs Ci-St	Formed in warmer air lifted over colder air (warm front).	20,000 to 25,000 ft 6 to 8 km	High sheets that appear thin and transparent. May produce halo around sun or moon.	None
ALTO-CUMULUS	Ac	High lifting of warm front or wave or slow overturning of layer.	Around 10,000 ft 3 km	Like cumulus puffs only higher and packed together in a a layer (due to auto-convection of lifted layer).	None
ALTO-STRATUS	As	Formed in a warm front or a cooled layer.	Around 10,000 ft 3 km	Pale sheets blurring the sun. May have occasional gray streaks. Does not cause Halo.	None
NIMBO-STRATUS	Ns	Formed from stratocumulus in warm front or in cooled layer.	Usually below 6,500 ft-2 km	Darker than stratus. May appear wet. Blots out sun. Often visible rain falling.	Steady rain
STRATO-CUMULUS	Sc St-Cu	Break-up of stratus due to a decreased stability; lifting in warm front; thermal clouds spreading out (overdevelopment).	Usually below 6,500 ft-2 km	Gray and dark cloud spread out in puffy layer. Often some blue sky. Rounded masses or rolls.	None
STRATUS	St	Warm front lifting or cooling of air layer.	Below 6,500	Gray, low sheet covering a large area with a fairly uniform base.	Occa-sional drizzle
CUMULUS	Cu	Localized lifting convection currents (thermals).	Typically 2,000 to 14,000 ft 600 to 4,000 m but may reach 22,000 ft in high mountains	Like puffs of cotton or wool. Separate clouds with cauli-flower type tops.	None
CUMULO-NIMBUS (Nim-bocumulus)	Cb Cu-Nb CuNim	Lifting of unstable or humid air over mountains or due to cold front passage. Also excessive build-up of thermal currents.	Surface to 75,000 ft 25 km	Dark and towering with large billows. The upper reaches often exhibit smooth areas which may form an anvil head.	Heavy to violent

Altostratus clouds with cumulus on the right.

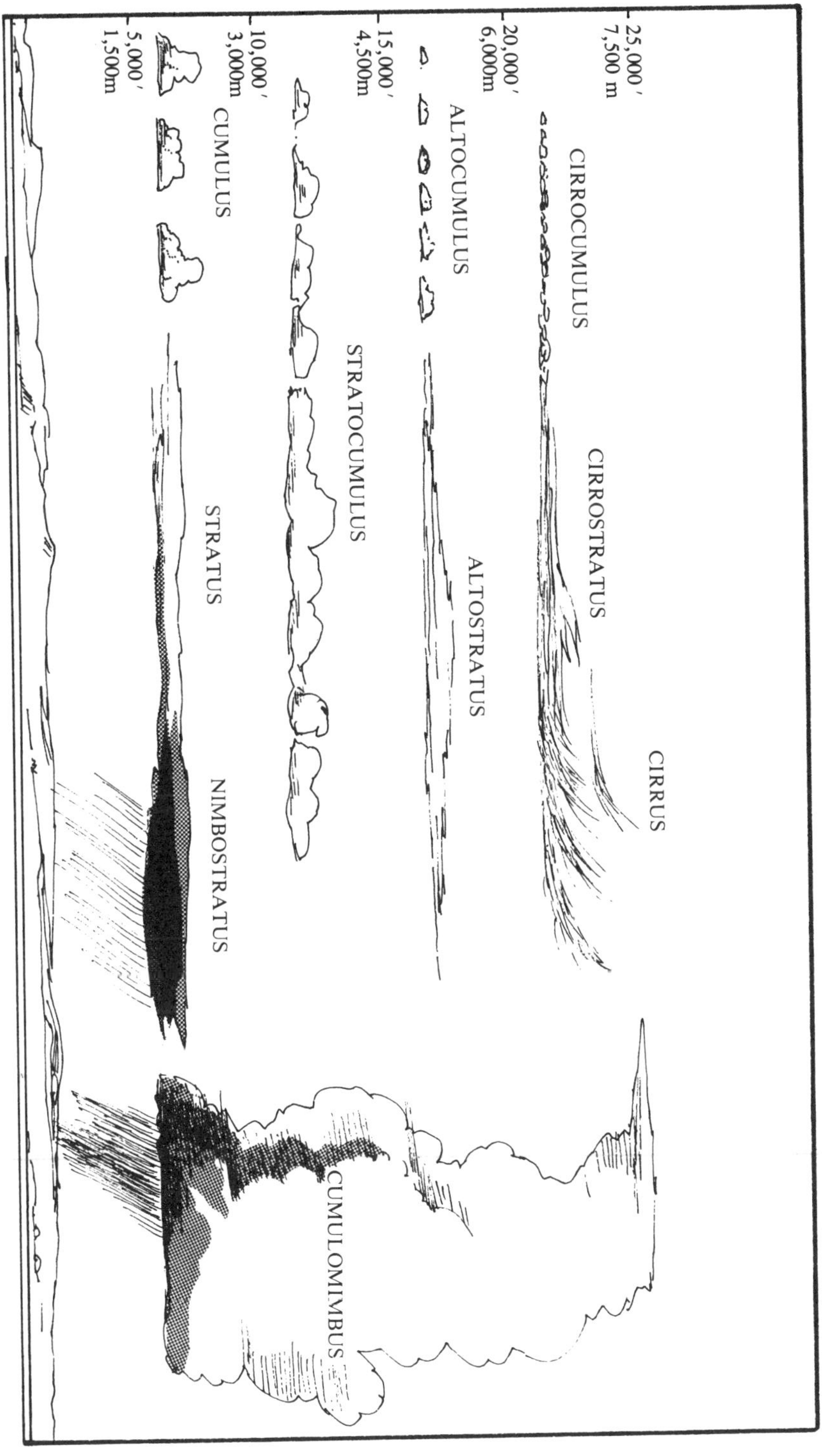

Figure 29 - Cloud Types and Altitudes

SEASONAL AND LOCATION VARIATIONS

Cloud heights and types vary with the season and latitude. Since we now know how clouds are formed we can readily figure this out. When the air is colder as it is in Polar regions or in winter, the relative humidity is higher, and the air is more nearly saturated. Cold winter air may feel dryer than summer air in the house, but outside the air is usually much nearer to its saturation point in winter. As a result any lifting creates clouds sooner and bases are lower.

In addition, greater heating of the earth's surface produces much more vertical convection causing condensation and rain which tend to remove water from the atmosphere thereby raising saturation points. As a result, the clouds tend to be much higher at the equator than the poles and clouds tend to be cumulo-type at the equator and stratus-type at the poles. Likewise clouds tend to be higher and more cumulo-type in summer and more stratus-type and lower in winter. Figure 30 shows cloud height variation over the earth with cloud symbols indicated.

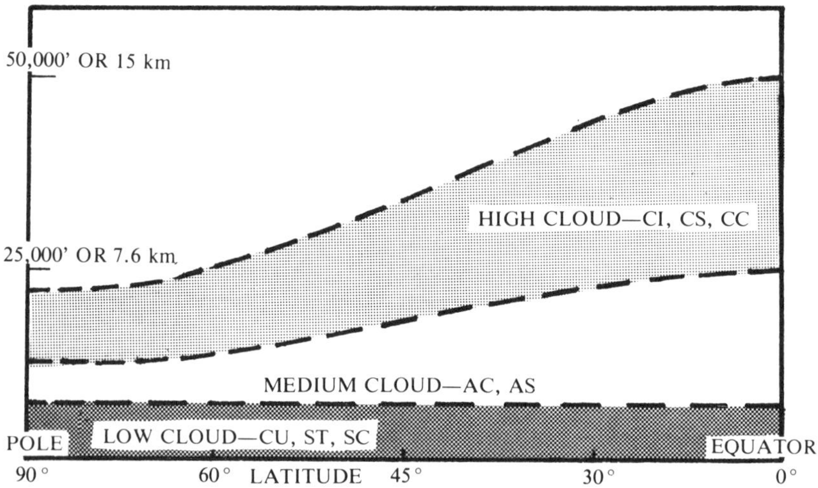

Figure 30 - Cloud Heights and Latitude

LESSER KNOWN CLOUDS

Cloud watching isn't as exciting as bird watching (of either type) perhaps, but there are enough different forms of clouds to provide interest to pilots, especially since they can give us further information about conditions. We'll list the various clouds and give their description as well as what they portend.

FOG—This cloud form is well-known and needs no description. It is found when warm, moist air from sea moves over the land (advection fog), or when the land radiates heat at night to cool a moist layer lying above it (radiation fog). Here are a few old-timey sayings that tend to be true about fog:

A summer fog for fair, a winter fog for rain,
A fact known everywhere, in valley and on plain.

Stratocumulus clouds on an unsettled day.

Often a fog burning off in the morning indicates a good thermal day.
When fog goes up, the rain is o'er,
When fog comes down, 'twill rain some more.
Evening fog will not burn soon,
Morning fog will burn 'fore noon.
and:
Fog that starts before the night will last beyond the morning light.

CAP OR CREST CLOUD—A cap cloud is formed over the top of a mountain when air is lifted as a general wind strikes the mountain or when upslope breezes due to heating slip up the mountain's sides and over the top to reach condensation level (see figure 31). A cap cloud often begins as a thin wisp over the mountain in the late morning and grows until late afternoon. These clouds may readily reach thunderstorm proportions in very unstable air.

The cloud bases of cap clouds often lower as the day continues (unlike

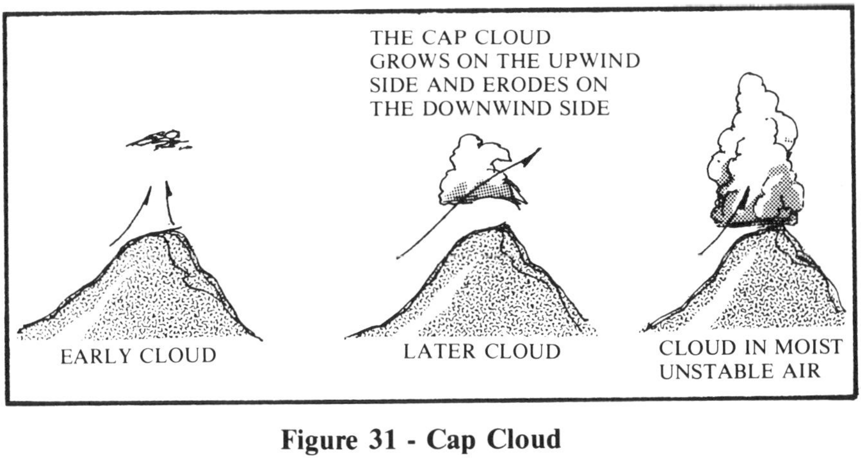

Figure 31 - Cap Cloud

cumulus clouds created by thermal currents whose bases often rises as moist air near the ground dries out) due to a continued supply of humid air, especially in seaside mountains. This lowered base can obscure the mountain and put a damper on soaring. Tropical islands with all their allure of warm breezes, warm seas and warm natives have one major drawback for pilots: they usually cloud up and rain by early afternoon.

A cap cloud is continually formed on its upwind side and continually erodes downwind. Thus it stays put over the mountain top and doesn't drift with the wind. For this reason it is not a good indicator of wind velocity although it may lean in the direction of the wind aloft. Also by carefully watching it build and erode you may be able to tell the wind direction at cloud height.

A cap cloud over Mt. Sherman in the Mosquito range of Colorado.

BANNER CLOUD—Another type of cloud that forms on a mountain crest is a banner cloud as shown in figure 32. Here rotor air blowing upslope on the downwind side of the mountain as well as drifting snow combine to create a cloud that streams out downwind from the crest. This type of cloud usually foretells high winds at the cloud level and should serve as a warning to sport aviation enthusiasts.

ROTOR CLOUD—When winds blow across a mountain chain or ridge it may form a roll of air downwind from the mountain like a tornado on its side. This is known as a rotor and is often found beneath waves. This cloud tends to stay in the same place and indicates very serious turbulence. The cloud is formed on the upmoving side of the rotor and erodes on the downward side. Rotor clouds are pictured in figure 33 along with wave clouds.

WAVE OR LENTICULAR CLOUDS—These clouds are known affectionately as lennies for their cross-section in very lens-like as shown in figure 33. They form when the air undergoes up and down undulations (waves)

Figure 32 - Banner Cloud

caused by the air blowing over hills or mountains. Wave clouds are also more or less stationary since they grow in the upward portion of the wave and erode in the downward portion. These clouds are oriented perpendicular to the wind but do not give much information about the wind velocity (since they are stationary) other than indicate at least a 15 mph (24 km/h) wind. We deal with waves in Chapter VIII.

LEE SIDE CLOUDS—In very humid conditions—often when it is raining or clouds are near a mountain top in layers—small broken wisps will appear on the downwind side of the mountain. This is caused by the upslope drift due to the rotor against the mountain. Lee side clouds are indicators of the wind's direction.

BILLOW CLOUDS—Sometimes clouds that appear like long ripples in water will show up high in the sky. These are billow clouds and they are formed when one air layer (warm) moves over another with enough velocity to create close waves just like in water. What distinguishes billow clouds

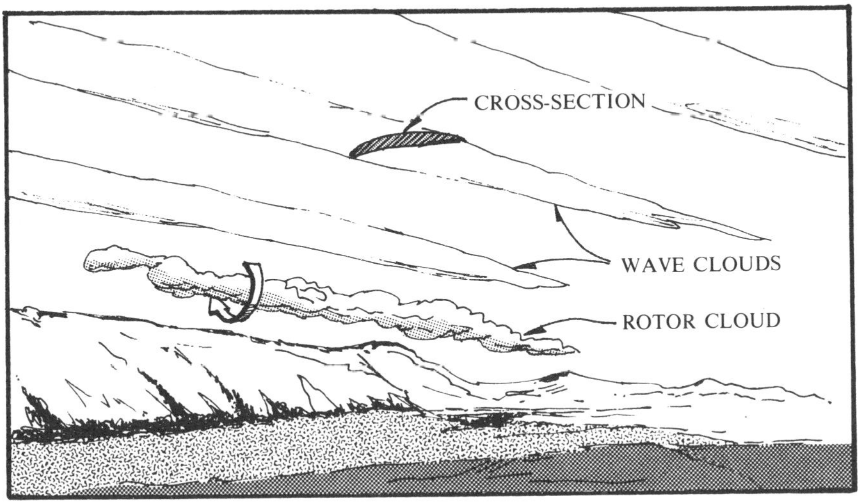

Figure 33 - Lenticular Wave Clouds

Long wave clouds. Note the double layers in the large cloud.

from wave clouds is their spacing. Wave clouds are much further apart, are often stacked one atop another and are frequently lower than billow clouds. Billow clouds move with the wind as shown in figure 34, but do not drift as fast as the wind in the upper air mass. Billow clouds often foretell a change in weather as they are frequently formed by an approaching warm front. When the billowing process produces long, close rolls of clouds they are called *Roll Clouds*.

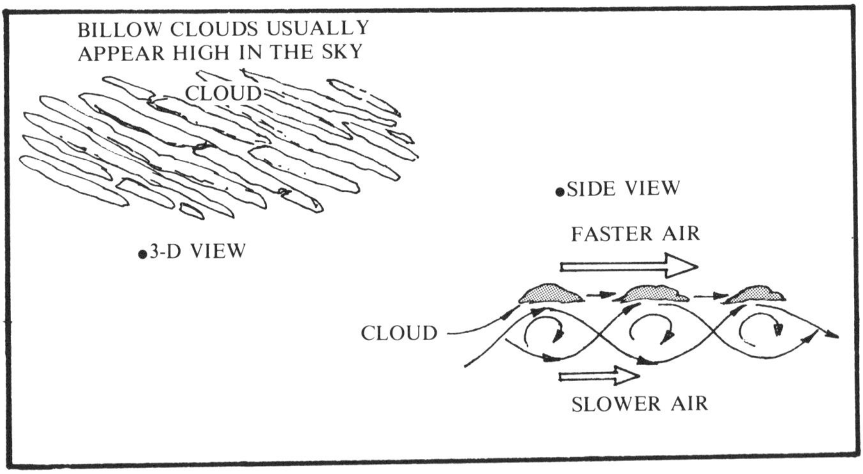

Figure 34 - Billow Clouds

MAMMATA CLOUDS—These clouds often appear under the shelfs of thunderstorms and look like udders hanging below the cloud. They are so named because they are someone's idea of mammaries. Mammata clouds indicate slight downdrafts below the cloud as they entrain cloudy air downward before it can evaporate.

Billow clouds extending from a band of stratus.

Mammata clouds and a rainbow in front of a thunderstorm.

PILEUS CLOUD—The top of a towering thunderstorm often pushes the air above it upward as it climbs rapidly to high altitudes. This pushed up air will often form a cloud that looks like a veil over the thunderstorm top as shown in figure 35. This is the pileus cloud and its real significance to us is that it indicates lift above the thermal cloud. A pileus cloud is one of the highest clouds and as such has served as a name for various sport aircraft.

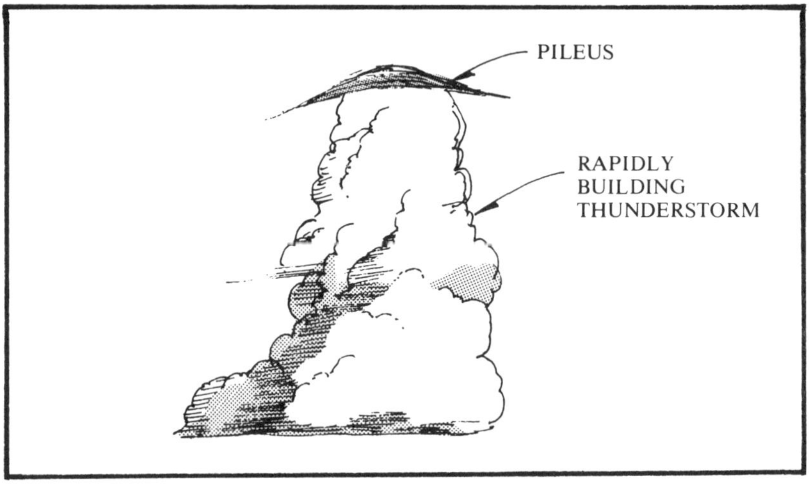

Figure 35 - Pileus Cloud

We can also identify a few other types of clouds based on their peculiar shape. These are *Altocumulus Lenticularis* which are simply high wave clouds that show some cumulus activity due to instability created by the wave lifting. *Altocumulus Castellanus* are cumulus type clouds connected in rows or spread out groups with high tunel-like structures. These clouds often portend thundery weather. *Fracto-Stratus or Fracto-Cumulus* are simply ragged clouds of a given general type (stratus or cumulus). They may be broken apart by high winds or irregular vertical motions.

Fractocumulus in strong winds. The tilt of the distant clouds indicate the wind direction.

WHAT THEY TELL US

Clouds are up there where we want to be. Because they are in the air environment they can tell us what the air is doing. From this we can often discern what the current conditions are as well as what is going to happen.

WIND VELOCITY

Clouds can generally tell us the wind velocity (speed and direction) at their height. However, as we just learned, certain types do not drift with the wind (cap, banner, rotor and wave clouds). Furthermore, stratus clouds will not demonstrate any drift if they are so undifferentiated or widespread that we cannot see them move. Another problem arises when cumulus type clouds are growing so rapidly that even their upwind edges appear to move outward. Finally, cumulus clouds formed on thermals arriving at altitude with a slow horizontal velocity acquired below may actually drift slower than the surrounding air due to the inertia of the thermal air (their mass can measure into the thousands of tons).

Given all the above exceptions, we still can get a very good idea of the wind velocity by watching the cloud drift. The best way to do this is to stand next to a building or tree and compare the cloud's position as it moves in time. Figure 36 shows a cumulus cloud in various wind velocities. If you are flying, there is no way of separating the cloud's drift from your own, but you can still observe the cloud drift and hence the wind velocity by watching the movement of the cloud *shadows* along the ground.

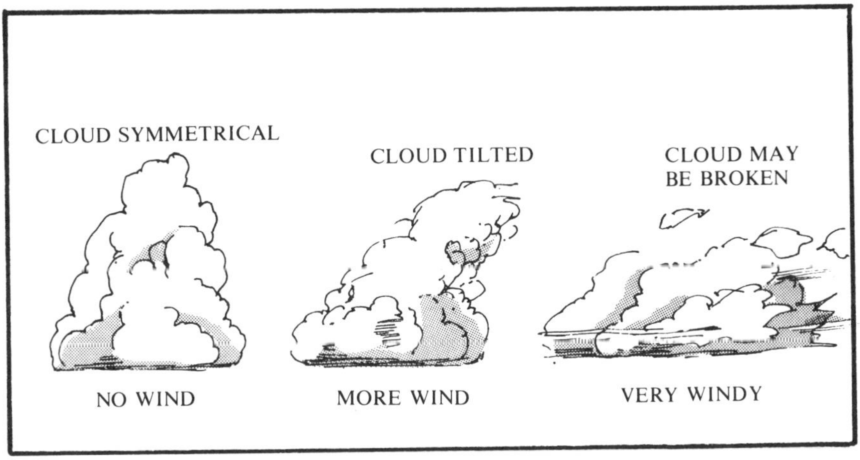

Figure 36 - Cumulus Clouds in Wind

When two or more layers of clouds exist there is often a parallax problem whereby their relative motion makes the high clouds look like they are moving backwards or drifting slower than they really are. This is illustrated in figure 37. Here the upper level clouds appear to be moving slower because they are much further from us. The way to overcome this problem is to use a building or tree to hold your eye in one place as mentioned above.

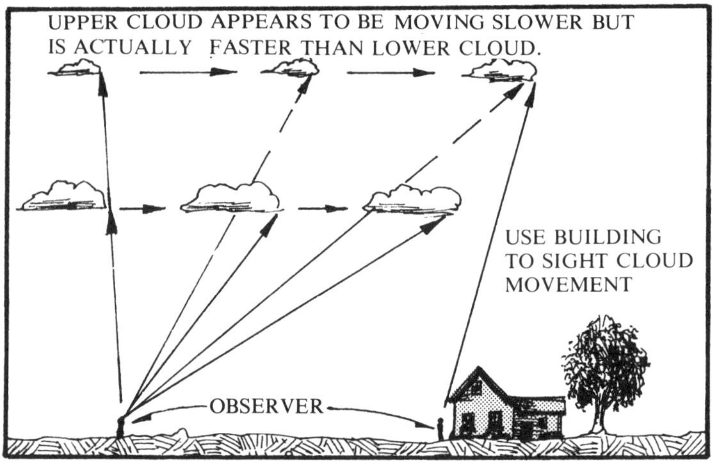

Figure 37 - Judging Wind Speed at Different Levels

WEATHER INDICATIONS

Clouds can give us an idea of what weather to expect. Lower clouds give us hints of what's in store several minutes to hours in the future, while higher clouds can predict what's happening in hours or days ahead. Every cloud has a message of some sort and it's not always about the wind.

Cumulus clouds that flatten out into a layer indicate that the air has stopped rising upon reaching a temperature inversion (warmer air). Such an inversion is often associated with an approaching high pressure system. This indicates clear weather ahead for the air sinks slowly in a high pressure system which clears out all clouds except isolated cumulus created by thermals able to rise through the sinking air.

Bands of clouds are frequently seen in the sky. They can be anything from different thicknesses of stratus layers to a line of towering cumulus to cloud streets. The latter is dealt with in Chapter X. A line of towering cumulus often preceded a quickly advancing cold front with all the attendant changes brought by the advancing cold mass. Great turbulence and a wind direction change should be expected.

High bands of clouds are often good indicators of future weather. Long bands of high cirrus associated with the jet stream parallel this high river of wind (see Chatper V) and often foretell the future direction of the surface wind. When upper level clouds move in a direction markedly different from lower winds (say 90° or more) it generally means the wind on the surface is going to change—usually to that of the upper level. If the band of clouds in the jet stream is stationary, the weather is unlikely to change for the next twelve hours or so.

If high cloud is moving away and the sky is clearing, the system that created the cloud has probably passed and better weather is on the way. On the other hand, if a layer of stratocumulus cloud is approaching with little or no surface wind the line of the advance indicates the upper wind direction with the surface wind blowing beneath it at about 20° to the left or counterclockwise as we go down (clockwise in the southern hemisphere).

When bands of cirrus in its various forms are followed by thickening clouds and lowering clouds, there's a good chance that a warm front is on the way and will be in the area within 24 hours. The speed, strength and attendant severity of the front can be predicted by noting how fast the clouds increase and are moving.

There are other causes of cloud bands in the air such as thermal rolls along a ridge or mountain chain as well as a long area of convergence. In general, moving bands of clouds are the important signs of changing weather.

SIGNS OF LIFT

Soaring pilots are always looking for lift and besides other gliders and birds climbing as well as dust devils, cumulus clouds are a glider pilot's friend. Of course, the cumulus cloud must be of the lower variety or they won't be based on ground thermals and are thus not readily usable. Even when cumulus clouds are thousands of feet above a pilot it often pays to move under them for thermals tend to feed in multiples and the whole extent of the lift can be rather spread out and reach to the ground.

Clouds based on wave lift and convergence lift are also good indicators. I have witnessed several competition pilots cross a five mile valley to get under a flat, long convergence cloud. They were rewarded with a flight in solid lift that continued for tens of miles.

When extending oneself by gliding long distances to find lift under a cloud it is obviously important to know what type of cloud it is and how active it is. In general, less active clouds tend to be flatter although they also tend to have less sink around them. In Chapter IX we explore the nature of thermal clouds in more detail.

SIGNS OF TURBULENCE

Turbulence is a mixing of the air. Because a cloud is borne on the air it can often indicate what amount of turbulence we should expect. Figure 38 shows various clouds in turbulent conditions.

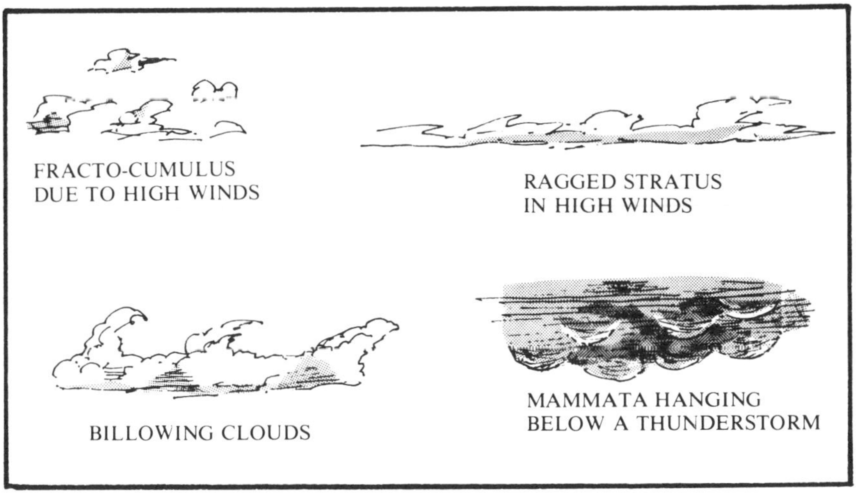

Figure 38 - Cloud Signs of Turbulence

Cumulus clouds indicate turbulence caused by thermals. The amount of cloud boiling in a cumulus head and how ragged it is help determine how strong the lift and how strong the wind is. Strong conditions generally relate to strong turbulence. Any time clouds are torn apart we should expect turbulence.

Other clouds such as rotor clouds and cumulonimbus thunderheads are signs of *extreme* turbulence. Stratus clouds in general are signs of gentle conditions, but when they are formed where two different layers of air are mixing or shearing, they can indicate turbulence. Stratocumulus clouds caused by mixing or overturning layers should be expected to be turbulent. Watch the details and fine edges of clouds for signs of turbulence.

RAIN SIGNS

Clouds can take on many hues, depending on how they pass the sunlight. When it is illuminated low a cloud can produce awe-inspiring sunsets. This lends us the old adage which holds up to modern science: "Red sky at night, sailor's delight; red sky in the morning, sailors take warning."

The darkness of a cloud is often dependent on how it is illuminated—when the sun is in back of it, it absorbs the light and appears darker. However, in general we can tell a cloud's load of moisture and thus how apt it is to produce rain by noting how dark it is, especially at the base. The darker a cloud becomes, the more likely we are to encounter drops from heaven.

Clouds Tell Us...

Wind velocity—Note isolated cloud drift, lean and raggedness.
Weather—Note changing wind directions at different levels and changing types of clouds. Cloud bands often indicate changes.
Lift—Note the presence of cumulus, wave or convergence clouds.
Turbulence—Note cloud types and raggedness
Rain—Note cloud shading and changes in shade as well as build-up in size.

FLYING IN CLOUDS

This book is not a book about how to fly. However, certain characteristics of clouds that we have learned here should be pointed out to all pilots contemplating entering their depths. When clouds form they release latent heat. This makes them more unstable which often results in greater turbulence. Taking a flight on a small airplane will demonstrate the difference in turbulence above, in or below the clouds. In the clouds is usually the roughest with below next. We should mention the fact that turbulence associated with rotor and thunderstorm clouds can tear an airplane apart.

A greater danger than turbulence in clouds is disorientation or vertigo. Because the normal visual references are not available in clouds, the eyes and sense of balance are not in agreement and total spatial disorientation can occur. The only way clouds can be safely flown on a continuous basis is by an instrument rated pilot with a minimum of a turn and bank indicator and a compass (preferably a gyro type since magnetic compasses are not accurate in turns).

SUMMARY

Flying above a cloud with the sun at your side produces a strange shadow of you and your craft on the cloud surrounded by a bright halo tinted in rainbow colors. Pilots know this spectacular sight as a "glory." It is caused by the normal shadow process and the reflection and refraction of light by the cloud.

The glory is but one of the visual beauties that clouds offer uniquely to pilots. There are also billows and canyons to play in, light pillars and silver linings to fill our eyes. All of this and more is brought to us by clouds in the sky.

Clouds also tell us much about the nature of the sky for they are offspring of moisture originating at ground level and carried aloft by many processes. They are altered in character by the surrounding air in which they pass their life cycle. It seems that every pilot should have a more than average interest in clouds for by their nature they predict how joyful a flight will be and they can greatly alter the tenor of that flight for better or for worse.

We cannot really touch the clouds but we can learn their lessons and join them for a brief spell in the great expanse of sky.

Light and varied winds aloft create confused cirrus patterns and generally mean fair weather.

CHAPTER IV

The Big Picture—General Meteorology

When the public thinks of the weather they think of the TV weather personality, a sunny day for the weekend or perhaps that winter storm the paper predicts. They certainly aren't intrigued by an undulating jet stream or a front wafting across the country, although their interest may be piqued by the occasional hurricane and a tornado or two. But pilots need more than an armchair understanding of general meteorology in order to predict changes in conditions and expected lift patterns.

Meteorology is just a fancy name for weather. We can think of general meteorology as weather changes that take place on the scale of hundreds of miles (or kilometers) and may require a day or days to manifest themselves. By comparison, micrometeorology—small-scale effects—occurs on the order of tens of miles or less (80 km or less) and lasts less than a day. While most of our study of the air's behavior concerns micrometeorology, in this chapter we explore general meteorology.

Of course it takes many heavy books to completely cover the range of general weather, so we won't presume to create consummate weathermen with one simple chapter. However, we can readily provide a summary of the necessary ideas so that a pilot can understand the processes pertinent to flight. Note: in Chapter XII we cover sources of weather information and the art of making predictions with the success of a weatherperson.

MOVEMENT OF THE ATMOSPHERE

From space our humble planet looks like a big blue billiard ball. The sun smiles down on it from nearly 93,000,000 miles (148,800,000 km) away. Yet through this considerable void of space much warmth is transmitted to heat the earth's surface. The equatorial areas of the earth garner most of the sun's warming radiation simply because these areas face the sun more directly. The warm tropical areas heat the atmosphere that lies above them while the cold polar areas cool the overlying air.

As we have seen previously, relatively warm air rises as convection currents while cool air tends to settle. As a result we should expect a general circulation about the earth as shown in figure 39. Note that the rising air is centered at the point of the sun's most direct rays which may range from 23.5° north latitude to 23.5° south, depending on the time of year.

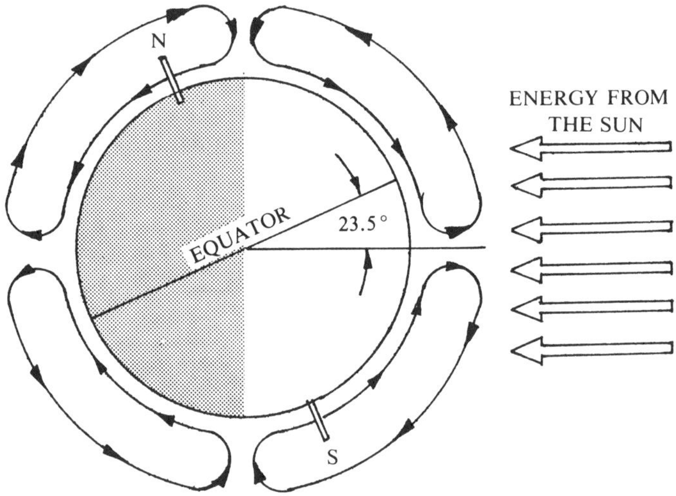

Figure 39 - General Circulation

In the real world, however, the circulation of the atmosphere is not so simple because of Coriolis effect. As we learned in Chapter II, the Coriolis effect causes all free flowing or falling objects to tend to turn right in the northern hemisphere and left in the southern hemisphere. As a result, the circulation of air is interrupted as shown in figure 40.

A great volume of air rises in the equatorial regions then moves either north or south aloft due to the accumulation of air in the upper levels. Coriolis effect is zero at the equator and strongest at the poles, so at first it doesn't alter this poleward flowing air, but by the time the air has covered about 30° in latitude it has been turned by 90° and again accumulates aloft. Here some of the air sinks to the surface to separate and flow north and south, and some of it renews its poleward journey aloft. The air moving south at the surface is again turned to the right (in the northern hemisphere) to create the westward flowing trade winds. The air moving north also turns right to create the prevailing westerlies that flow eastward in the middle latitudes (see the figure).

The air aloft continuing north eventually loses heat through radiation, contracts and sinks in the polar regions. This air moves down over the polar caps cooling further, turning right to form the polar easterlies, and progressing to meet the westerlies in conflict at about 60° latitude. Being much lighter, the warm and often moist air in the westerlies is forced up over the cold polar easterlies and helps to feed the supply of air at the poles.

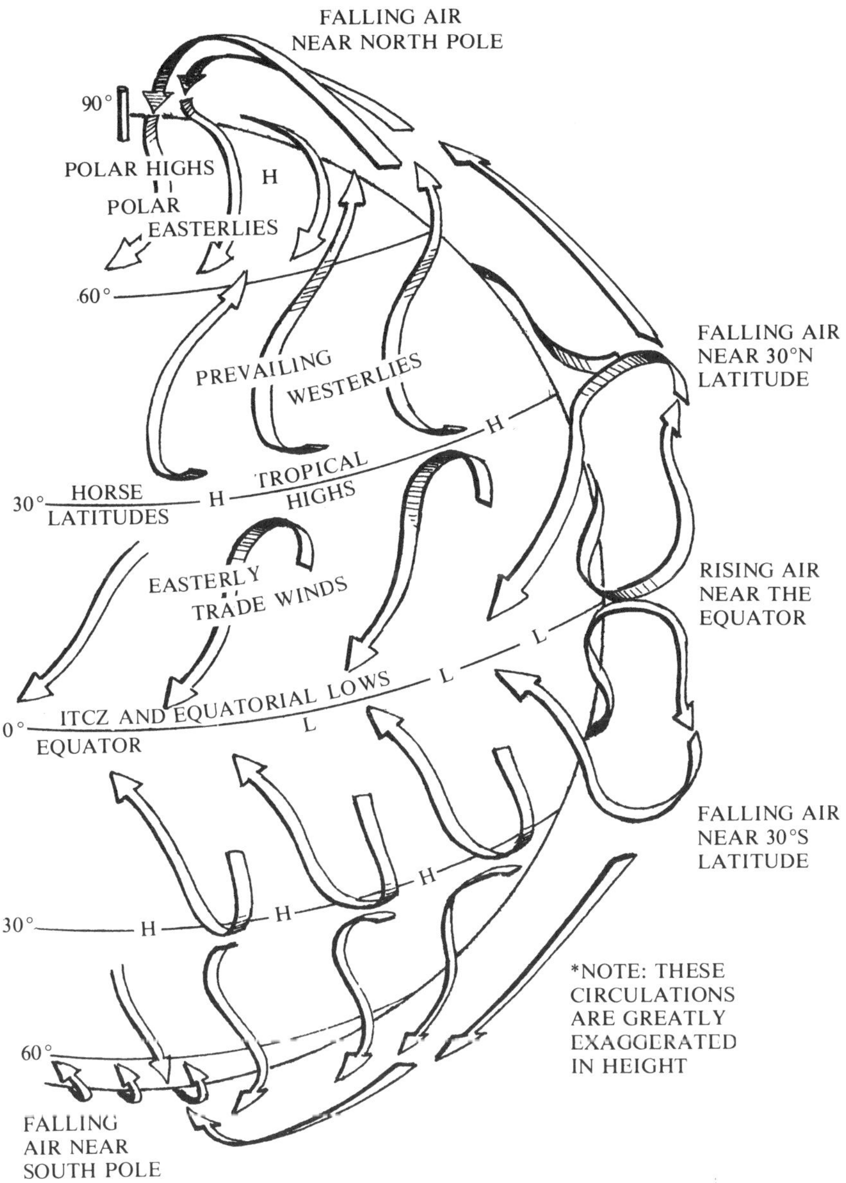

Figure 40 - The Circulation Model

The continual pile-up of air and thus pressure at the polar caps causes sudden random outbreaks of "polar waves" moving southward in the northern hemisphere (and northward in the southern hemisphere) as far as 25° latitude in extreme cases. This breakout relieves the polar high pressure and brings cool air well into the warmer climes.

Near the position of 30° latitude not much wind is moving near the surface. A band of high pressure known as the *tropical highs* circles the globe at this level in both hemispheres. This area is called the horse latitudes because early sailors would find themselves becalmed in the tropics and have to lighten their loads considerably. Overboard went ballast, personal effects and especially such superfluous things like horses brought to colonize the New World. The lightened ship would then hopefully respond to the soft breezes leaving the floundering horses behind to name a region of quiet sea.

Near the equator a band of low pressure exits known as the equatorial lows. This area is also called the *Intercontinental Convergence Zone* (ITCZ). Again not much surface wind blows here because it mostly moves vertically. The ITCZ is well-known as the doldrums, so named by long ago sailors bored with sitting and watching the horizon for signs of winds day after day.

One glimse at a satellite view of weather patterns on earth indicates that the above discussion is also a simplification. However, this model serves as a reasonable starting point and illustrates well the continuous shoving match that occurs between the polar highs with cold air masses and the tropical highs with warm air masses that is the cause of constantly changing weather in the mid-latitudes or temperate zone.

AIR MASSES

When a body of air—usually 1,000 miles (1600 km) or more across—has assumed a uniform temperature and moisture content, we call it an *air mass*. The temperature and humidity of an air mass determine its nature. It may be cold if it originates in the polar regions or warm if it comes from the tropics wearing sunglasses. If it comes from the sea and has picked up moisture it is known as a maritime air mass. If it comes from the land and is dry it is known as a continental air mass. Figure 41 shows the four types of air masses as listed in the chart below and their typical paths across North America and Europe. Note that in Australia the air masses are mainly maritime whether polar or tropical.

Air Mass Types

Continental Polar—Cold and dry. Originating over land.
Maritime Polar—Cold and humid. Coming from the sea.
Continental Tropical—Warm and dry. Originating over land.
Maritime Tropical—Warm and Humid. Coming from the sea.

We will spend some time discussing the origin of air masses for their background greatly affects the weather they bring us. For example, a moist (maritime) air mass tends to be lighter since water vapor is lighter than dry air. A relatively moist mass will tend to ride over another mass it encounters of similar temperature. Also, if a saturated air mass is lifted by riding over a mountain chain or colder mass, cooling may occur to the point where the water vapor condenses to form clouds and possible rain. This condensation process releases latent heat into the air and thus the air may be warmed considerably if it continues over the mountains to be compressed at lower altitudes (see Appendix III). This action accounts for the vast strip of bone-dry territory that exists east of the coastal ranges from North to South America as well as northern Africa.

A cold air mass barreling down from the north generally passes over warmer ground which heats it from below causing instability as shown in figure 42. On the other hand, a warm air mass sliding north into cooler areas often becomes more stable as the bottom layers are cooled as shown in the figure. The latter statement is not a firm rule; in many cases warm air masses bring with them and take up so much humidity that they become unstable and may even exhibit thunderstorms.

In figure 43 we show a side view of the atmosphere from the pole to the equator demonstrating how air mass movement invades the mid-latitudes both from the north and south. This idealized cross-section should be related to figure 40. Note the rising of the tropopause toward the equator

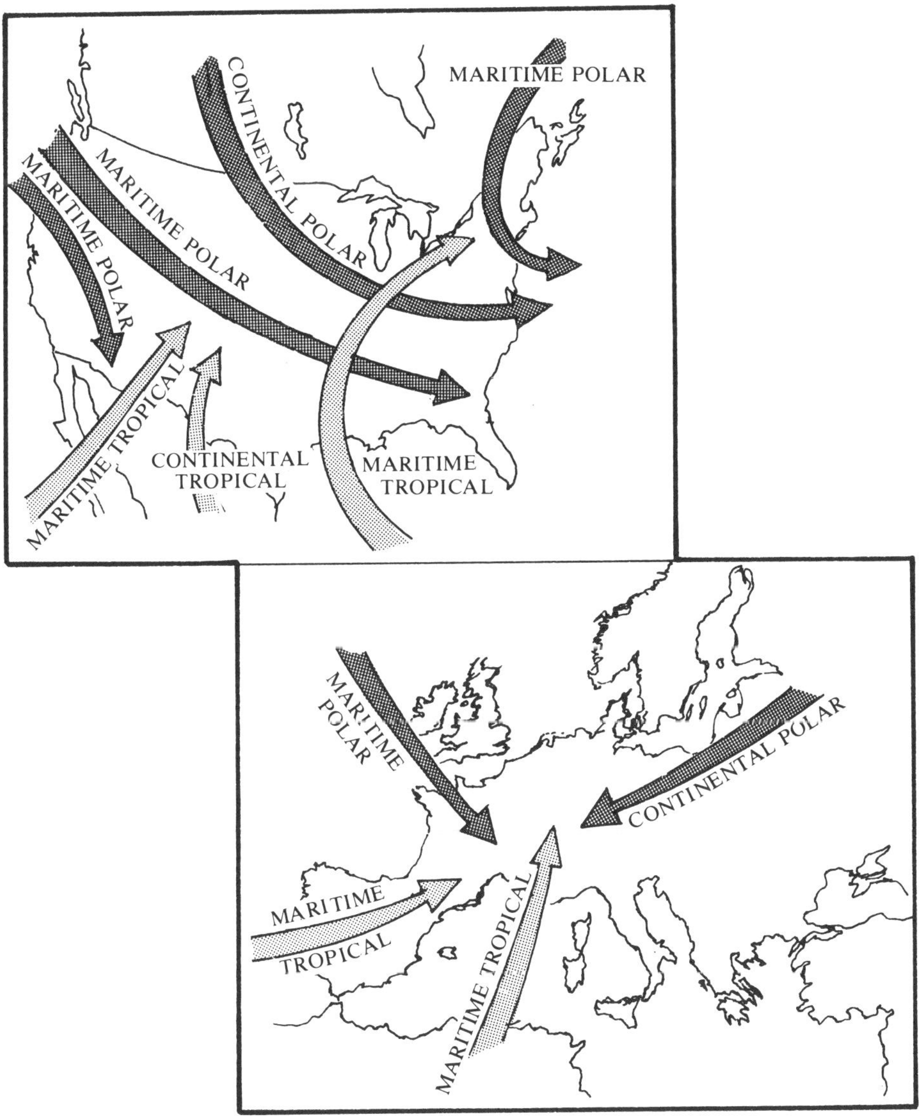

Figure 41 - Air Mass Movement

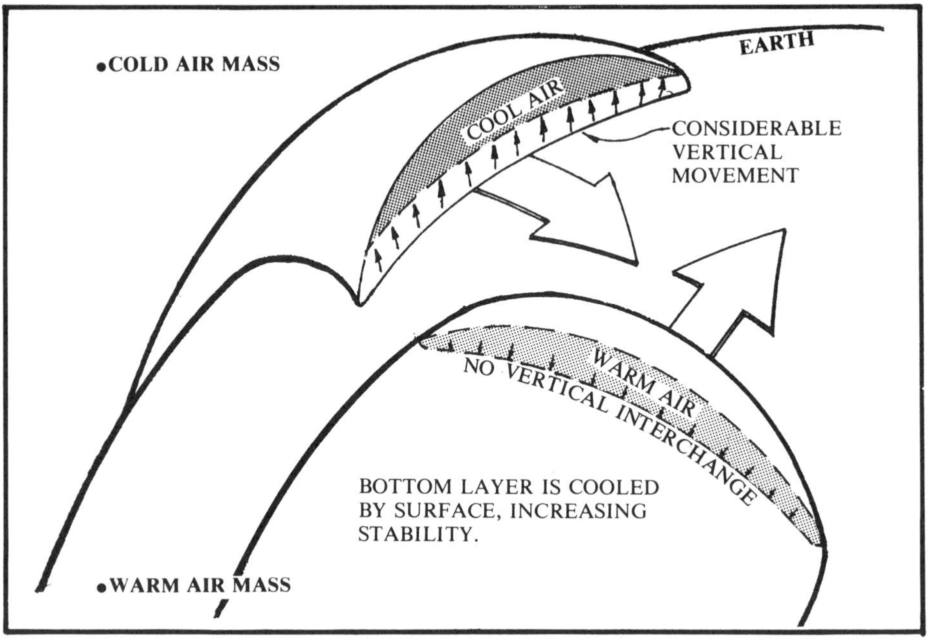

Figure 42 - Air Mass Stability

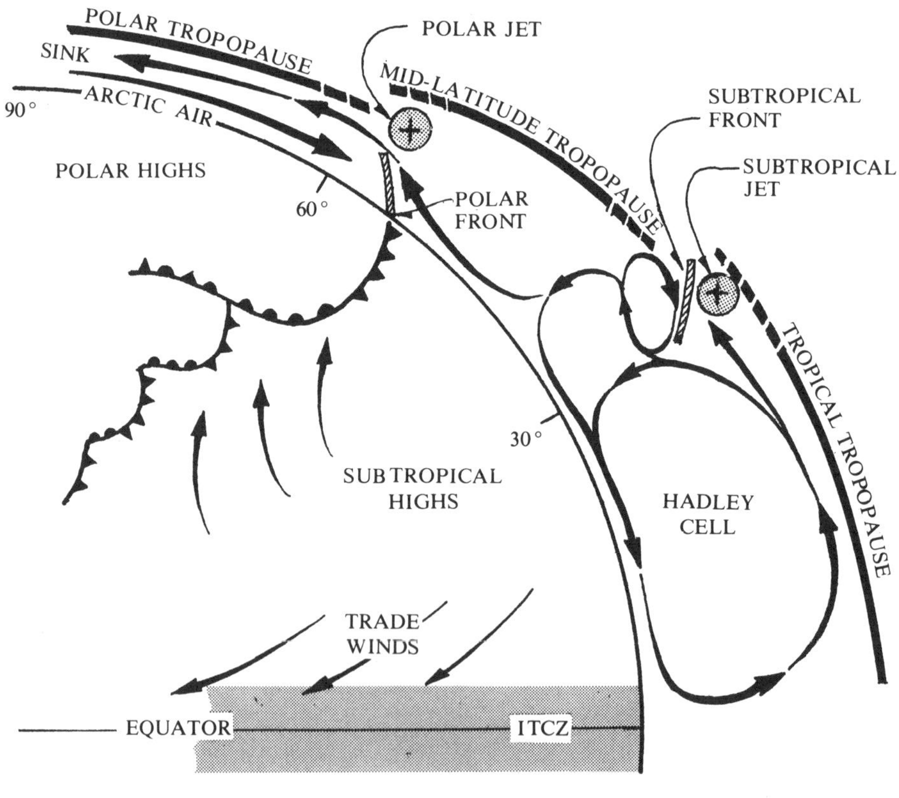

Figure 43 - The Lower Atmosphere in Cross Section

and the *Intertropical Convergence Zone* (ITCZ) where much of the circulation arises. Also indicated are fronts and the jet streams that flow west to east (into the paper) which we discuss in the next chapter.

THE MEANING OF FRONTS

Most of us have seen fronts on the TV weather marching across the country. These fronts herald changes in weather and usually are accompanied by a spell of clouds and rain. We pilots are most interested in the behavior of fronts for they are important predictors of good and bad flying weather.

A front is simply the boundary between a cold and a warm air mass as shown in figure 44. If the colder air is advancing, the front is known as a *cold front*. If the warmer air advances it is a *warm front*. Figure 45 shows how fronts are depicted on a surface weather map.

Sometimes an air mass moves in then stops when pressure in front of it builds up. The frontal boundary then is known as a *stationary front* and is shown in the figure. The most important difference across a frontal boundary that determines where the front lies is density which is related to the temperature (a pressure change also occurs). Air masses of different densities don't mix readily in a similar manner to oil and water. Consequently a stationary front can maintain its identity and persist for days.

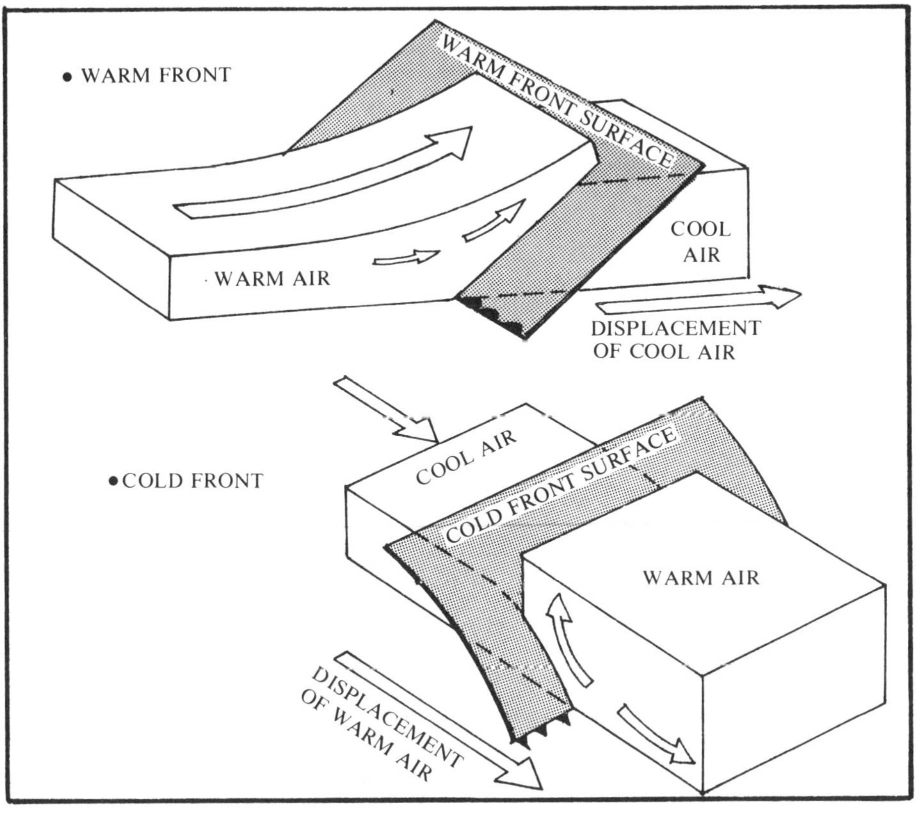

Figure 44 - Warm and Cold Fronts

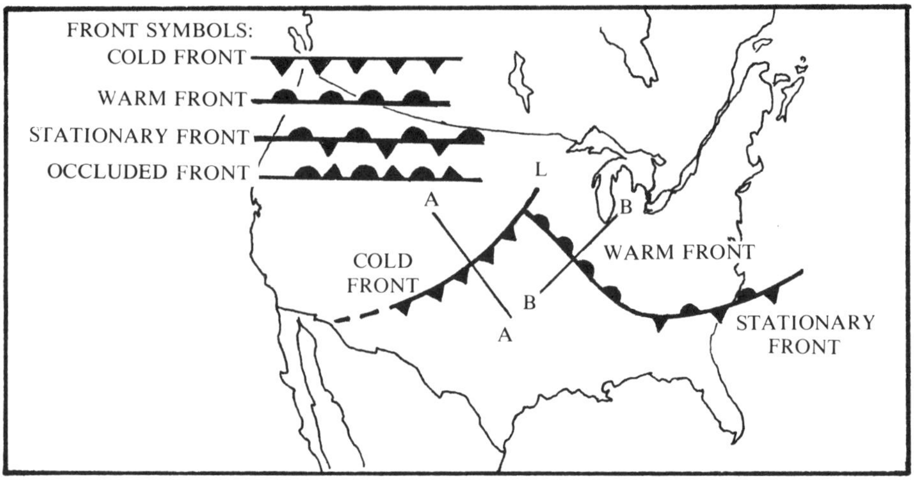

Figure 45 - Fronts on a Surface Map

FRONT CHARACTERISTICS—COLD

A cold front generally comes from the north and moves southerly in the northern hemisphere or from the south moving north in the southern hemisphere. This front is at the leading edge of a cold, often dry air mass.

A sectional view of a cold front through line AA in figure 45 is shown in figure 46. Here we see two possible situations: the cold air replaces either unstable air or stable air. In the former situation the unstable warm air is lifted by the plowing cold air and forms convective clouds. Often

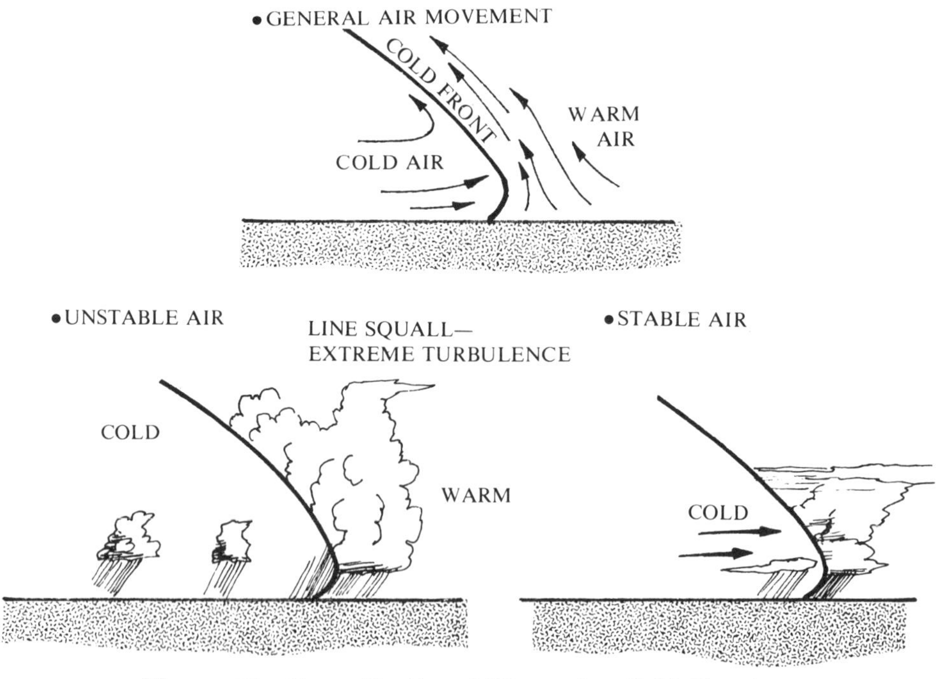

Figure 46 - Cross-Sectional View of a Cold Front

thunderstorms or squall lines accompany this type of frontal activity.

Squall lines are continuous thunderstorms that precede a front by as much as 50 to 300 miles and extend generally parallel to the front. The extremely violent weather contained in a squall line is covered in Chapter XI. The presence of a squall line may extend the period of bad weather associated with the front considerably.

Cold fronts tend to be energetic compared to warm fronts and may travel at speeds up to 40 mph (64 km/h), especially in winter when the air is more dense. The faster moving fronts exhibit more violent weather but winds drop off sooner after their passage. The slope of a cold front is from 1/30 to 1/100 which creates strong updrafts as it moves vigorously forward, lifting the warm air. The slope depends on the temperature contrast between the air masses and the wind speed across the front.

In stable conditions more stratus type clouds may form before and behind a cold front as it progresses. In this case, a lingering shower may slow the clearing of the sky after the frontal passage, but the passage itself is accompanied by less violent conditions.

Pilots generally look forward to the passage of cold fronts, especially in the warmer months. The reason for this is a cold front usually brings clear or cumulus studded skies, great visibility, thermal lift for those who soar and denser air to stimulate the engines of those who use power.

FRONT CHARACTERISTICS—WARM

A warm front is often a pilot's nemisis. Such a front can present an unpleasant package of cloudy skies, high humidity, haze, excessive heat and rain that lasts for days. Only if the air mass is dry or it's the dead of winter do we grudgingly welcome a warm front.

To see how a warm front works, see figure 47 (This is a cross-section through line BB in Figure 45). You will note that the warm front rides up

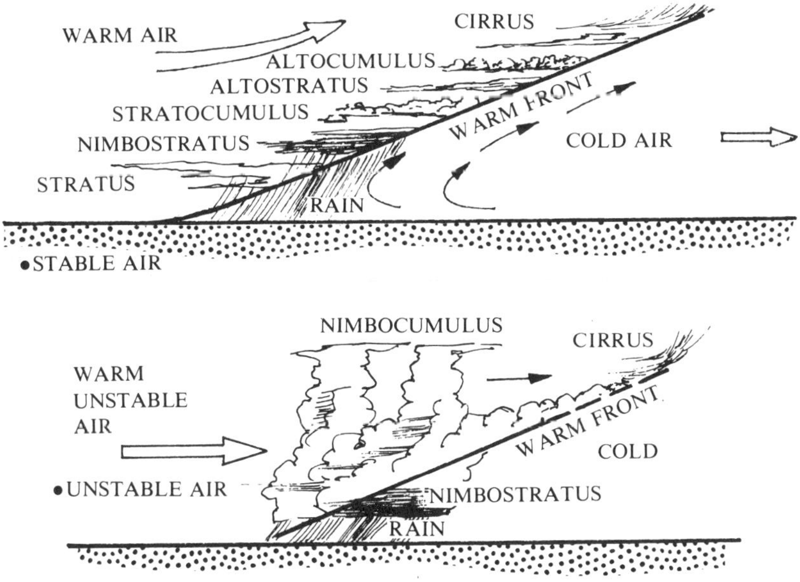

Figure 47 - Cross-Sectional View of a Warm Front

over the cooler air it is replacing. Because warm fronts tend to move slower than cold fronts—15 mph (24 km/h)or less—and introduce less dense air, they tend to ride up over the cooler air in a gradual manner as shown in the figure. The slope of a warm front is from 1/50 to 1/400, considerably less than that of a cold front.

The shallow slope of a warm front results in a wide area of cloud cover, often spreading over a distance of 1,500 miles (2,400 km). Also, because of this expanse of clouds and the front's slow advance we can often predict the approach of a warm front a day or two ahead of time by noting a gradual increase and lowering of clouds starting with cirrus and progressing to cirrostratus or cirrocumulus as indicated in the upper figure.

The figure shows two types of warm front: those bringing either stable or unstable air. In the first case we should expect long periods of steady rain and generally smooth conditions except perhaps near the frontal boundary. In the unstable case we should expect bursts of heavy rain intermingling with steady drizzle as well as dangerous turbulence associated with thunderstorms (see Chapter XI). In either case the time during which a warm front passes is often best spent indoors doing something constructive such as reading a weather book.

Here is a summary of fronts:

> *Cold Fronts*—Pass in a matter of hours unless preceded by a squall line or trailed by another front. They usually bring dry, cooler air with great visibility and unstable conditions.
> *Warm Fronts*—May take a day or several days to pass with their attendant clouds and rain. They bring warm, humid and often stable air trailing behind them.

FRONTAL ACTION

We can think of a cold front as the leading edge of a wave of cool air busting out from the polar regions. Because of the generally westerly wind that assails the temperate zones where all this frontal action takes place, the fronts themselves tend to eventually drift toward the east (in both hemispheres). The time lapse action of ideal frontal activity is shown in figure 48. We use a rectangle to represent a hypothetical continent or land mass.

At first the figure shows a wave of cold air moving down over the continent. The cold front on the surface is shown as it progresses. It is extending from a low pressure system indicated by the L on the map and driven by a high pressure system marked H. We explain these matters below.

As the front drops down over the land it is eventually opposed by high pressure at lower latitudes so that it slips off toward the east. The western portion of the cold front then stops its expansion and also begins to move east as a returning warm front. The original cold front continues to the east and may die at sea or eventually lose its identity by warming if it lingers over land.

A new low pressure system with a new cold front is seen to be following the same pattern of the original front. Thus we have the standard process of cold front followed by a warm front followed by a cold front and so on, especially

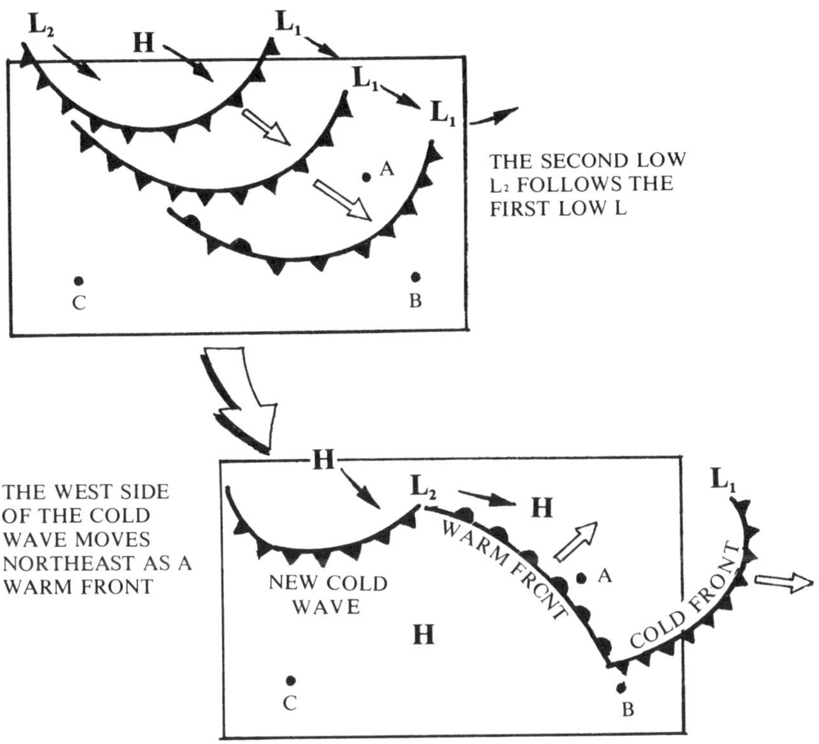

Figure 48 - Frontal Movement

for a resident at point A. A citizen at B may experience cold fronts only in a serious winter while the lucky person at C may never see a front.

Once again we have greatly simplified matters, for irregularities always occur in the atmosphere, but this model will help you understand the more complex frontal movement when you see it unfold or depicted on weather maps.

VARIATIONS IN FRONTS

We have already seen that fronts can accompany stable or unstable air masses and they can be fast or slow moving. They can also vary their behavior according to the terrain or pressure systems they encounter. High mountain chains can partially block a front and complicate its pattern as it passes. The movement of pressure systems can stall a front and render it stationary until it loses its identity, moves back as the opposite type front or regains impetus to continue on. Sometimes a cold front may stall and flatten out to lie nearby in a west-east direction, possibly moving up and down to create bad weather for days for the grumbling nearby inhabitants (see figure 49).

On other occasions, the eastern portion of a cold front may push vigorously forward to bulge up so that the front appears from the west or even southwest (in the northern hemisphere) as shown in figure 50a at point A. This is sometimes called a *back door front*, although a true back door cold front appears as in 50b when a cold front comes down from the

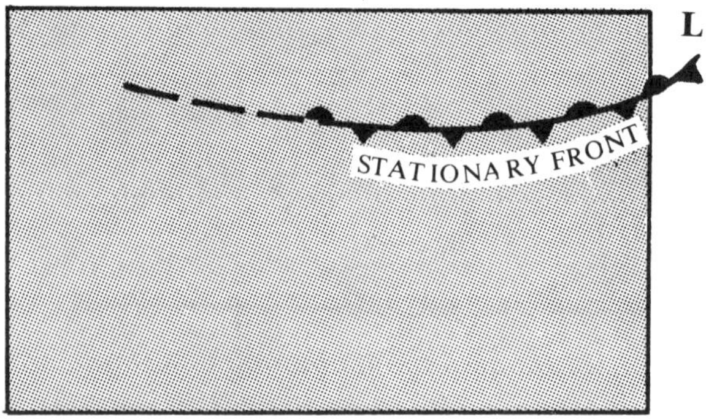

Figure 49 - Stationary Front

northeast (again in the northern hemisphere). This action is common in the northwest areas of the North American continent as well as Europe. A back door front usually brings a relatively shallow layer of cool, unstable air, so the thermals associated with such a front are limited in height.

Another important cold front variation is an *upper level front*. This occurs when cool air is lifted over mountainous areas and meets colder air on the other side. In this case the cool air does not descend but produces frontal activity aloft as shown in figure 51. This upper level disturbance can create severe weather including thunderstorms and tornados.

The next variation to identify is a *secondary cold front*. There are often perturbations or oscillations in the vast glob of cold air at the poles as if it were a great mound of shaking jello. This can cause wave after wave of

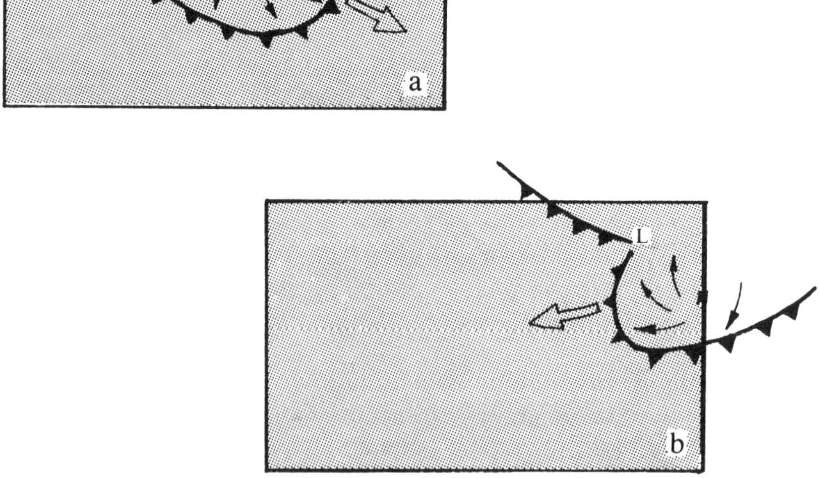

Figure 50 - Back Door Fronts

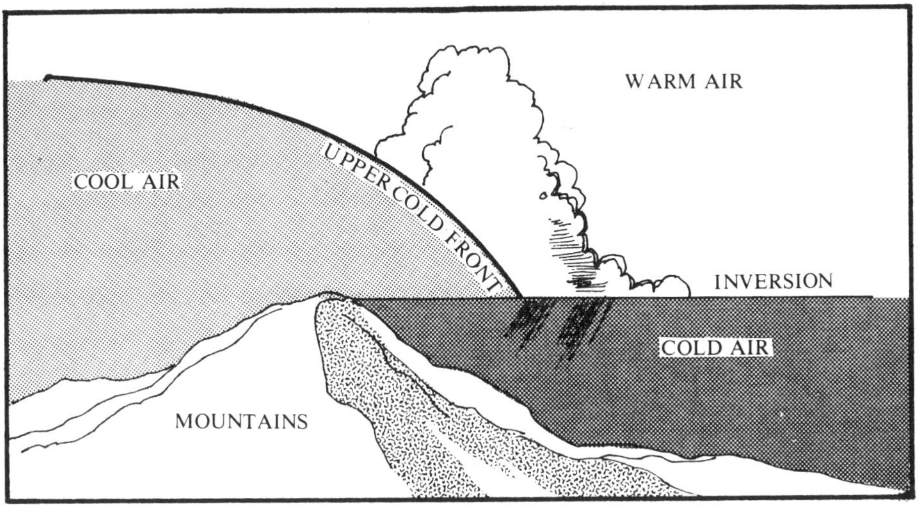

Figure 51 - Cold Front Aloft

cold air to break into the warmer latitudes. Fronts following close on the heels of a preceding front are called secondary fronts (even though there may be several) and appear as in figure 52. Each successive wave tends to bring colder air and needless to say such a march of fronts can disturb the weather for days.

We should mention *arctic fronts* which are simply strong polar fronts from the north maintaining their identity well into the temperate areas. Finally, we look at occluded fronts.

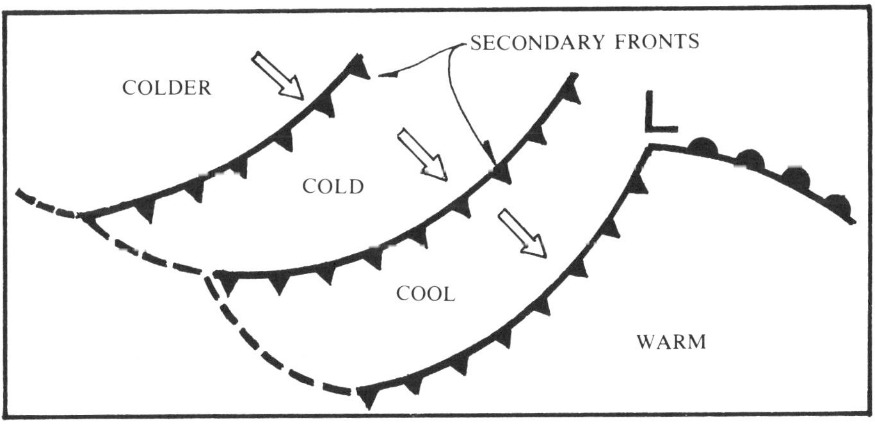

Figure 52 - Secondary Fronts

OCCLUDED FRONTS

When a cold front catches up to a cool wave that preceded it , we have a situation known as an *occluded front*. This is shown in figure 53 with the shaded area representing the occluded portion. Note the wind directions as shown by the arrows and the positions of the three different air masses. This state of affairs is known as a *cold occlusion*.

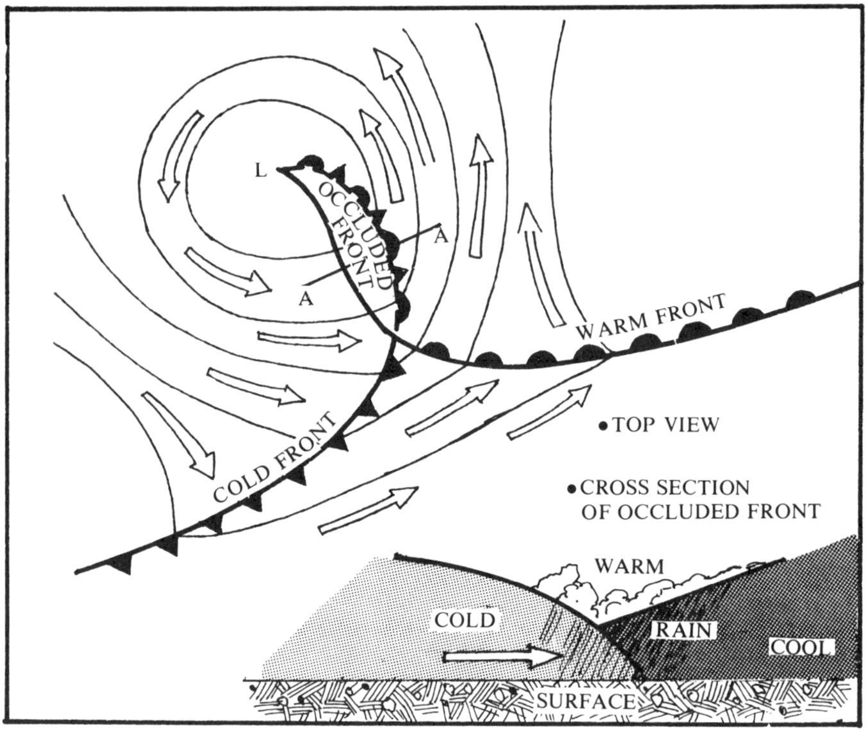

Figure 53 - Occluded Front

A *warm occlusion* can occur when a succeeding cold front is not as cold as the front it catches. This is common along western coasts where a maritime cool air mass meets a cold continental mass that has plowed under a warm mass. The cross section of such a warm occlusion is shown in figure 54.

A cold occlusion will exhibit clouds and winds aloft similar to a warm front as it approaches, but behave similar to a cold front in terms of winds and weather as it passes. Conversely, a warm occlusion exhibits cold front weather in the upper level cold front region then appears as a warm front as it passes at the surface. In either case occlusions can cover a vast area with clouds and rain for they generally occur when a weather system has

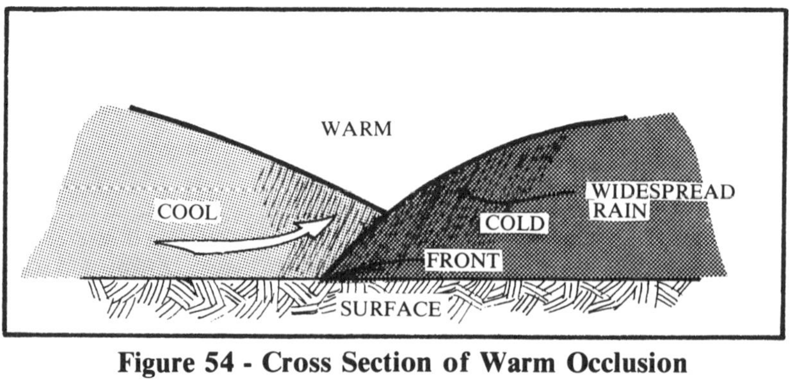

Figure 54 - Cross Section of Warm Occlusion

slowed or stagnates. Often such occlusions stay in one place drizzling and dampening every pilot's spirits until they decay and lose their identity. Coastal rainy seasons such as that along the Pacific Northwest in North America are caused by a succession of warm occlusions intruding from the Pacific Ocean.

SEASONAL CHANGES

We have already discussed how the tilt of the earth combined with its orbit around the sun changes the position of maximum heating on the earth. The result is a change in air mass movement and frontal patterns as winter melts into summer and summer yields to winter again in any given area.

In figure 55 we depict the general winter and summer patterns of cold and warm front passage throughout the world. The main features to note

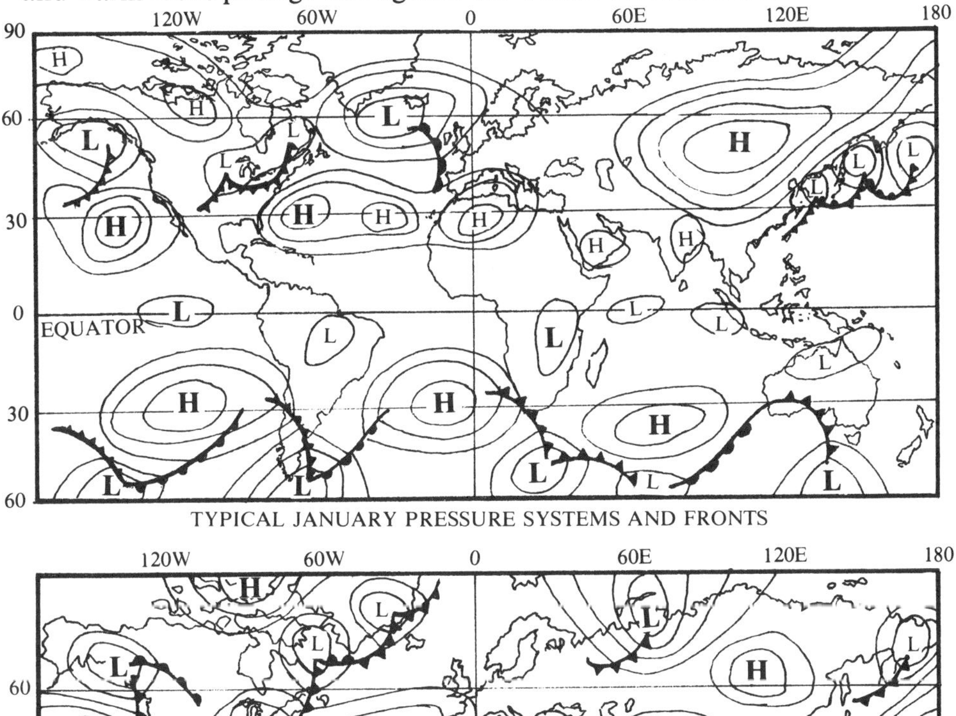

TYPICAL JANUARY PRESSURE SYSTEMS AND FRONTS

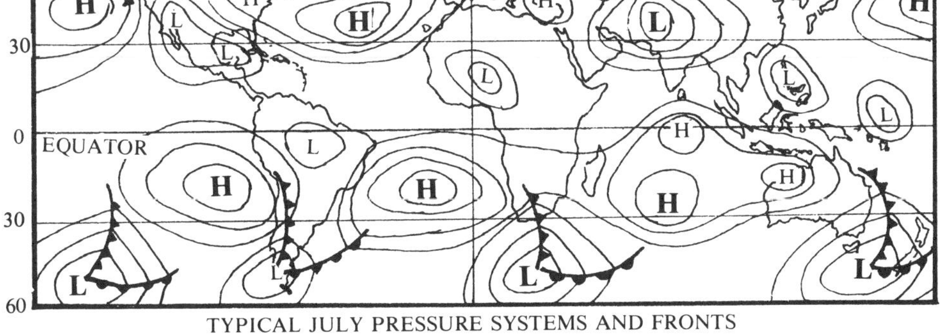

TYPICAL JULY PRESSURE SYSTEMS AND FRONTS

• NOTE THE LARGER THE PRESSURE SYMBOL, THE STRONGER THE SYSTEM.

Figure 55 - Worldwide Fronts and Pressure Systems

are the general north and south limits of the fronts in the different seasons as well as the different positions of the pressure systems.

Just as the birds tend to desert the temperate climates in winter then return to regale us with their songs in spring only to subdue their activity in the heat of summer then again become active in fall, so too does the weather seem to operate. We have higher relative humidities, lower cloud bases, colder air and more frequent frontal passage in winter. As spring arrives cold fronts and warm fronts alternate on a more regular basis. Good flying weather often exists after these fronts pass. In summer warm fronts with hot, hazy, often stable air dominates the moister regions, putting a damper on soaring flights while deserty areas become very unstable in the extreme sunshine. Finally in fall regular cold fronts return to the temperate zones and the cycle begins anew.

HIGH PRESSURE SYSTEMS

We know that the presure we detect in the air at any level is dependent upon how much air is piled up above us. The more air aloft, the more it weighs and presses on our bodies and instruments. We have also seen how the complexities of atmospheric circulation on the earth results in the piling up of air in certain general areas. This action creates a relative high pressure system at the earth's surface. If we moved under this high our barometer would read higher than at any surrounding area.

The build-up of a high pressure system can be thought of as the prime mover of the air masses and fronts at their boundaries. Imagine this: air moving in aloft towards the north pole settles down and cools over the cold snowy ground. This air contracts and becomes more dense, building up more and more pressure at the surface as more air moves in aloft. Finally, this northern mass of heaving air can no longer be contained by the pressure of air to the south so it breaks out in a sudden wave invading the south as a cold front preceding a cold air mass.

Figure 56 shows a satellite's view of the north pole and the cold air mass poised to invade the south. You may notice the high pressure systems are generally located over the land masses above 60° latitude. Also the cold air extends further south over the land. This is because land cools more readily than water, creating colder air above itself. This illustration, also shows the principle paths of invasion of the cold air when it breaks out into a cold front in the northern winter. Note that the cold fronts crossing Europe are often back door fronts. A similar map can be drawn for the southern hemisphere (see figure 57), but the lack of significant land masses from about 40° to 70° latitude results in a more uniform polar air mass and more regularly spaced highs and frontal outbreaks.

A band of high pressure systems also extends around the globe—above and below the equator—at 30° to 40° latitudc. Thcsc highs are centered over the sea area that is cooler than the land masses in the tropics, especially in summer. These tropical highs are the main systems that oppose the polar highs and prevent cold fronts from reaching the tropics. They in turn send warm fronts into the temperate areas in summer. Figures 40 and 43 can help you imagine where surface highs should appear.

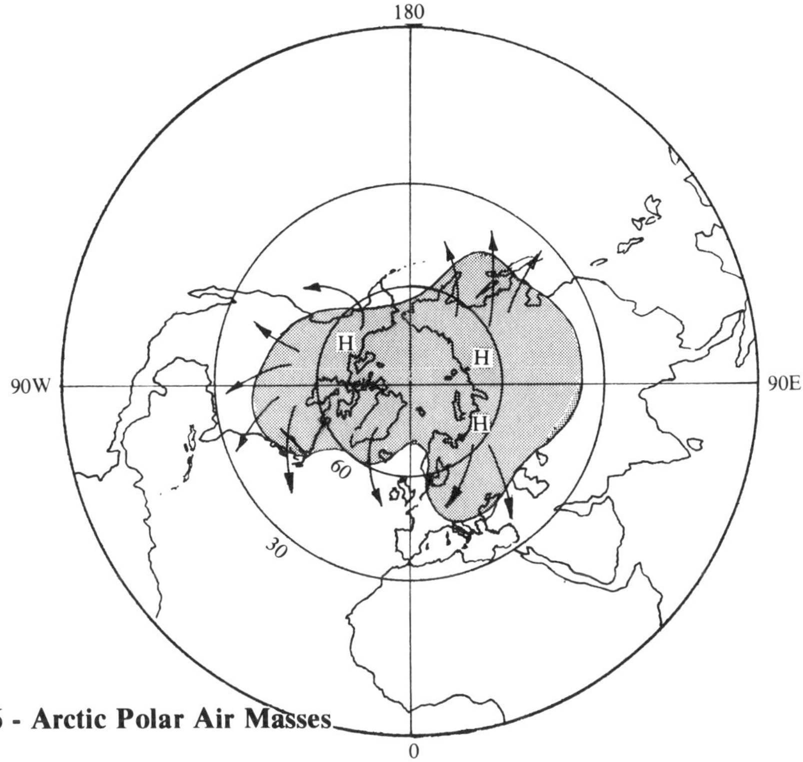

Figure 56 - Arctic Polar Air Masses

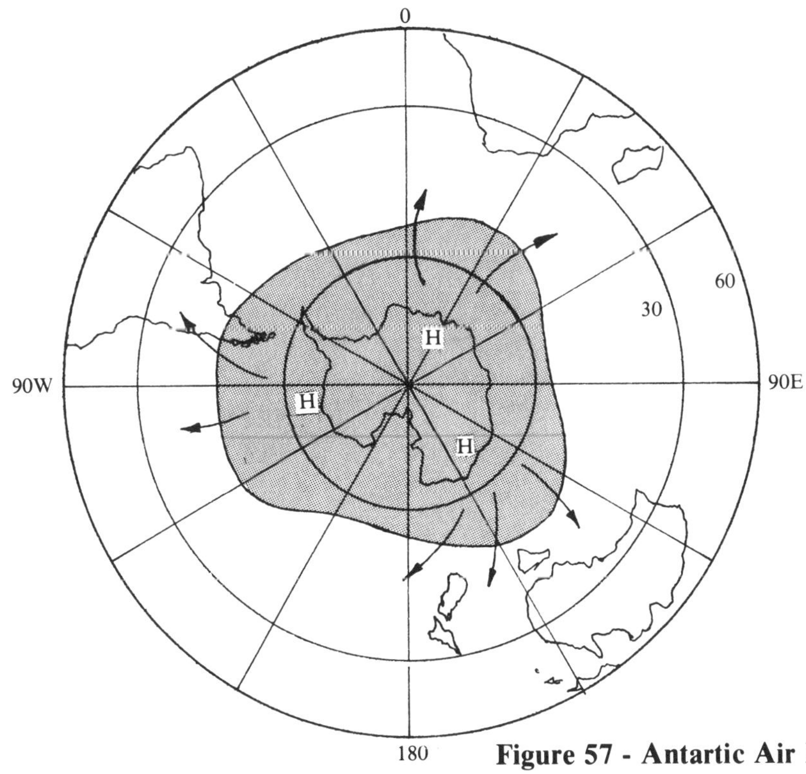

Figure 57 - Antartic Air Masses

LOW PRESSURE SYSTEMS

While we can readily understand how highs are created by an abundance of air aloft, it is a bit more difficult to imagine air being taken away from an area in the upper atmosphere to create a surface low. But this is exactly what happens. To see this we need to picture the general circulation at altitude—say 18,000 feet (6,000 m)—as in figure 58. Here we have pictured the north pole and the general wind flow at altitude.

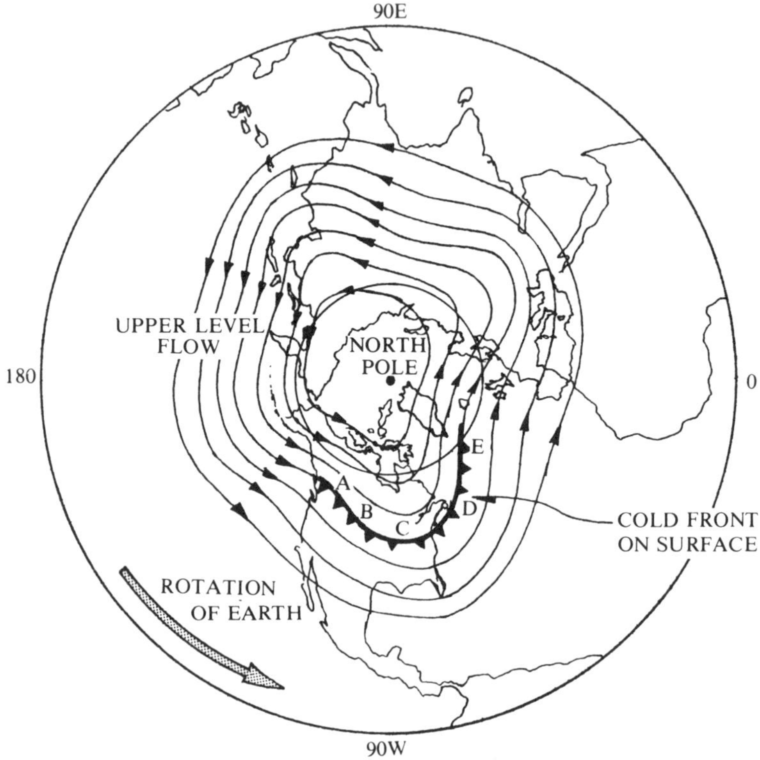

Figure 58 - Upper Level Circulation

Except for a deep layer of westward flowing air that reaches into the troposhere in equatorial regions, the upper air flow consists of a broad band of eastward flowing wind in both hemispheres known as the *circumpolar westerlies* (see below for more details). These westerly winds do not flow in a neat circle, but undulate with the northern air mass to the tune called by the pressure systems.

Now look at the figure and imagine the earth as a rotating disc. As an object spinning with the disc moves away from the center it tends to slow down (scientists call this conservation of angular momentum). As it moves toward the center it tends to speed up. You can experience this yourself by going around on a playground merry-go-round and leaning toward or away from the center. You will experience a force increasing or decreasing your rate of rotation respectively. A spinning skater similarly changes his or her rate of rotation by distributing the mass of his or her arms inward or outward. The wind behaves the same way. When it moves south it gets further away from the earth's spin axis so it slows down. As it moves north it acquires more relative speed because it moves closer to the spin axis.

Looking again at the figure we start at point A and follow the airflow and see that the air gradually decelerates until we reach the slowest moving air at point C. Continuing around the path we see the air beginning to move north and accelerating until we reach the fastest moving air at E. In fact, all the places where the flow is closest to the poles contain the fastest moving air and vice versa.

Proceeding one step further we ask ourselves what happens at points B and D. Think of this: if you are driving on a crowded single lane highway and a car up ahead suddenly slows down you find yourself and the drivers near you suddenly bunched up behind this slower vehicle. In front of the slow driver traffic thins out as the other cars pull away from him.

The same thing happens to the air molecules as they follow their undulating path around the earth. When a portion of the air is slowed, the air behind it tends to bunch up, while the air in front of this slow flow spreads out. By studying the figure you can see that air at B will be crowded or bunched while air at D will be thinned.

The crowding of air at B is called *convergence* (meaning coming together). In effect the air actually piles up at this point in the upper levels causing an increase in pressure at the surface or a ridge of high pressure.

The thinning of the air at point D is known as *divergence* (meaning separation). The effect of this thinning is an area of low pressure at the surface below D. Thus we see that the actual cause of low pressure cells on the earth's surface is contained in the peculiarities of the upper air flow. In addition, we have found another source of surface high pressure areas.

Cold fronts on the surface tend to break out and extend below a wave aloft. The fronts themselves extend from the surface lows produced by the mechanism indicated above and exhibit a slight trough of low pressure along their length. This low trough affects the wind flow as we shall see in the next chapter and can be used to detect the passage of a front. By watching your barometer or altimeter you can see falling or rising pressure. Falling pressure means the approach of a front (both cold and warm) or low pressure system while rising pressure indicates a receding front. A change from falling to rising indicates the frontal passage. Barometers marked with stormy-rain-change-fair-dry use the pressure change related to fronts and pressure systems to give these predictions based on current pressure readings. In Chapter XII we learn how to use the barometer to predict the weather in more detail.

LOCAL LOWS

Low pressure systems can be impressively huge affairs covering half a continent. But lows can be much smaller and form from a mechanism other than varying upper level flow. For example, an area that is heated more than its surroundings will heat and expand the air above it so that air aloft flows away and a *local low* is formed. This is an important phenomenon that helps create the sea breezes as well as other local flows and will be explored in Chapter VII.

On a slightly larger scale a persistant summer low known as the California low is created over the American Southwest desert. Also a series of lows along the eastern face of the Rocky Mountains or along the southern slopes of the Swiss Alps are formed in the same manner with the addition of blockage of the wind by the mountains which prevents the low from dissipating by drawing in air at the surface.

The Rocky Mountain low in particular creates an initial flow from the Mississippi basin all across the plains from east to west at the surface in the summer months (see figure 59). This flow is turned northward by Coriolis effect so that it creates the prevailing southerlies on the Great Plains and sucks a considerable burden of moisture from the Gulf of Mexico to feed the famous thunderstorms and tornados of Dorothy's Kansas and surrounding states.

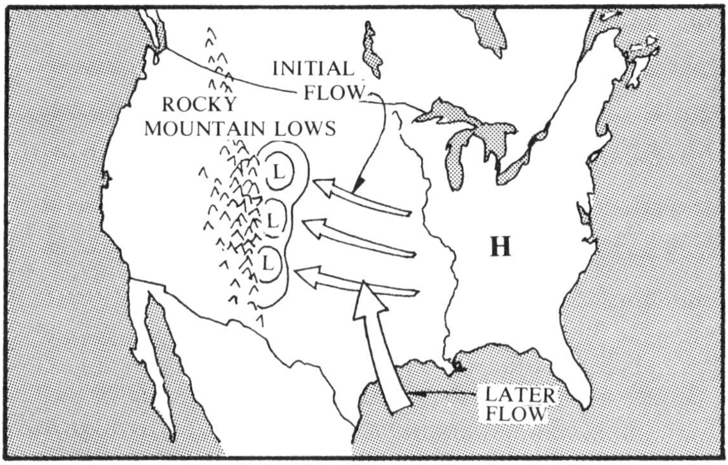

Figure 59 - Rocky Mountain Lows

Likewise the low near the Alps pulls in Mediterranian moisture to drop showers on northern Italy. Later we will look at tropical weather and see a further production of local lows.

HIGHS AND LOWS TOGETHER

By now you probably know that highs and lows exist side by side in the atmosphere and they interact to produce winds and weather. Let's see this in more detail. Figure 60 shows the general relationship of high and low pressure systems in terms of how they are formed. In summer the highs tend to initiate over the cooler sea while the lows are born over the land. The opposite is true in winter.

Since a high pressure wants to squeeze the air out below it we would naturally expect the air to flow from high to low pressure. This is exactly what it does initially. We can picture this as a surface contour like in figure 61. Here a fluid flows from the highest point to the lowest point. However, in the free air Coriolis again enters in to complicate the picture. In the northern hemisphere the flowing air turns to the right so that as it moves outward from the high it ends up circulating clockwise (when view-

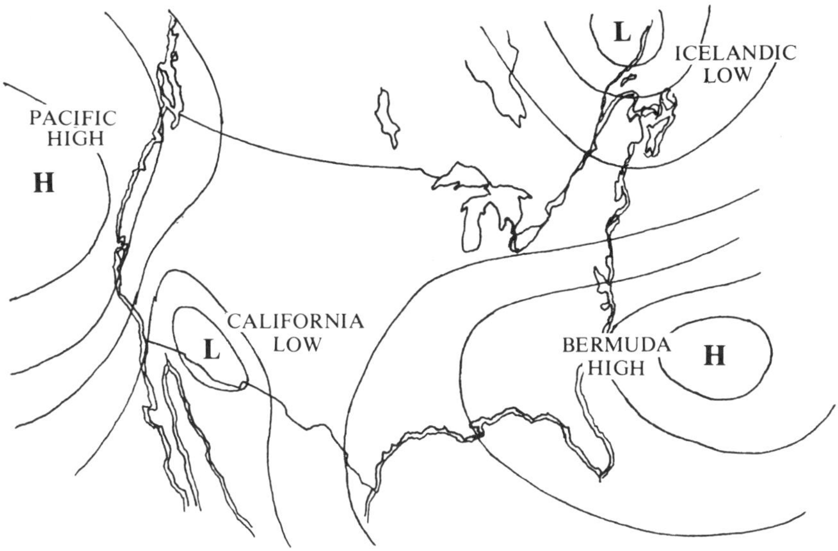

Figure 60 - Summer Pressure Systems

ed from above). In the southern hemisphere Coriolis effects causes a left turn so the flow around a high becomes counterclockwise.

The air moving inward towards a low is likewise turned right in the northern hemisphere. This action combined with the inward flow results in a counterclockwise flow around a low. In the southern hemisphere the flow around a low is clockwise. We summarize these important flows below:

Circulation Around Pressure Systems		
	Northern Hemisphere	Southern Hemisphere
Highs	*Clockwise*	*Counterclockwise*
Lows	*Counterclockwise*	*Clockwise*

This is an important concept for pilots in order to deduce wind direction from the surface maps. To aid your memory, all you need to know is that

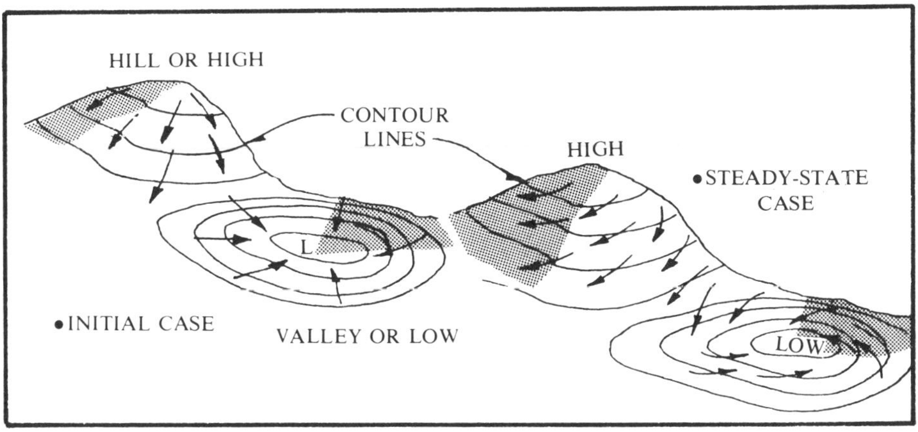

Figure 61 - Flow Around Pressure Systems

air turns right in the northern hemisphere so as it flows outward from a high it moves clockwise. From this we remember the flow is opposite around a low and everything is reversed in the southern hemisphere.

When lows and highs are produced by heating effects (and not by circulation aloft) they set up a mutual support system that appears as in figure 62. Here we see a relative high on the ground with air flowing to the low. Aloft we see a relative high over the low that sends air flowing over the high to reinforce it. This flow will continue as long as the heating differences last.

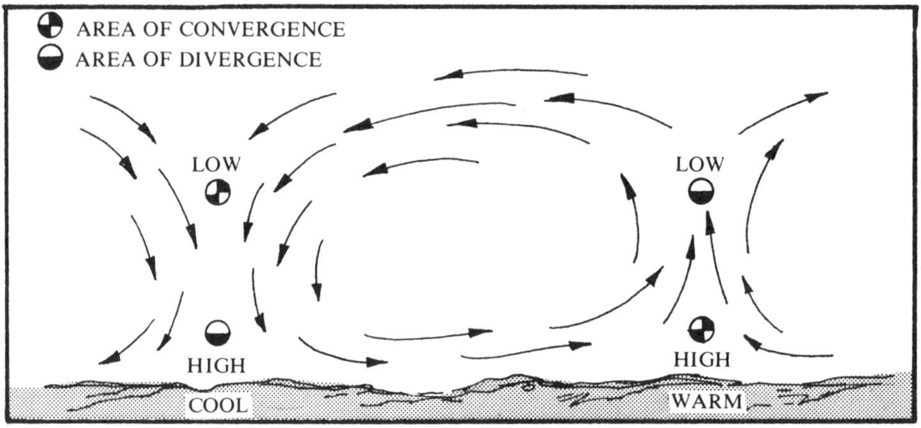

Figure 62 - Circulation in Lows and Highs

An important point to note in the figure is that the air is *subsiding* or descending above and around the high while it is *ascending* over the low. This is the case in pressure systems in general, no matter what creates them. This too has great importance to pilots for:

The air around a high pressure *is descending which results in compression, warming, reduced relative humidity and increased stability.*

The air around a low pressure *system is ascending resulting in expansion, cooling, increased relative humidity and decreased stability.*

The descending air at the high is moving downward only a few centimeters (about an inch) a minute at the most, but this is enough to produce the effects outlined above and create clearing skies generally associated with a high pressure system. The irony is that the subsiding air in the high produces an increasingly more stable air mass and is the main source of the inversion that frequently exists in the non-desert areas of the temperate zone. Yet post cold front, high dominated weather is what brings the most lower level instability and thermal production in these same areas. This action is a result of clear, cold air allowing plenty of sunshine through to heat the ground and produce lower level instability. Nevertheless, if a high lingers for days in an area it gradually increases the stability of the air mass so that eventually thermals are suppressed.

The ascending air around a low gives rise to much cloud cover and precipitation. It can also exhibit such instability that thunderstorms are formed. In general, the approach of a low is viewed with a jaundiced eye by pilots in humid areas, for no one likes to fly in the rain.

ISOBARS

It's really difficult to talk about pressure systems without talking about isobars. Isobars are simply lines connecting points of equal pressure on a weather map. The word comes from the Greeks. *Iso* means equal or same; *bar* refers to weight or pressure.

Just like the contour lines on hills or valleys circle these features on a topographical map as in figure 61, so too do isobars appear on a map. The isobars in figure 63 circle the high and the low and take more complex shapes away from the pressure systems. Because the isobars outline the highs and lows just like topographic lines do hills and valleys, we often refer to long areas of high pressure as *ridges* and lows as *troughs* or *depressions*. Low pressure systems are also known as *cyclones* and highs as *anticyclones*.

Isobars depicted on a weather map are usually in 4 millibar (mb) steps or 2 mb steps if the pressure change is weak and more detail is needed (see

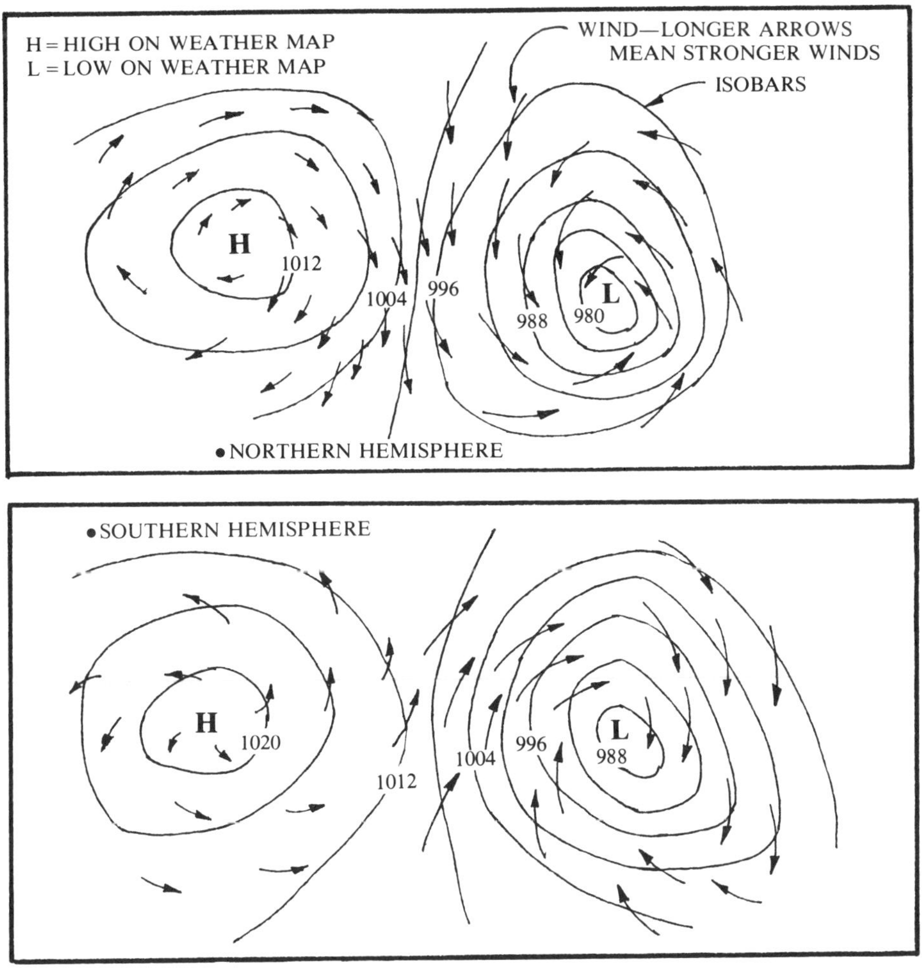

Figure 63 - Isobars Around Pressure Systems

figure 63). The closer the isobar lines are spaced, the greater the pressure difference in a given distance perpendicular to the isobars. Thus we say there is a steep pressure gradient (meaning change with distance) exactly analogous to steep terrain depicted on a topograhic map.

The greater this pressure difference, the greater the force driving the wind. Thus we can correlate wind speed with the spacing of the isobars the rule is: *the closer the isobar spacing, the stronger the wind.* The following chart provides a guideline to expected isobar spacing at the surface for a 15 mph (24 km/h) wind. Note that this changes with latitude due to Coriolis effect since Coriolis effect always opposes the pressure gradient force.

Isobar Spacing for a 15 mph (24 km/h) Wind

Latitude*	Spacing (Miles)	(Km)
60	144	230
55	153	245
50	162	260
45	176	282
40	195	312
35	218	349
30	248	397
25	297	475
20	364	582

*A given Isobar spacing produces stronger winds at lower latitudes.

In hilly or mountainous terrain the isobar spacing must be less for a 15 mph surface wind due to friction effects. The important point is to know your area to get an idea of how many isobars in a given distance present usable winds and at what isobar density do the winds get unmanageable.

ISOBARS AND THE WIND

Initially we can see that the air gets an impetus from the pressure systems to move perpendicular to the isobars. However, as explained before, Coriolis effect soons turns the wind so formed and it lines up with the isobars. This lining up is caused by a balance between the pressure gradient force, Coriolis effect and centrifugal force as explained in Appendix II. In figure 64 we see a weather map of the pressure systems at 18,000 feet (6,000 m) over the North American continent. The arrows show the wind direction—clockwise around the high, the opposite direction around the lows and always parallel to the isobars.

On the surface the story is somewhat different. The good uniform nature of the highs and lows and simple flows aloft are more complicated at the surface due to differential heating effects as can be seen in the lower drawing. Also friction of the air flowing over rough terrain results in a slowing of the air so that Coriolis effect is reduced. The consequence is that the wind crosses the isobars at the surface as shown previously in figure 63. The angle of crossing will be as little as 10° over water where there is little friction to as much as 40° or 50° over rough or mountainous

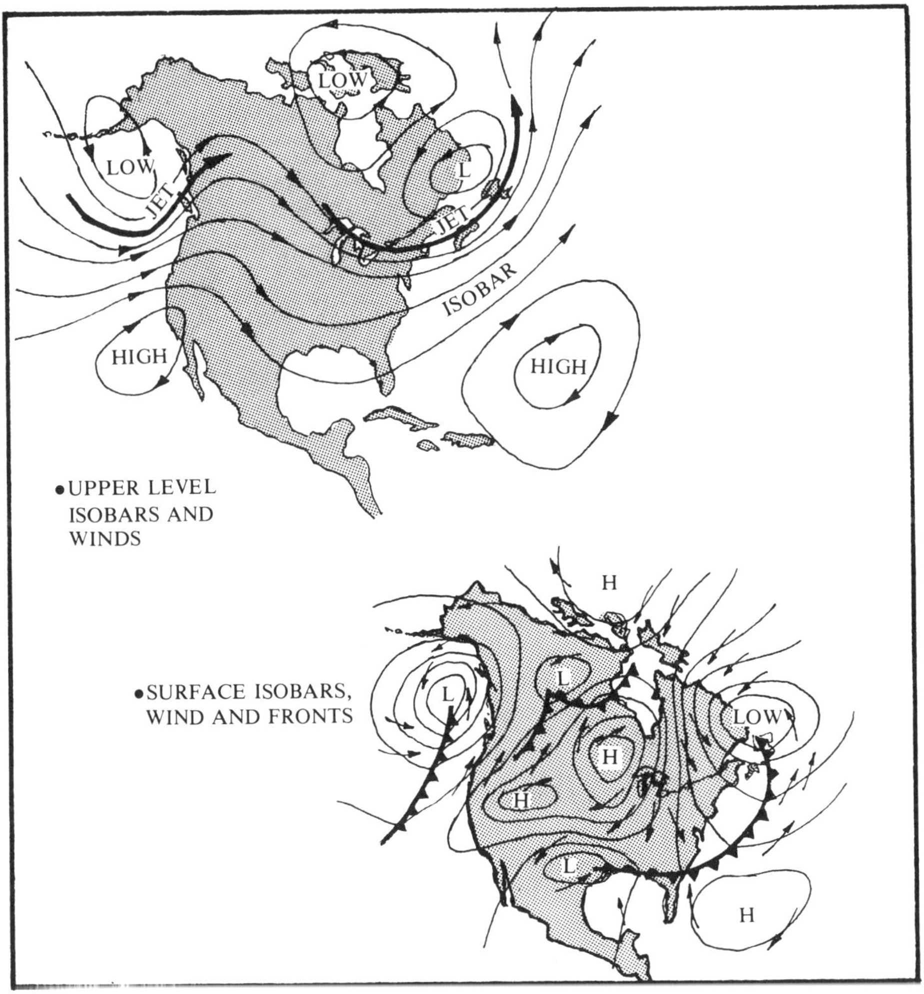

Figure 64 - Circulation at Altitude

terrain. This flow across the isobars is greater in winter than in summer and greater in latitudes closer to the poles since cooler air is denser and experiences more friction. With this knowledge we can predict surface wind velocities (speed and direction) given a surface pressure chart.

In both a high and a low pressure system the wind increases with increased pressure gradient. Also around a high the centrifugal force is added to the pressure gradient force while it is subtracted around a low. Thus we should expect higher winds around a high, but in fact the opposite is true for the isobars tend to spread out further in a high while they tend to contract in a low due to air circulation aloft, creating a greater gradient (tighter isobars) and higher winds. Also, the winds tend to get lighter as you move toward the center of a high or it moves over you while they usually get stronger toward the center of a low.

CIRCULATION ALOFT

At this point we can get a clearer picture of general circulation at altitude. To begin we should realize that we can draw an isobar map at any height to get a picture of how the pressure is distributed and thus how the winds blow. Weather stations send up instruments on balloons to detect pressure at different altitudes to enable a central processor to draw these maps.

Typical weather charts readily available besides a surface map are the 850 mb (about 5,000 feet or 1,600 m), 700 mb (about 10,000 feet or 3,300 m), 500 mb (about 18,000 feet or 6,000 m), and 300 mb (about 30,000 feet or 10,000 m). These heights are not exact, for in fact these are constant pressure charts that show the contours of a given pressure rather than the changes in pressure. For our purposes this is the same.

The circulation at altitude loses the surface details and tends to only exhibit the large-scale features. Completely closed circulations—highs and lows—decrease in frequency with altitude to be replaced by ridges of high pressure or troughs of low pressure. A typical upper level circulation with these features is shown in figure 65. Notice the ridge running down the middle of an upward bulge in the isobars causing clockwise rotation of the air. The troughs trail down from low pressure systems and extend where the isobars take a dip toward the equator, creating counterclockwise circulation. This model is for the northern hemisphere with opposite circulation applying south of the equator.

We have already seen that the general westerly winds circulating aloft dip down in ridges and troughs (see figure 58). These are known as *long waves* and there are usually from three to seven of them in existance around the globe. Superimposed on them are *short waves* which are the disturbances shown in figure 65.

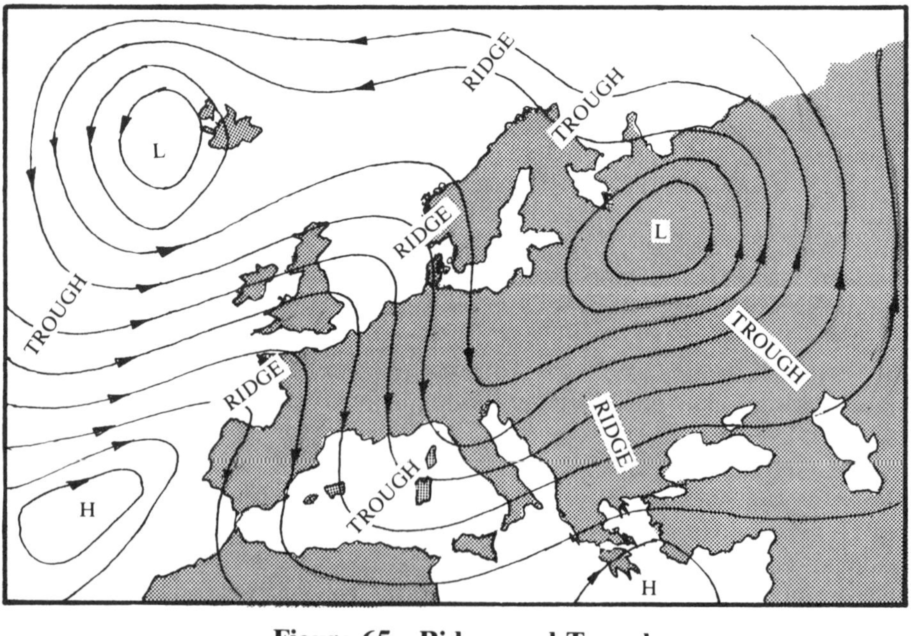

Figure 65 - Ridges and Troughs

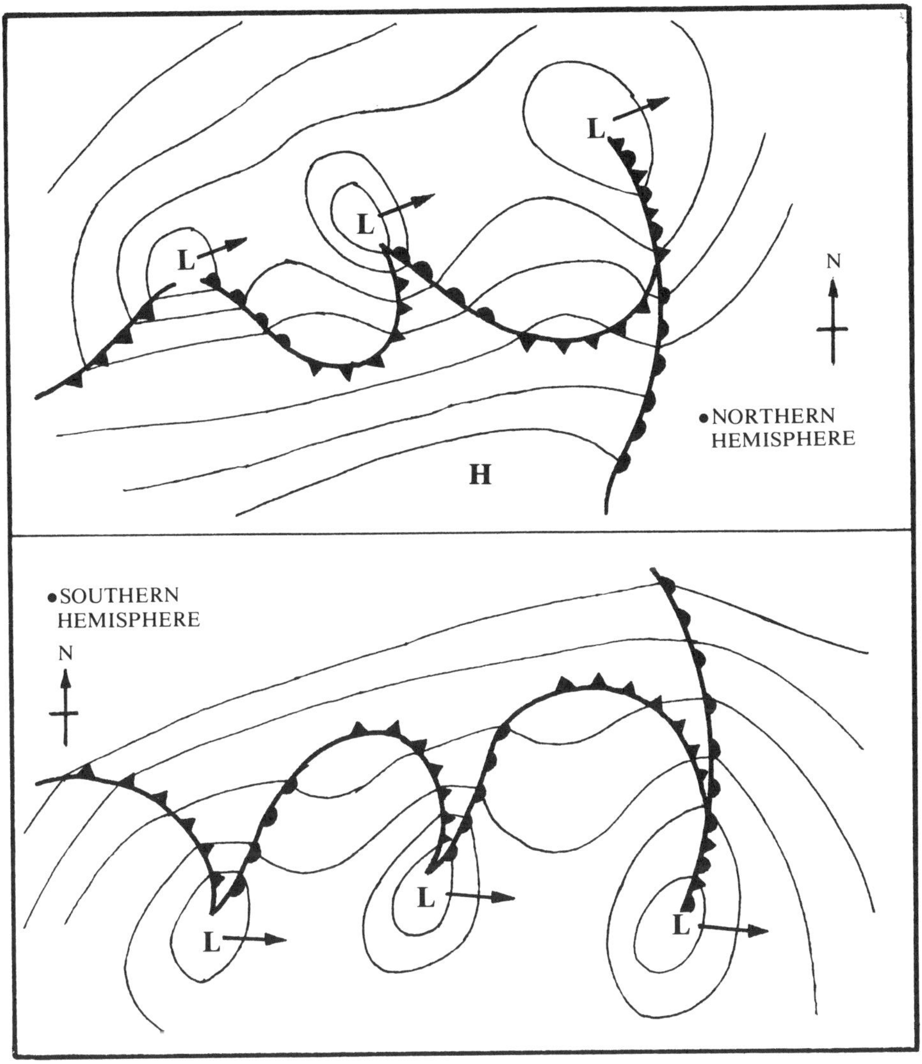

Figure 66 - Surface Waves

Long waves tend to move slowly, drifting eastward or remaining stationary for a few days. The airflow around them can carry tropical air far to the north or polar air far south. Short waves are faster moving oscillations proceeding through the long wave pattern. The surface situation below a series of short waves is shown in figure 66. The waves induce a series of lows on the surface that rapidly move up the general circulation pattern.

A TROUGH ALOFT

When an upper level trough is formed it creates clouds in the rising air just like a singular low pressure cell. Usually these clouds will be in high broad bands unless the trough deepens and reaches the ground in which case the clouds thicken and offer the dubious gift of precipitation typical of a ground based low. Understanding the mechanism of upper level disturbances—troughs—can allow a pilot to predict the winds and weather before and after their passage. In general, lifting and rain preceeds an upper level trough with southwesterly winds aloft and west winds at the surface (northwest aloft and west on the surface in the southern hemisphere). Then northwest winds occur aloft with surface winds remaining westerly (southwest and west in the southern hemisphere) and clearing takes place as the ridge passes. This process is shown in figure 67. Note that the wind turns in a counterclockwise direction with altitude before the trough and clockwise with altitude after the trough passes (for both hemispheres).

Upper-level troughs usually catch up to the main surface storm toward the east and meld into it by developing a complete circular pattern itself.

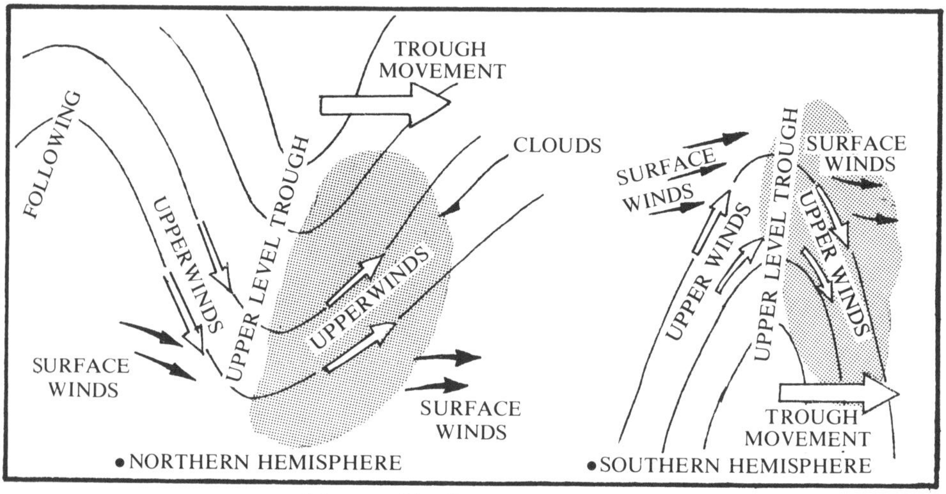

Figure 67 - Upper Level Trough

Altostratus with billow clouds indicating unsettled weather.

THE JET STREAM

In figure 43 we have indicated the existence of the jet stream at the boundaries of the air masses. We will look at how this is formed in the next chapter, but here we simply identify the jet stream as a flow of fast moving air—over 200 mph or 320 km/h in some cases—from west to east. The polar front jet stream rarely circles the globe but is found in segments 1,000 to 3,000 miles (1,600 to 4,800 km) long. The tropical jet stream tends to be weaker, higher and shorter.

The polar jet stream is a good indicator of surface weather. When it is lying flat across the country as in figure 68 there is very little surface activity and generally good weather lies south of it (north in the southern hemisphere). When it dips down as in the second part of the figure it indicates that a cold front and the formation of a low will soon follow. This jet stream dip usually precedes the front by several hours to a day, although a strong front may catch and pass the position of the jet stream eventually as shown.

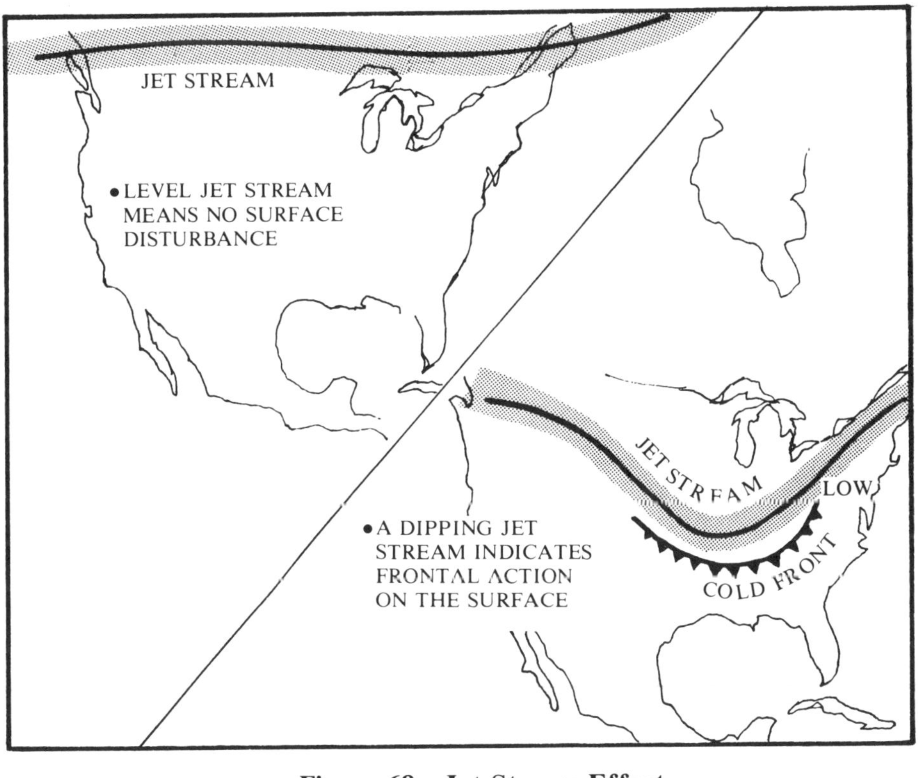

Figure 68 - Jet Stream Effect

STEERING WINDS AND PRESSURE PATTERNS

Even when an actual concentrated jet doesn't exist, the upper winds are known as steering winds for they help direct the movement of pressure systems and thus fronts. A low typically forms below the poleward flowing portion of these winds as we have seen previously. This low then rides slowly along below this flow to eventually stagnate and dissipate in the northeast extremes of land masses in the northern hemisphere or in the southeast below the equator.

Highs are likewise directed by the steering winds but they drift in a more southeasterly direction in the northern hemisphere and northeasterly in the southern hemisphere. Figure 69 shows the typical pattern of high and low pressure system drift in summer and winter.

Due to the seasonal change of these steering winds, at certain times of the year they are absent from some areas. This allows dominant pressure systems

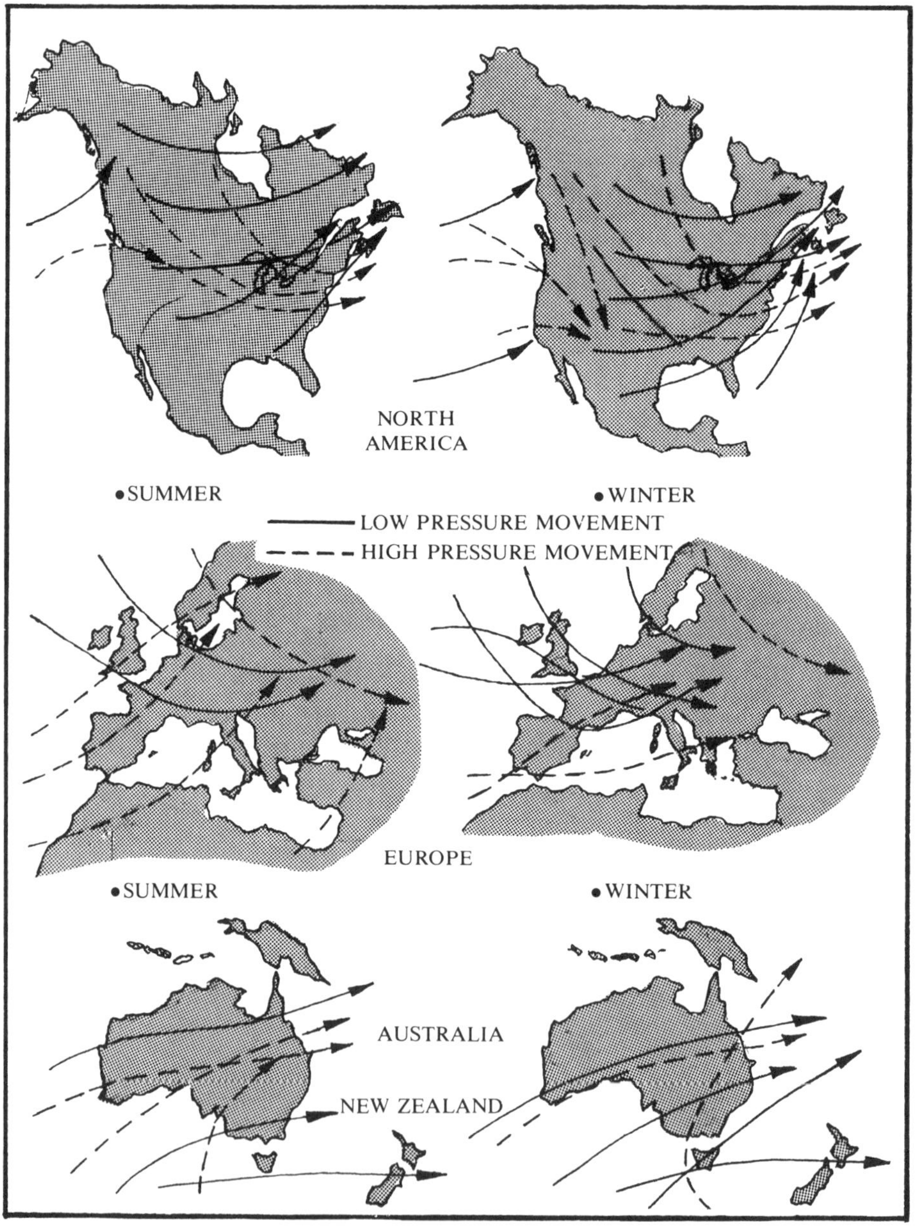

Figure 69 - Pressure System Movement

to set up and persist for days or weeks. For example, in the summer when the jet stream moves north in the northern hemisphere a steady high off the coast of Florida known as the Bermuda high pumps warm humid air into the southeast United States. Similarly a standing high off the coast of Portugal known as the Azores high keeps Spain and western Europe well supplied with warm air that helps populate the famous beaches.

Besides the two well-known systems above, the northern summer sees a Pacific high, a California low and a strong Mid-East heat low. The winter brings on the Canadian high, the Siberian high and the Icelandic low.

TROPICAL WEATHER

Below latitudes of about 20°, Coriolis force becomes insignificant so that winds no longer parallel the isobars but cross them at ever increasing angles, even at altitude. Near the equator the air flows generally perpendicular to the isobars. In addition, fronts and large pressure systems rarely visit the tropics, but diurnal effects, seasonal changes, small-scale waves and tropical cyclones do.

Diurnal effects refer to changes that take place on a daily basis, usually as a result of the sun's heating. These effects can be a major source of weather activity and wind since the tropical sun is so intense. Heat lows (see figure 62) form over drier and barer areas while relative highs appear over cooler areas (lakes and forests). Mountains and other topographical features also lead a hand in altering the daily air flow. The important point to realize is that the weather tends to be repetitive and consistent in the tropics from day-to-day in a given season.

Another diurnal effect is the atmospheric tides. Just like the oceans of water, the ocean of atmosphere exhibits a high and low tide twice a day. This oscillation affects the pressure on the surface in the range of 2 to 3 mb. If we consider local noon to be when the sun is directly above our longitude, then the time of highest pressure due to tidal action is 10:00 and 22:00 hours with lows appearing at 4:00 and 16:00 hours. These lows and highs move from east to west and alter the wind flow as they go.

Seasonal changes in the tropics are due to the earth's tilt and are related to the north and south migration of the Intertropical Convergence Zone (ITCZ). The ITCZ is labeled in figure 40 and consists of the band of equatorial heat lows with easterly flowing winds converging on the area of greatest heating. Figure 70 shows the typical ITCZ in January and July with the typical wind patterns. Note that these winds are mainly easterly except where the ITCZ migrates far from the equator. When these winds blow off the oceans they bring seasonal rains to the nearby land areas.

The band of tropical highs that typically reside around 30° latitude frequently exhibit waves that propogate from east to west. These are known as *easterly* or *transverse waves*. They bring weather disturbances in the form of towering cumulus clouds every four to six days to tropical areas in the local summer and autumn.

Some of these waves deepen and become more serious matters. In this case they are known as *tropical depressions*. If the pressure continues to fall and winds increase to over 33 knots (38 mph or 61 km/h) the system is

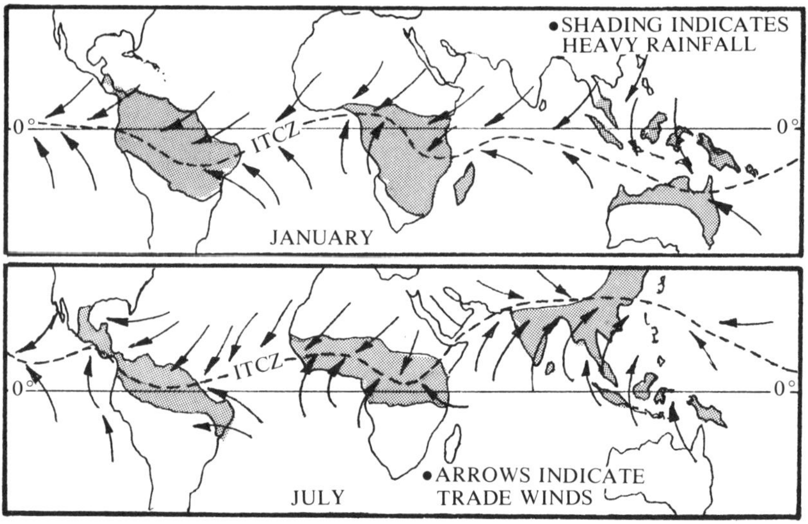

Figure 70 - Intertropical Convergence Zone Positions

known as a *tropical storm*. If the low continues to build and wind speed reaches 62 knots (72 mph or 115 km/h), we have on our hands a full-blown *tropical cyclone* which is variously known as a hurricane, typhoon or cyclone depending on the region.

Tropical cyclones are huge affairs covering hundreds of miles. They are fed by the latent heat of evaporation that exists in the humid air over the ocean and is released when this humidity condenses into clouds. They are giant heat engines and appear in cross section as shown in figure 71. Here

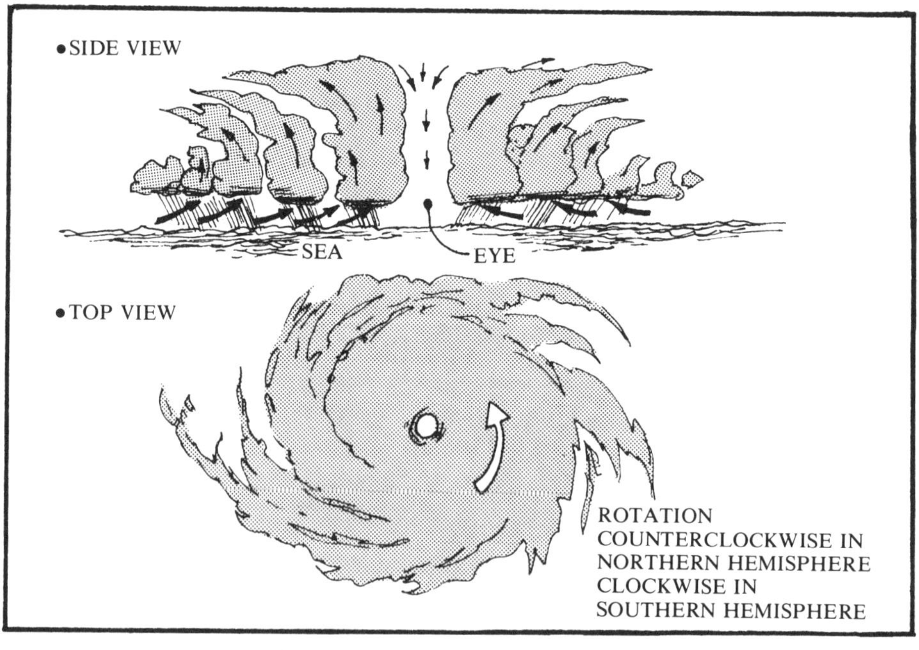

Figure 71 - Views of a Hurricane

the famous "eye of the hurricane" shows as a calm central region while the rest of the cyclone rages around in a mad swirl—counterclockwise in the northern hemisphere and clockwise in the southern hemisphere.

Tropical cyclones tend to follow a westward path then angle poleward as they are steered by the upper winds. They may remain far at sea and eventually turn into typical large-scale lows as they enter the temperate zones. Otherwise they slam into coastal areas bringing damaging tides, floods and winds which can exceed 100 mph (160 km/h). In this case, the cyclone quickly dies because the ground cuts off the supply of warm, moist air that fuels the snarling behemoth. Even so, such a storm hitting the coast can cause heavy rain and clouds over a third of a continent for several days after its demise. Needless to say, pilots near seaboards attractive to tropical cyclones are not pleased when they are approaching.

ISLAND WEATHER

Many of us dream of living on a tropical island for a spell, but despite the allure of gentle ocean breezes, blue crystal water and lithe natives, the flying weather isn't always ideal. Certainly an island situated in the trade winds or the Pacific westerlies as is Hawaii receives the blessing of constant, smooth, dependable winds which are fine for soaring. But most islands are little weather systems unto themselves.

Certainly islands do not lack a source of moist air—usually warm—for they are surrounded by it. When the land heats up in the day it forms a local heat low that sucks in air from all sides (in the absence of a general wind) that converges and rises over the island forming a cloud. This cloud continues to grow and may reach thunderstorm size with a subsequent heavy shower. An island acts like a mountain forming a cap cloud, but with all the supply of moist air the island doesn't need to be high to form clouds. If the island is mountainous it creates a cloud all that more readily and is often clouded over by early afternoon. Figure 72 shows an island in the afternoon in calm and windy conditions. If you plan to visit mountainous tropical islands, expect to do your flying early.

Other large islands out of the tropics such as Vancouver Island, Long

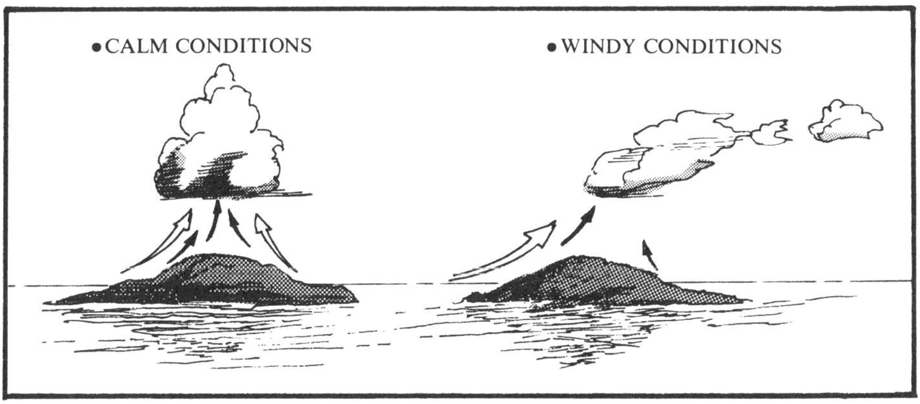

Figure 72 - Island Conditions

Island, the Azores, Ireland, Japan, and so forth exhibit sea breeze effects around their coast lines which combine with other local effects and large scale wind movement. We mention these island winds in the next few chapters.

SUMMARY

The changing, moving, seething atmosphere above us is a complex fluid that pulsates to the cyclic beat of the sun and the urge of gravity. To fully understand its ways and means we must start with the big picture, look at more and more details until we are observing very local effects then relate these back to the overall scale of things once again. A weatherman's understanding takes a lifetime of study. We can only hope to grasp the salient points immediately, but as we continue our experience in the air our personal observation coupled with our elementary knowledge of the sky's behavior pays off in dividends of understanding and practical flying excellence.

The basic concepts are these: Solar heating creates world-wide circulation that is altered by Coriolis effects and limited by gravity. This circulation creates imbalances of temperature distribution and pressure that cause outbreaks or waves of cold air from the poles and warm air from the tropics trying to reestablish equilibrium. The leading edges of these waves are known as fronts which have their own identity and associated weather.

Winds are created and driven by pressure differences over a distance which are known as pressure gradients. Aloft the pressure patterns are larger, less complex and more spread out giving rise to generally stronger winds that uniformly follow the pressure isobars. At the surface friction causes the wind to cross the isobars at an angle less than 45° usually. This knowledge helps a pilot predict the wind direction and strength at any level, given the appropriate weather maps.

Add to the above understanding the knowledge of the movement of fronts and pressure systems as well as the possibility of secondary disturbances and a pilot is well-armed to predict what the next day or two will bring in terms of flying conditions. That's as much as the average weather station is willing to do with any certainty.

CHAPTER V

Wind Patterns

Because we depend on the air for our very life's breath we have developed a subconscious awareness of its changes. With our vision we note clouds drifting on high while our senses of hearing and touch can directly detect the air's movement.

We call the moving air wind and we describe this movement variously as a breeze, drift, eddy, gale, gust, howl or zephyr. We also have names for special winds like chinook, foehn, harmattan, khamsin, levanter, mistral, Mono, Santa Ana and sirocco indicating our awareness of their cause and effects.

We pilots have a great incentive to seek out the benign moods of the winds. Certain breezes can allow us to ferry across vast expanses of sky while strong blows can turn our flights of fancy into episodes of terror. For this reason it is important that we understand wind patterns.

As we have discovered previously, wind can be created on the large scale and blow for long distances. It can also be created locally and change quickly which is the subject of a later chapter. In this chapter we investigate the behavior of winds on the larger scale.

MEASURING THE WIND

We have discovered that wind is the wandering of air from place to place. Whether this air is rushing or merely drifting depends on the amount of pressure driving it. What we feel as wind is the push of air molecules against our body. If they are moving faster, more force is felt because the molecules have more momentum as they strike us. In fact, the force of the air varies with the square of the airspeed. In a similar manner our wings experience a force from the air moving over them. The limit to safe flying comes when the speed of the wind approaches that of our minimum flying speed so that we can no longer maneuver with respect to the ground. Since different aircraft have different minimum flying speeds and controllability, they can safely handle different wind maximums.

On the surface we measure the wind with an *anemometer* (anemo means wind in Greek). The official height to measure surface winds is 10 meters or 33 feet above the ground. This means the measurement is above some of the ground friction and turbulence. Various types of anemometers or wind speed indicators such as a spinning cup or impeller type (see figure 73) measure true wind speed because the vane moves along with the wind velocity. Other types such as a pitot tube or floating ball type vary their reading according to air density (dependent on air temperature and altitude) and read relative wind speed since they detect dynamic pressure of the wind. They read low at altitude or in hot, humid conditions. That's not as bad as it sounds, for usually when we are interested in the wind speed we want to know its effect on our flying craft and this depends on dynamic pressure of the air. A true 15 knot wind at 10,000 feet will have less effect on us than a true 15 knot wind at sea level.

You can use your airspeed indicator as an anemometer—they are iden-

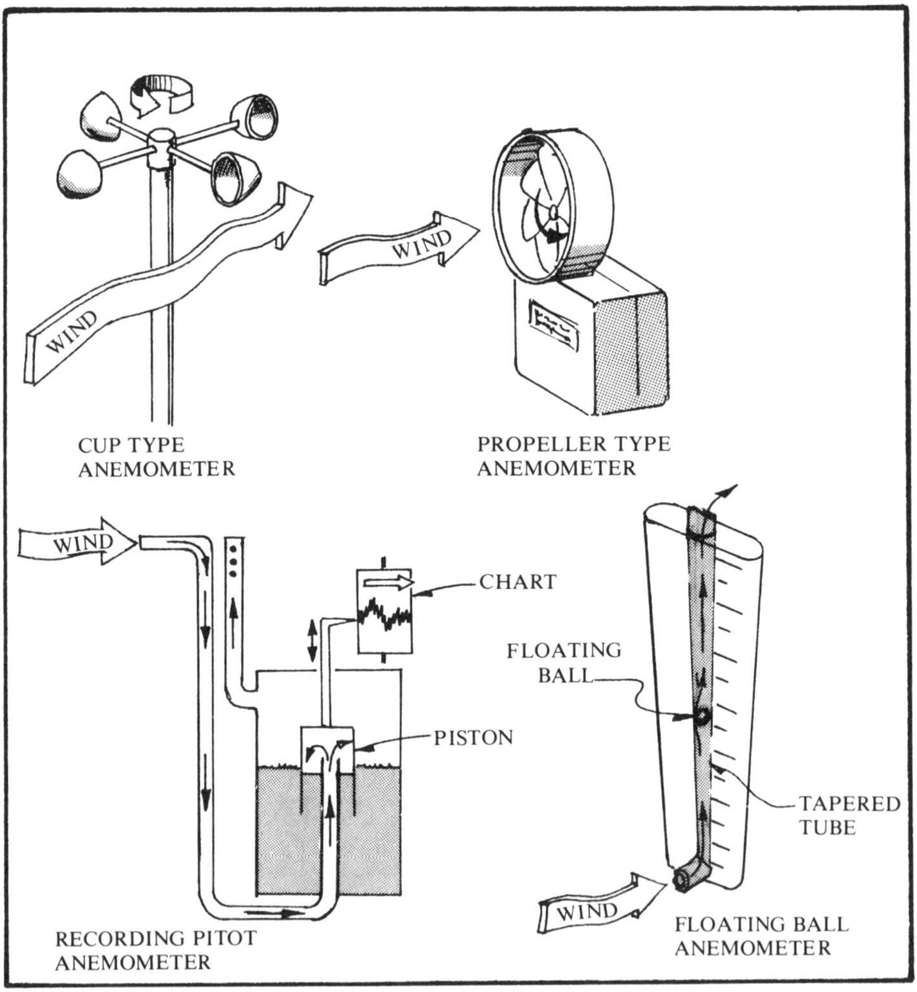

Figure 73 - Wind Speed Indicators

tical except for the mounting system perhaps. Just note the type of system you have so you know if you are reading true or relative wind speed. (Incidently, an airspeed indicator reading relative velocity is not disadvantageous for although it reads low at altitude your aircraft is flying faster in this less dense air. Thus your relative airspeed indicator will always read the same at stall and other cardinal speeds no matter what the air density.)

Weathermen and seamen have classified winds over the years according to their strength. The chart below is a guide to official wind measurements and terms:

Wind Speed Classification

Weather Classification	*Speed Knots**	*Km/h*	*Beaufort Force*	*Terms used in General Forecasts*
Calm	Less than 1	Less than 2	0	Calm
Light Air	1-3	2-6	1	Light Winds
Light Breeze	4-6	7-12	2	
Gentle Breeze	7-10	13-19	3	
Moderate Breeze	11-16	20-30	4	Moderate Wind
Fresh Breeze	17-21	31-39	5	Fresh Wind
Strong Breeze	22-27	40-52	6	Breezy
Moderate Gale	28-33	53-61	7	Strong Wind
Fresh Gale	34-40	62-74	8	Gale
Strong Gale	41-47	75-87	9	Severe Gale or Storm
Whole Gale	48-55	88-102	10	
Storm Gale	56-63	103-117	11	
Hurricane	64-71	118-142	12	

*One knot equals 1.15 miles per hour

WATCHING THE WIND

If an anemometer is not immediately at hand we can still get a good idea of wind speeds by watching its effect on the environment. The handy chart on the next page is a good guide.

These wind effects relate only to wind speed. We also need to know the wind direction when we are flying. In the air we can detect the movement of wind lines in grass and trees to surmise the direction as well as speed. Flags, smoke and clothes on the lines all give useful information of wind direction and speed with flags being best, followed by smoke, then clothes. Flags trail in the downwind wind direction then flutter, slowly wave, flap and snap as the wind increases (be cautious of relying on large, heavy flags in light winds). Smoke lies flatter as the wind speed increases.

Bodies of water can be a reliable source of wind information once you learn how to read it. Large water expanses may show ripples or waves that move in the direction of the wind. Be cautious not to confuse current effects with wind effects. Spend some time watching water surfaces from the shore and note how they react in different winds. Off the coast of California the Pacific cliffs afford an excellent vantage point for wind watching.

Wind Table

Wind Velocity	Effects On Environment
calm	Smoke straight up. No movement in vegetation.
0-3 mph (0-5 km/h)	—Smoke straight up. Leaves begin to rustle.
3-5 mph (5-8 km/h)	—Smoke leans, twigs move.
5-10 mph (16-29 km/h)	—Smoke leans about 45°. Small branches and grass begin to move.
10-18 mph (16-29 km/h)	—Smoke lies about 30° up from horizontal. Whole branches begin moving. Grass waves. Clothes move on a line.
18-25 mph (29-40 km/h)	—Smoke lies flat. Large branches wave. Grass ripples and clothes wave. Dust swirls begin.
25-35 mph (40-56 km/h)	—Large limbs and medium trunks swag. Clothes flap. Dust and snow blow readily.
35 and over (56 km/h)	—Larger trees sway, cars rock. Difficulty walking into wind.

Frequently shear lines—the meeting of different volumes of air—can be observed in the interplay of different wind lines on the water.

Small ponds and lakes can be used in quite a different manner. Because they are contained in a basin of earth, they are always lower than the surrounding ground. Hence they block the air on their upwind side. The result is a calm mirror-like surface near the upwind shore and a sheet or streams of ripples further downwind as shown in figure 74. The stronger the wind, the more widespread is the disturbed area. From some positions you cannot detect the difference in the smooth and rippled surface due to the sun angle, so change your position to be sure to get a true reading if you are relying on wind lines. This skill should be in every pilot's repetoire for oftentimes ponds are the only wind indicators available.

One interesting feature that pilots in wooded areas can observe is the progression of wind lines up a tree-covered mountain. As gusts of wind blow up the slope it turns up the leaves so that the lighter sides show, indicating where the strongest flow exists. This feature can allow you to tell wind direction on the mountain as well as the presence of thermals or gusts and how far down the mountain the stronger winds extend. All this information can be obtained in the valley to gain insight into soaring prospects before trekking up the mountain. This is an important guide to pilots in Europe and the eastern U.S.

PREVAILING WINDS

Any given point on the earth experiences winds from all directions, but some given directions are much more favored than others. Over a long

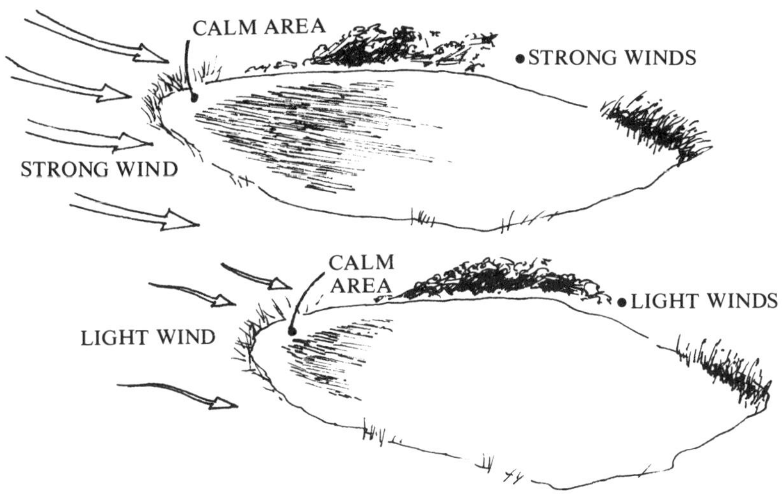

Figure 74 - Wind on Water Effects

period of data collecting we can draw a "wind rose" for an area that depicts the frequency of wind direction. One such wind rose is shown in figure 75. Here we see that southwest winds are the most common and would be considered the prevailing winds (the peak representing the wind is downwind from the rose center). South wind is also well represented. West then northwest winds are frequent with a bit less southeast then a small dose of northeast winds. East winds tend to be the least frequent.

This wind rose example is actually quite similar to the situation on the eastern half of the North American continent and parts of Western Europe. Later we'll see why these wind patterns arise. Other areas have different prevailing winds and frequencies. The important thing is to recognize the tendency for winds to blow a great percentage of the time in a few preferred directions. This recognition is important in order to know how to orient airstrips or locate flying sites.

WIND FLOW NEAR THE SURFACE

At this point we know why the wind blows but we need to point a clear picture of how it reacts with the terrain and behaves in the lower levels

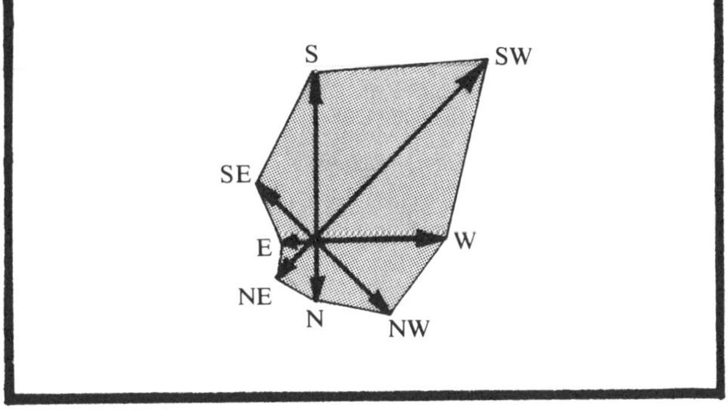

Figure 75 - Wind Rose

where we fly. The best way to imagine the wind flow and currents is to watch moving water. When a stream bed is smooth water exhibits uniform steady flow up to very fast speeds. When the stream is studded with rocks, logs or other obstacles, the flow becomes quite irregular, even at slow stream velocities. The water prefers to flow around rocks rather than over. When it is forced to go between two rocks it wells up and flows faster. When it does flow over a rock it will exhibit ripples or waves downstream. Behind many rocks it breaks into eddies or oscillations.

The airflow reacts to irregularities in the terrain in the same manner. It would rather go around an isolated mountain than over. It is channeled by valleys and forced to flow parallel to mountain chains and ridges. It wells up and moves faster when it is forced through a gap. It also exhibits turbulent eddies (swirls) and waves which are the subjects of later chapters. The main thing to learn from this mental exercise is that the terrain greatly alters the wind flow near the surface by forcing it to alter its path to pass by ridges, hills, tree lines and mountains.

Besides altering the wind's path, all these obstacles in the way also slow the wind near the surface. This is called the wind gradient (change with height) and is covered a bit later.

In addition to the mechanical effects outlined above, heating effects greatly alter the surface flow. This subject is left for Chapter VII, but here we need to understand that near the surface is where the wind exhibits most of its irregularities. The chart below helps define the behavior of the air in the lower layers.

Wind Layers

Approximate Height Above the Highest Surface Feature	*Nature of the Airflow*
	Free atmosphere. Here the air viscosity is not significant because it doesn't react with solid objects. The movement of the air is due only to the pressure gradient and Coriolis effect.
1,500 to 3,000 feet (500 to 1,000 m)	
	Region of transition. Here the effects of surface friction influence the wind structure. Sea breezes extend into this layer. Coriolis effect and density changes (heating effects) are major forces in this layer.
150 to 300 feet (50 to 100 m)	
Surface Boundary Layer	Region of continual shearing action. The wind structure is determined mainly by the nature of the surface and temperature changes with altitude. Valley and mountain winds exist within this layer.
(Top of Terrain)	

We have already seen in the last chapter how the surface friction effects slow the wind and cause the flow to cross the isobars whereas in the free atmosphere the wind follows the isobars. Now let's take a closer look at these surface patterns.

WINDS AROUND FRONTS AND PRESSURE SYSTEMS

A typical cold front and low pressure system is shown in figure 76. The drawing is three-dimensional and depicts a low at the surface with a cold front trailing south (northern hemisphere drawing). The low and the front extend up into the atmosphere.

The high pressure systems in the cold and warm sectors are shown with light lines representing the surface isobars. The larger shaded arrows represent cold air pushed by the high in the north while the dark arrows are the warm air driven by the warm sector high. Both sets of arrows tend to flow toward the relative low pressure along the front.

The cold flow curves under the warm flow and presents west to northeasterly winds once the front passes. The action of the warm air is most interesting. As it reaches the front it is lifted over the cold air and turns right, eventually achieving a southwest to northwest flow aloft.

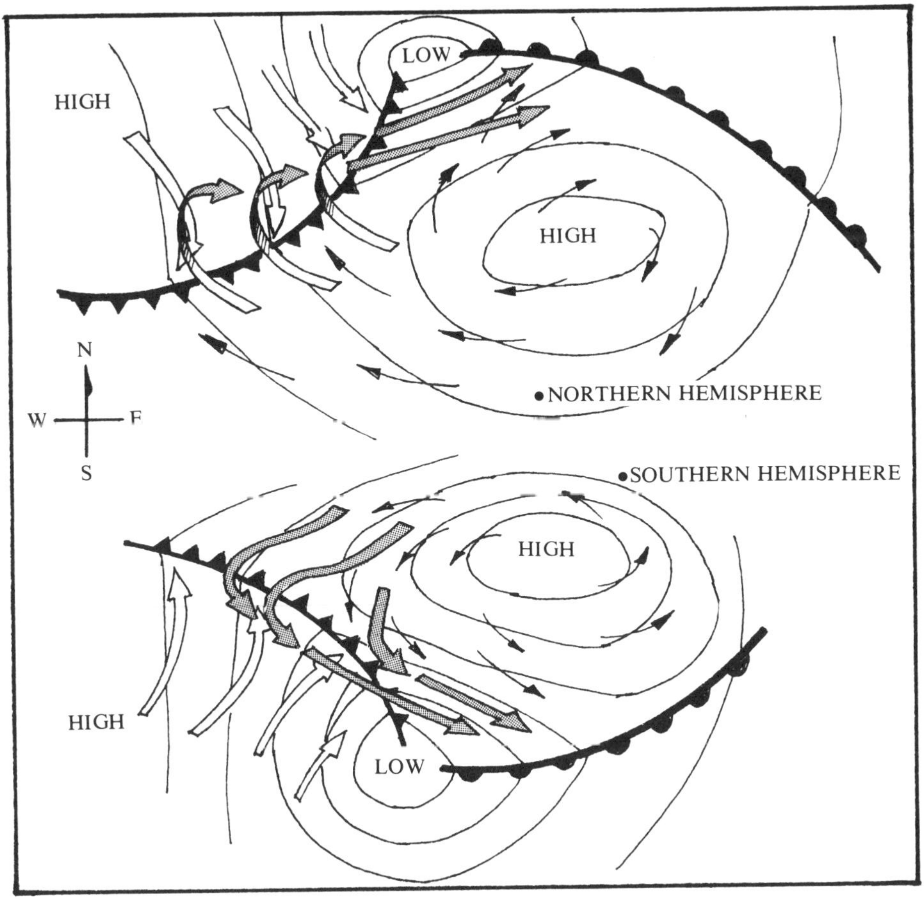

Figure 76 - Wind Flow Near a Cold Front

Frequently when a cold front approaches an area from the northwest in the northern hemisphere, the surface winds will be south to southeast while aloft the upper clouds can be seen to drift out of the northwest. Pilots have experienced thermaling up in the southeast flow before a front and drifting toward the northwest originally but gradually changing drift direction until the thermal heads back southeast in the northwest flow aloft. This can take place in a few thousand feet (1,000 m or more) of climb. Pilots have expressed surprise when confronted with this phenomenon and concern for the possibility of shearing action aloft. However, the observations described are a normal state of affairs in pre-frontal conditions. Of course very near the front there can be shearing action (see the next chapter) and severe weather, but the conditions we describe can precede the cold front by a day or two.

In the southern hemisphere the reversal of Coriolis effect and the arrival of cold fronts from the south creates a situation shown in the lower figure. North or northeasterly winds proceed the front while west to south winds follow its passage.

As we have seen in the preceding chapter, the western extension of a cold front stops its progress and often returns back as a warm front. Near this front we can draw a similar three-dimensional pattern to fathom the wind flow. This is shown in figure 77. As the cold front spreads southward it will be preceded by south to southwesterly winds then present north to northeasterly winds as it passes (northern hemisphere). Once it stops and heads back to the northeast as a warm front the winds in the cold sector turn clockwise to become southeasterly to south, changing to southwest as the front passes. As the warm air pushes north it flows over the cool air

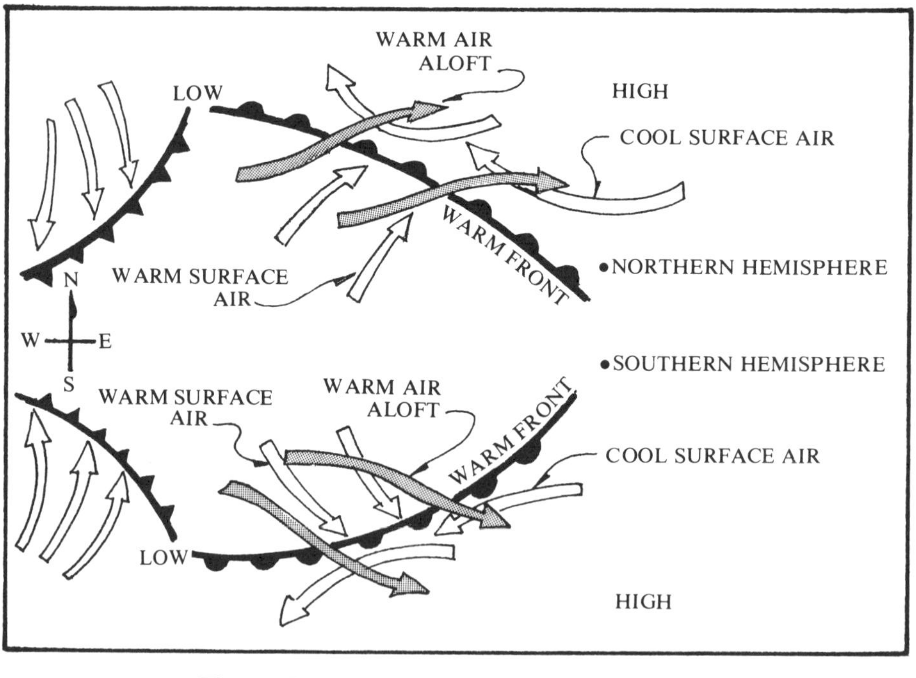

Figure 77 - Wind Flow Near a Warm Front

and turns to the right to flow westerly aloft.

In the southern hemisphere the situation appears as in the lower figure. Here we see northeast to north winds proceding the warm front while northwest winds follow it.

MOVING WEATHER

We have learned that weather generally moves from west to east in the temperate zones of both hemispheres. This allows us to picture some typical situations so we can learn what weather and wind to expect given a rough idea of the position of weather systems.

A RIDGE OF HIGH PRESSURE PASSES

In figure 78 we show a high pressure ridge in the northern hemisphere. Assume your initial position is at A. As the system moves east the portion along the line AA′ passes you. First you experience northwesterly flow that gradually turns more west and diminishes. When the ridge maximum (dashed line) passes over you the winds may be calm and skies clear. Then they increase and turn southwest to south as the ridge departs east. Skies may again cloud up as the next low pressure approaches.

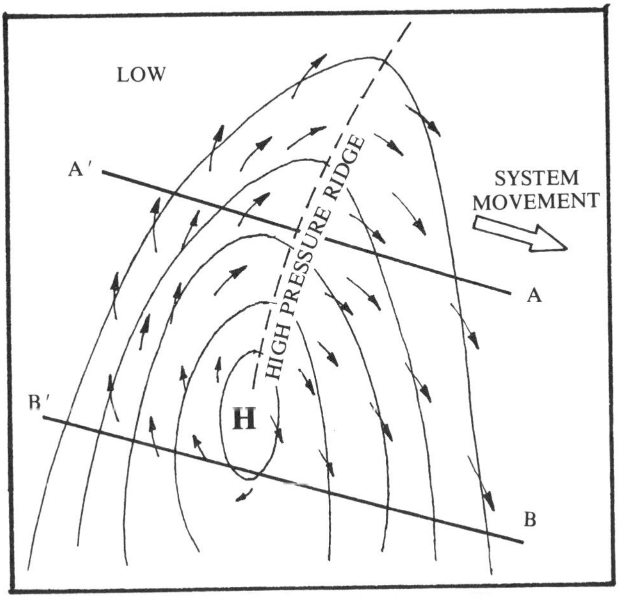

Figure 78 - Wind Flow Near a Passing Ridge of High Pressure

In the southern hemisphere the process is the sáme except initial winds will be southwest changing to north or northwest.

Now imagine that you are located at B. As the system moves east you experience the conditions along line BB′. In this case an initial northwesterly wind turns clockwise to become northeast, east, southeast and eventually south. In any case it will be quite light or negligible in the weak pressure area below the high. In the southern hemisphere the winds turn counterclockwise from their initial southwest flow in this situation.

A LOW PRESSURE SYSTEM PASSES

In figure 79 we have a low pressure system and the typical fronts that trail from it in the northern hemisphere (note that the warm front shown is the western portion of the previous cold wave that is moving back north as a warm front). Now assume you are stationed at point A and the northeasterly drifting low brings you conditions along line AA′. First you experience a southeasterly drift that turns counterclockwise and strengthens to east, northeast then northerly.

Although you don't encounter any fronts in this area (except possibly an occluded front that has wrapped well around a mature low), there is plenty of rain and clouds that increase and lower as the wind turns east. This weather will continue until the winds turn northwesterly which may take twelve hours or more. If the low passes well to your south (hundreds of miles) you may miss the rain and only see high clouds in the southern sky.

If you are positioned at point B the moving system will present the weather along line BB′. In this case an initial westerly wind will turn counterclockwise toward the southwest to southeast, increase in speed then turn a bit clockwise to southwest (if it is not already southwest) as the warm front passes. Then the winds remain steady until the cold front passes and a sharp veer to the northwest occurs. Finally the wind weakens and turns more northerly or westerly depending on the position of the next high.

The weather associated with the event described above is simply warm front weather followed by clearing between the fronts followed by cold front weather. The warm front normally brings clouds and rain for a day or days while cold fronts usually pass in less than a day.

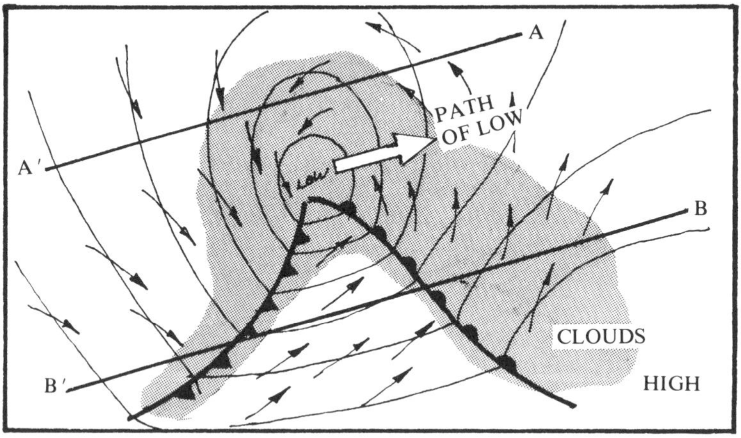

Figure 79 - Wind Flow With a Passing Low - Northern Hemisphere

In the southern hemisphere the situation is shown in figure 80. If you are stationed south of the low track at A you will experience northeasterly winds turning more east then continuing clockwise to end up southwest. The weather and wind strength will be similar to that as described along AA′ above.

If you are at B your original westerly winds will turn northwest to northeast as the warm front approaches then turn northwest between the fronts

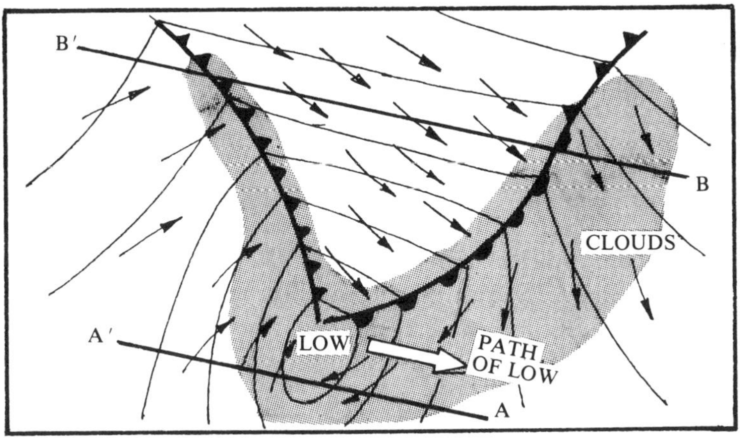

Figure 80 - Wind Flow With a Passing Low - Southern Hemisphere

and quickly change to southwest as the cold front passes. Again the weather will be as described for the northern hemisphere.

Use this general outline of wind and weather behavior to understand the flows in your area related to the position and movement of weather systems. Observe the wind given the current weather map to gain experience predicting the wind you desire for excellent flying. As any pilot or sailor can tell you the predictions based on the data aren't always right but we greatly enhance our chances for good flying when we learn when to expect it.

AREAS BEYOND THE FRONTS

The above explanations are fine for the flying public that lives where fronts are frequent visitors, but many pilots live beyond the reach of such weather systems. Also some areas require modification to our simple model. Figure 81 presents typical sea level weather maps for the months of January and July respectively. The most frequent pressure systems and frontal activity is shown.

Starting with the North American continent we see that our ideal fronts assail the North and continue across to the East. The northwest coast from Alaska down to northern California experiences moist cold fronts from the ocean and the back sides of the continental cold fronts which often combine to form an occlusion. The winds experienced in these areas are those explained in relation to the respective fronts.

The southwestern United States escapes the marauding fronts normally and is dominated by a heat low in summer and a pacific high in winter. The winds tend to be northwesterly in winter and weak or easterly in summer (from southeast to northeast depending on location). In these areas local heating effects including the seabreeze along the coast are often the most important determinants of the wind's direction. It is not uncommon for a pilot to encounter completely opposite winds in a flight of only a score of miles due to the production of local winds.

Turning to the European continent we see that the influence of the Icelandic low tends to pull frontal systems around it so that cold fronts often hit the British Isles and western Europe traveling from the west or

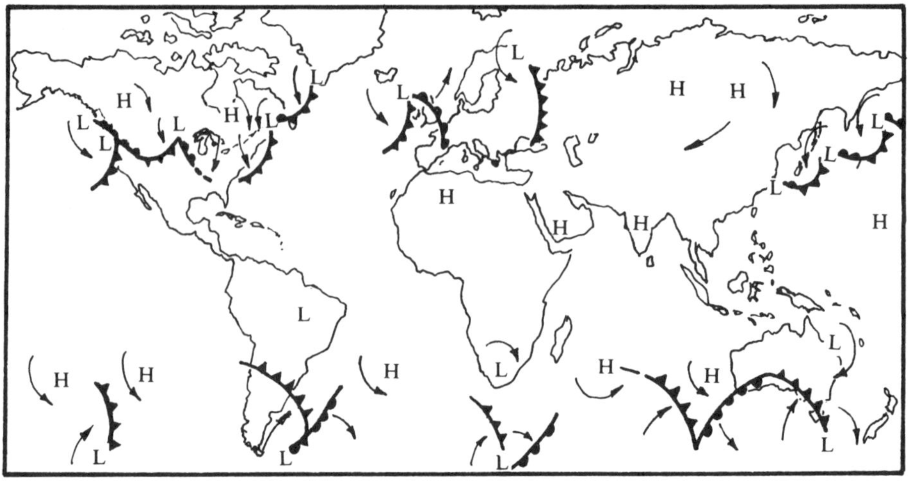

•JANUARY SURFACE CONDITIONS

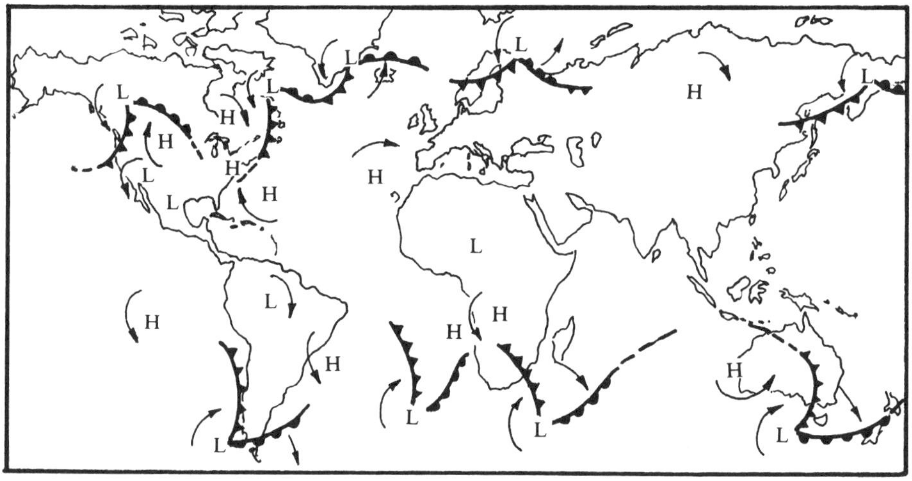

•JULY SURFACE CONDITIONS

Figure 81 - General Worldwide Surface Conditions

southwest. The positioning of these fronts and the fact that they travel over the warming effects of the Atlantic ocean change their nature. We have to rotate our expected winds up to 90°counterclockwise from the model given in the previous section when viewing western Europe. Cold fronts are often not as severe and carry much moisture to fall on the Alps as snow due to their maritime nature. Also, as we have seen, Europe can experience the back side of a continental cold front that offers much colder air to the continent in a manner similar to that experienced by the American Northwest.

Australia experiences classic frontal passage in its southeastern half. Naturally the east coast feels these fronts the most. The west coast of Australia is dry from lack of fronts due to the permanent tropical high that lies off the west coast. The north of Australia reaches to 15° latitude and experiences tropical weather with occasional cyclones and torrents of rain. Winds in Western Australia are light southerly or dominated by heating effects while those in the more populated eastern portion of the country are typical for the current pressure systems and fronts.

South Africa tends to be high dominated in its local winter and low dominated in summer from heating effects. Cold fronts cross the area in winter and bring tropical winds. At other times of the year heating effects including the seabreeze are important determinants of the wind as in southern California.

BACKING AND VEERING WINDS

We have seen winds change direction over a distance along the surface. They also change with altitude. Here we introduce two terms that describe these direction changes (whether with distance, height or time) and are frequently found in weather reports or sources of information.

A *backing wind* is one that turns or changes in a counterclockwise direction as shown in figure 82. A backing wind in the northern hemisphere is most common a day or so after a cold front passes and eventually protends the approach of the next front—usually a warm front.

A *veering wind* is one that turns in a clockwise direction as shown in the figure. The wind often veers immediately after frontal passage.

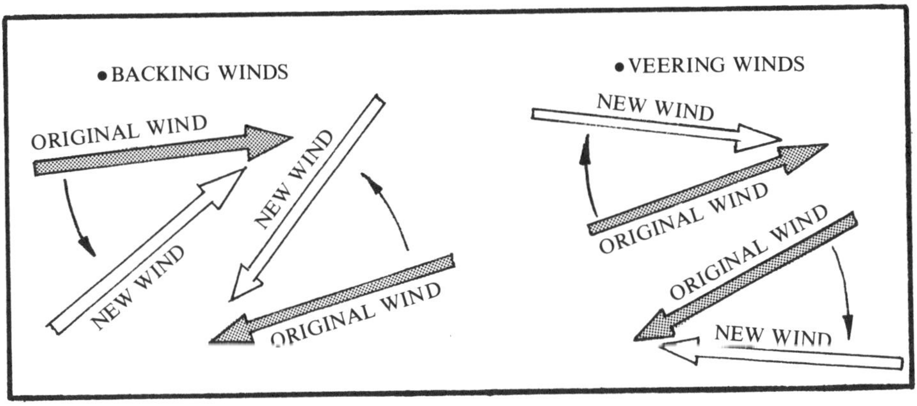

Figure 82 - Backing and Veering Winds

THE WIND GRADIENT

We have seen that the wind is slowed by surface friction as it nears the ground. We have also seen that this slowing causes the wind to cross the isobars at the surface while it follows them aloft. Let's investigate these matters in more detail so we know what winds to expect at altitude.

The wind gradient (again gradient means change) near the ground can be pictured as in figure 83. This graphical representation of the wind speed with height can be called the *wind gradient profile*. In uniform conditions (no thermals) over smooth ground the flow is non-turbulent and the largest change in wind speed occurs close to the ground. In rough terrain turbulence exists that tends to spread out the speed differences in adjacent layers so the flow changes speed more gradually with height as shown in the figure.

The problems related to wind gradient are well-known in aviation circles, for descending into such a wind can rapidly reduce airspeed. In addition, the presence of wind gradient means that the winds aloft are stronger than those on the ground. It is this matter that we investigate next.

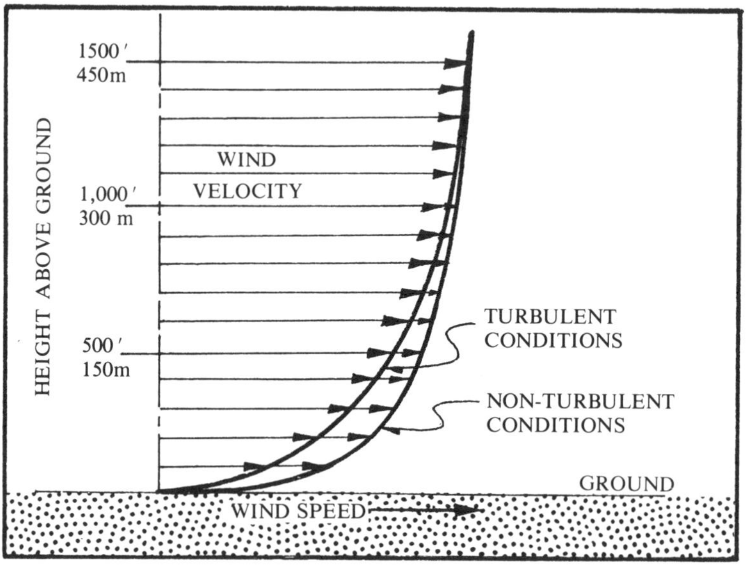

Figure 83 - Typical Wind Gradients

WINDS ALOFT

We have seen earlier in this chapter that winds from 1,500 to 3,000 feet (500 to 1,000 m) above the top of the terrain are no longer under the influence of the terrain's friction. Winds at this level are said to be in the *free atmosphere* and they reach their *free stream velocity*. What this means is by the time we reach several thousand feet above the local mountain peaks we encounter wind that is following the isobars and is flowing with the speed appropriate for the pressure gradient at that level. Wind that obeys the direction of the isobars away from the friction layer is said to be a *geostrophic wind.*

Since the wind at the surface crosses the isobars we can expect that it must change directions with altitude as it changes speed. This velocity (speed and direction) change is shown in figure 84. Here we see the wind turn clockwise (veer) as we rise from the surface to the level of the free stream where it is parallel to the isobars. Also we see it increase from 10 mph to 20 mph in the same height.

Also in the figure we show the situation over a smooth surface (water) and rougher terrain. In the first case the turning of the wind is much less because it is not slowed as much at the surface. In the second case the additional slowing results in more turning at the surface.

The turning of the wind with altitude is an important concept for pilots when following thermal tracks, plotting courses, conserving fuel or using soaring faces of varying mountains. It should be noted that unstable conditions with thermals tend to distribute the air up and down so that the average turning of the wind is less than it is in stable conditions. Also note that long valleys tend to turn the surface wind so that it flows parallel to the valley. This action may increase the amount of turning in the wind direction from the surface to higher levels.

The direction of turning with altitude up to the free stream velocity is usually clockwise (veering) in the northern hemisphere and counter-clockwise in the southern hemisphere. Above this level the wind will turn again to conform to the flow of the higher winds. In the temperate climates these upper level winds are usually westerly although in rare cases when the jet stream has moved far poleward some upper level meandering can occur.

The best way to tell the upper level wind direction is to watch the drift of high level clouds (use a stationary ground object as a reference). Invariably the free stream wind will turn to this direction, usually in a gradual manner. Looking at the figures 58 and 68 in Chapter IV showing the upper winds in relation to pressure systems and fronts will help you visualize what upper winds to expect. Note that upper winds almost always blow parallel to fronts. The greatest change (up to 180°) is found in winds that are easterly at the surface in temperate zones. All this knowledge is useful when working high thermals that drift with the wind.

We have seen that the wind above the surface influence increases to the free stream velocity. Above this what happens depends on the location of the upper level winds and especially the jet stream. Weather charts of the higher elevations depict these upper winds, and some weather broadcasts for public consumption also indicate the jet stream path. In the absense of

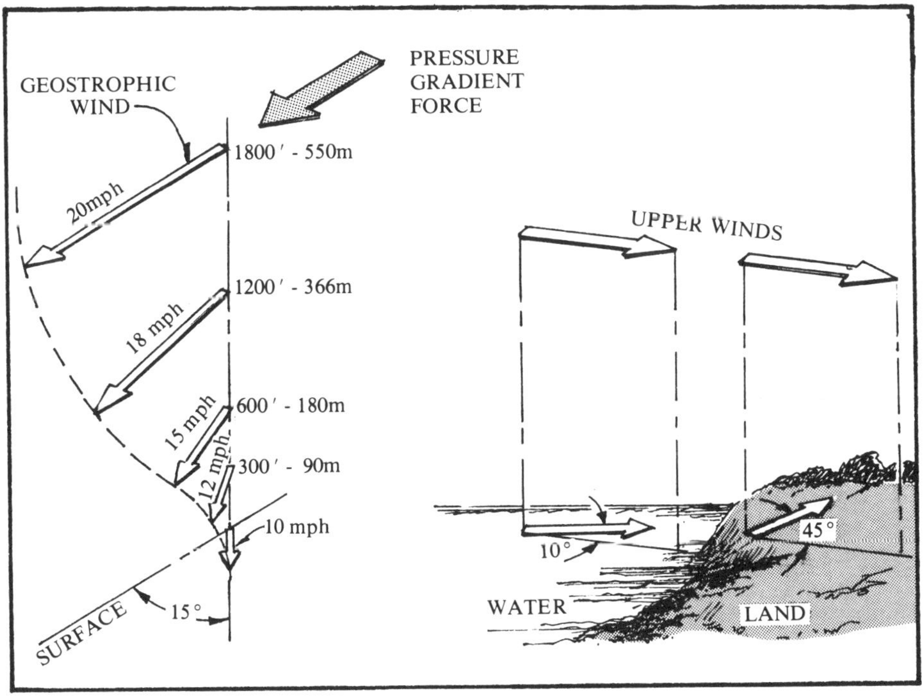

Figure 84 - Wind Velocity Changes with Altitude

Spreading contrail in high winds aloft.

these sources of information we can make the general rule that *winds increase aloft in a warm sector high and decrease aloft in a cold sector high.*

We should note that pressure systems at altitude may differ markedly from those at the surface. Also, the lower atmosphere may be layered with air masses as one moves over or under another. The layers often differ in temperature, moisture and motion. Thus we may find a change in wind speed and direction several times as we venture aloft. Generally the wind aloft indicates what the wind at the surface will become. The wind aloft changes first.

JET STREAMS

We have previously mentioned the jet stream several times for it is an integral part of weather in the temperate zones, but the curious thing is, it wasn't even known until large aircraft started flying high enough in World War II to feel its effects.

Actually, more than one jet stream commonly appears. One type is the *subtropical jet* shown previously in figure 43. These jet streams are quite high (above 40,000 feet or 14 km), located around 30° latitude and are shorter and weaker than the jet that lies in the temperate zone. They are of little consequence to sport pilots.

The *polar front jet* is located at the boundary of warm and cold air in the temperate zone as shown in figure 43. It is a current of fast-moving air in the polar westerlies and appears as in figure 85. In actuality, the jet doesn't appear with distinct layers as shown for there is a smooth transition from the jet maximum in the center to the lesser winds along the sides.

The fastest winds in the jet stream occur when the jet meanders farthest poleward while the jet slows down when it dips toward the equator. The reason for this is given in the discussion accompanying figure 58. The polar jet is located near 30,000 feet (10 km) up and can reach 200 knots over North America and Europe and up to 300 knots over Japan and New

Zealand where conditions for its formation are favorable.

Jet streams are caused by the strong temperature contrast between the polar and tropical air masses. They can be thought of as a combination of a convergence zone and a pressure gradient flow created by the temperature difference. This flow is induced toward the poles but turns right in the northern hemisphere and left in the southern hemisphere to circle the globe from west to east. Zones of strong horizontal temperature gradient, surface fronts and jet streams are typically found together.

The importance of the polar jet to sport pilots is twofold. First it helps move fronts and low pressure systems as explained in the last chapter. Watching the jet stream can give us a warning of the weather. Secondly it naturally means strong winds aloft. Flying under a jet stream is not necessarily dangerous, but one should expect higher winds to develop during the day as mixing takes place. Surface gustiness can be severe under a jet stream if fast-moving pockets of sink reach the ground.

The jet stream can often be seen on high because it frequently sports a long band of cirrus clouds. This band moves parallel to the jet stream but also may translate sideways as the jet meanders north and south.

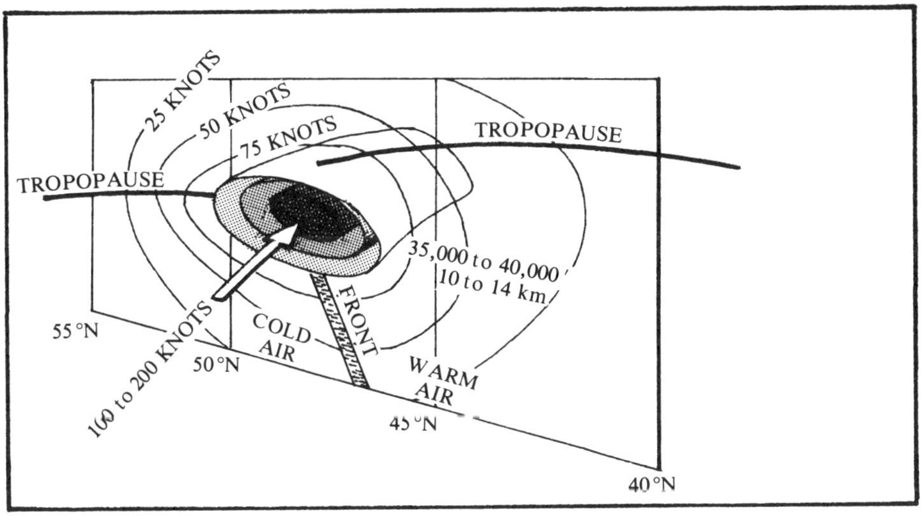

Figure 85 - The Polar Jet Stream

DIURNAL VARIATION OF THE WIND

The daily or diurnal variation of the general wind should be well known by anyone who has spent a couple decades on this planet. We know it usually picks up during the day and dies out at night. Figure 86 shows a typical 24 hour surface wind chart. Note that the maximum occurs in the early afternoon when peak heating and thermal exchange occur. The minimum is in the wee hours when the ground is coolest.

The occurence of these max and mins gives us a clue to the wind's variation. During the night the lower layer of the air is stable—a ground inversion occurs—so little vertical movement takes place. Consequently there is a large wind gradient resulting in little or no air movement near the sur-

Band of clouds parallel to the jet stream flow.

face. On the other hand, the sun's heating during the day causes an upward and downward exchange of air, bringing the higher velocity upper air down to the surface. The result is greater velocity surface winds when heating, instability and vertical currents are greatest in early afternoons. Figure 87 illustrates these matters.

A typical daily cycle starts out with still conditions in the early morning, then a gradual stirring of the air as the solar heating begins. The airflow will usually get a bit stronger and show a preferred direction as the morning continues. Once thermal activity deepens in late morning the exchange of air at different levels results in a rather rapid increase in surface winds. These winds often continue to build as thermal activity increases. Once solar heating begins to die thermals lessen and surface winds abate. Often this abatement can happen quickly. Winds continue to die into the evening

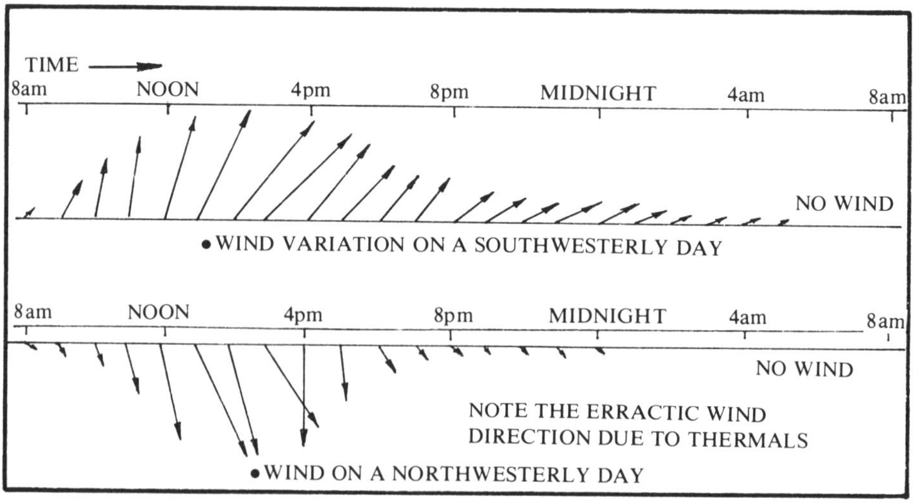

Figure 86 - Daily Wind Variation

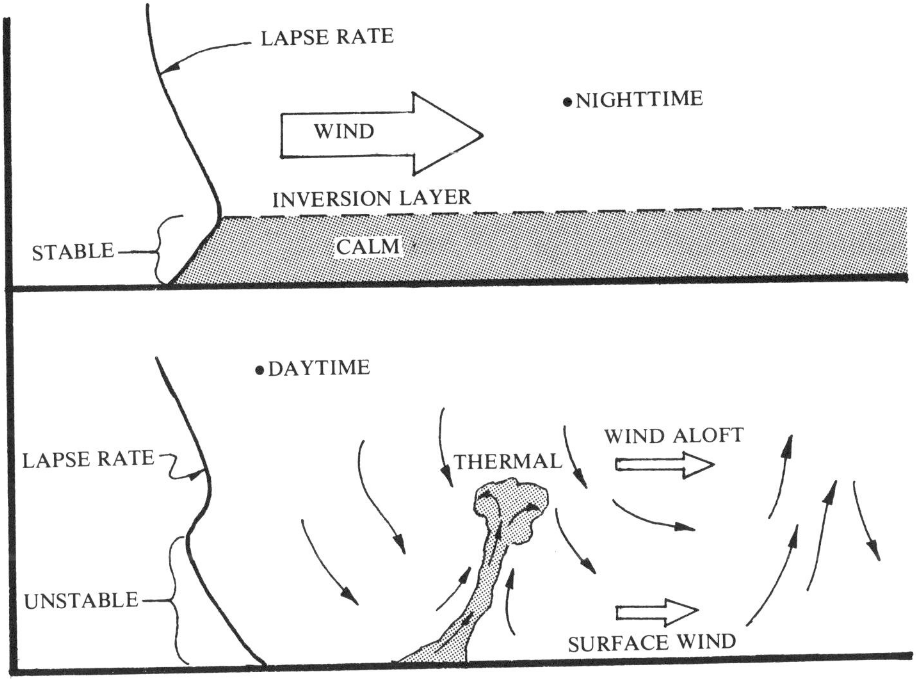

Figure 87 - Wind Brought Down From Aloft During Daytime

and night and are usually still in the wee hours. With the advent of a new day the cycle repeats itself.

There can often be exceptions to the above scenario. A very stable air mass will suppress thermals and may exhibit little or no surface winds. The hot oppressive days of summer in humid areas are of this nature. Another exception occurs when a front is in the area. In this case the surface wind may blow all day and all night with very little reduction in speed as evening falls.

LAYERING OF THE AIR

The air does not always consist of a single uniform air mass from the surface to the tropopause. Stratification or layering of the air can occur for several reasons. First we have already witnessed how a cool inverted layer next to the surface is created on clear nights. Also we have seen how a cold or warm air mass can move under or over one another. In mountainous areas cool winds sliding off the slopes can create layers. Also note the possibility of upper level air moving over cooler air held in the valleys of mountainous terrain. Finally we must mention the common occurence of an inversion layer in the subsiding air around a high.

All these layering possibilities are related to temperature discontinuities or changes. The important thing to note is that the wind profile is often tied in to the temperature profile (lapse rate). This is because air of different temperature has different density and air of different density does not readily mix. As a result, we often find a warm layer sliding by a cold layer with both of them maintaining their identities.

Figure 88 depicts several wind profiles and related temperature profiles. These profiles may change from day to night or they may be maintained for days if they are created by air mass incursion. Note that turbulence is often associated with the movement of two adjacent layers as we see in the next chapter. This turbulence can create its own temperature inversion by mixing the cool air above it down and the warm air below it up, creating an isotherm. At times the warmer air will contain enough moisure to form a cloud when it is lifted and a stratocumulus layer is formed. This cloud type is a good indicator of stratified air and turbulence at its level.

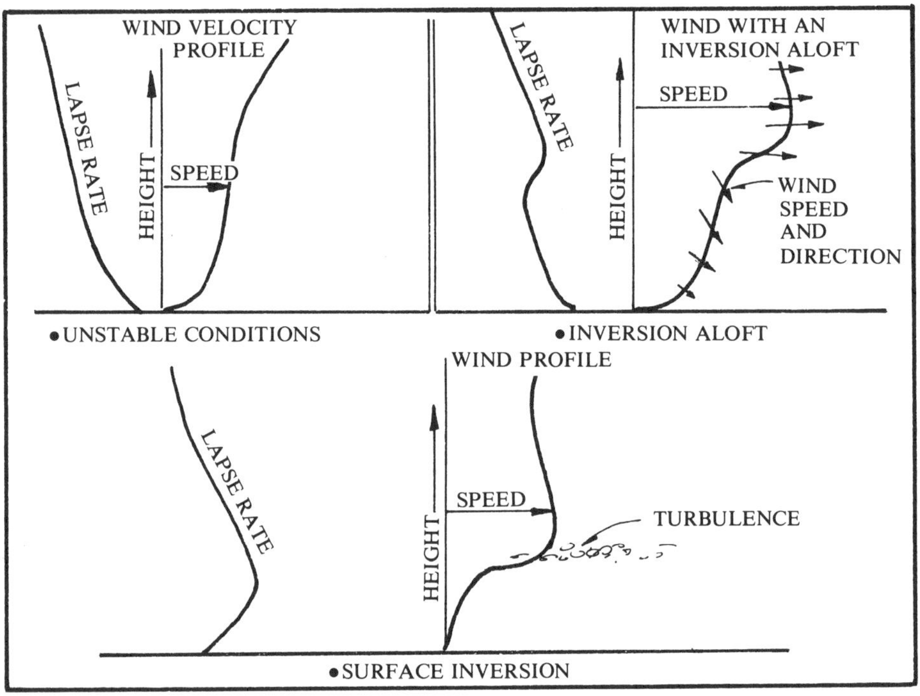

Figure 88 - Wind and Temperature Profiles

Balloonists traditionally fly in the early morning or late evening to avoid strong ground winds. However, as soon as they acquire altitude they enter winds that provide a free drift. By adjusting their height to enter layers of air moving in different directions, balloonists can greatly control the whereabouts of their flight. On many occasions when the wind sloshes around a valley from local heating effects a balloonist can go for a round trip and end back at his starting point by finding the layer of air moving in his desired direction.

LOW-LEVEL JETS

Low level jet currents aren't related to the polar jets, but are likewise caused by a temperature gradient. These low jets usually occur at night as a result of dense, cool winds moving from higher elevations to lower ground. The result is a reverse wind gradient—that is, the wind decreases with altitude. Actually, the traditional wind gradient is there with the

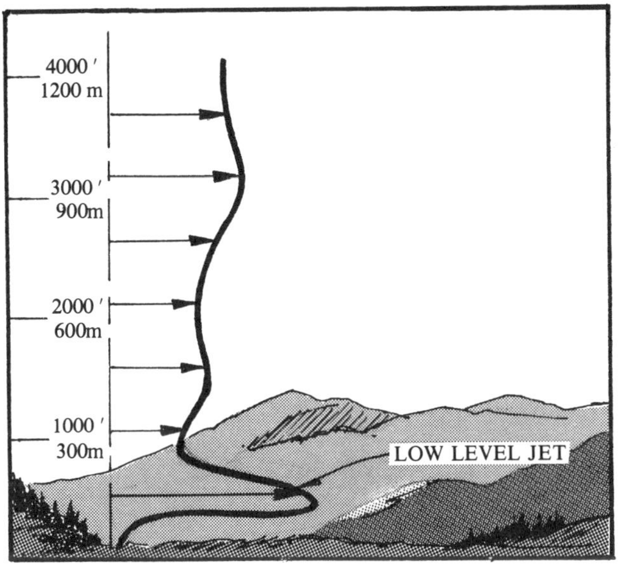

Figure 89 - Low Level Jet Wind

reverse gradient above it as shown in figure 89. As indicated, these low level jets can be quite intense and may contain turbulence.

During the day low level jets occur less often because they are "mixed out" by the up and down motion of the air. However, we have previously mentioned the wind that flows across the Great Plains in the USA (see figure 59). This flow can properly be termed a low level jet. It reaches its maximum during the day's maximum heating.

SPECIAL WINDS

We opened this chapter with a reference to some of the winds that have become so famous or infamous that they have acquired their own names.

Many of these winds are *foehn* winds which is a generic term that means the wind is dried and heated by compression as it drops over a mountain chain to a lower elevation (see Appendix III). The term foehn comes from the famous winds of Switzerland that flow down the steep valleys. A similar generic type wind is called a *bora* which refers to a falling wind that is cooler than the air it replaces. Bora winds are normally only found in polar areas such as Alaska and Scandinavia.

There are in effect two types of foehn winds. The first occurs when a cold, dry high pressure air mass stagnates near a restrictive mountain area. In this case the mountains block the lower flow and air aloft will spill over the tops of the mountains if a low pressure area exists across the mountain barrier (see figure 90). These foehn winds push out the previous air mass and commonly reach speeds of 40 to 60 mph (64 to 96 km/h) with a maximum of 90 mph (144 km/h) having been reported. These winds may last for several days with gradual weakening and occasional abrupt stops and starts. This type of wind is most common in the fall, winter and spring when pressure systems are strongest.

The second type of foehn wind consists of a deep layer of moist air that

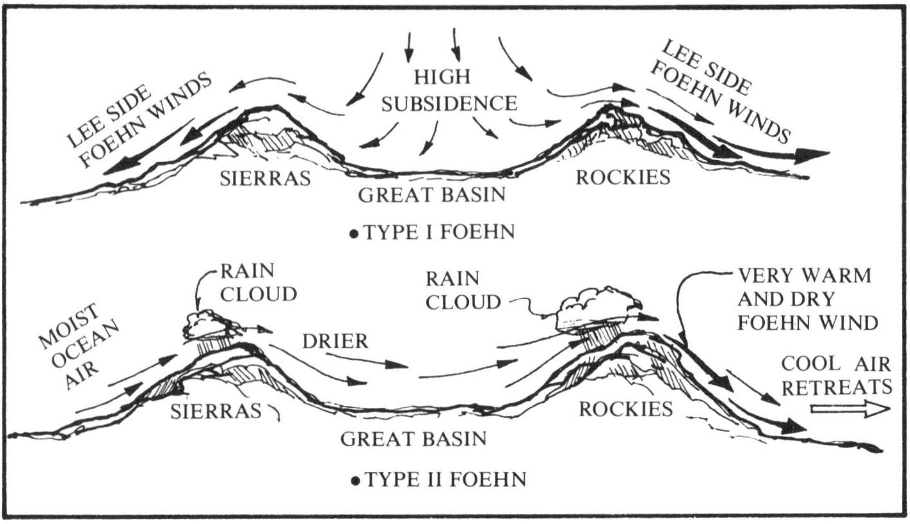

Figure 90 - Types of Foehn Winds

is forced across the mountains by larger scale circulation. This air is warmed and dried by the manner explained in Appendix III to descend on the lee side of mountains in strong drying gusts.

Foehn winds may act like a warm front and push the cooler air away on the lee side of the mountains if the pressure gradient is favorable. If this action doesn't occur, the foehn may only reach the higher elevations or reach the surface intermittently. The foehn can arrive as a front of great extent or as a relatively narrow current cutting through the lee side air, depending on the topography and pressure pattern. In any case when the foehn does reach an area it raises the temperature and drops the humidity sharply.

The action of foehns is often related to waves (see Chapter VIII) in the atmosphere. There is some evidence that strong foehn winds reaching the surface on the lee slopes are always wave related. As the wave changes wavelength or amplitude it changes where the foehn touches down. The well-known *foehn gap* of the Alps is caused precicely by this mechanism. As the air undulates up and down the stratocumulus cloud layer that often accompanies the foehn is dissolved in the down flowing portions as shown in figure 91.

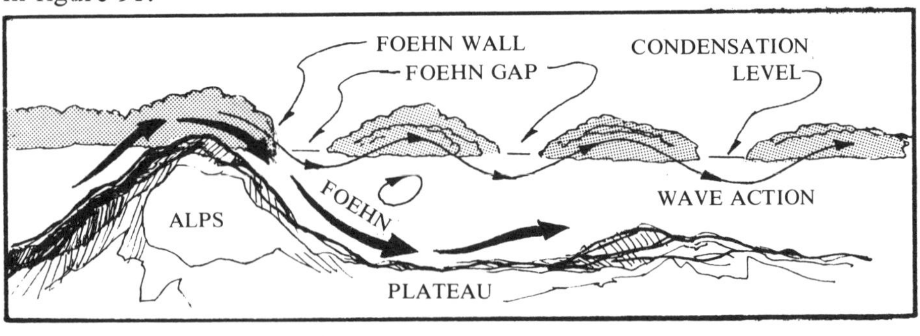

Figure 81 - The Foehn Gap

NORTH AMERICAN WINDS

Figure 92 shows the western half of the North American continent. Most of these special winds exist in the west for high mountains are an integral part of the foehn mechanism.

The *east winds* of the Cascades occur with a high far to the north. These winds may be weak and ride over the incoming cool marine air from the west or at times strike all the way to the valleys in strength. The *north* and *Mono* winds of Northern California are caused by a high sitting in the Great Basin between the Rockies and the coastal ranges.

The *chinook* winds blow on the eastern slopes of the Rocky Mountains from Alberta down to Colorado. They may produce quick wintertime thawing and have been known to raise the air temperature 30° to 40°F (17° to 22° C) in a few minutes.

In southern California the foehn winds are known as *Santa Anas*. These winds also develop with a high in the Great Basin and a favorable low along the southern coast. If the Santa Ana is weak it may affect only the higher elevations with the sea breeze and local slope winds being predominant. The sea breeze air may be returning Santa Ana air which has had only a short trajectory over the ocean and is drier than normal marine air. At other times a strong Santa Ana drops to the surface and rushes through the passes with strong northeasterly velocities. A strong Santa Ana wipes out the normal daily seabreeze and upslope/downslope patterns and brings turbulence to the area as it rolls over the mountains and hills.

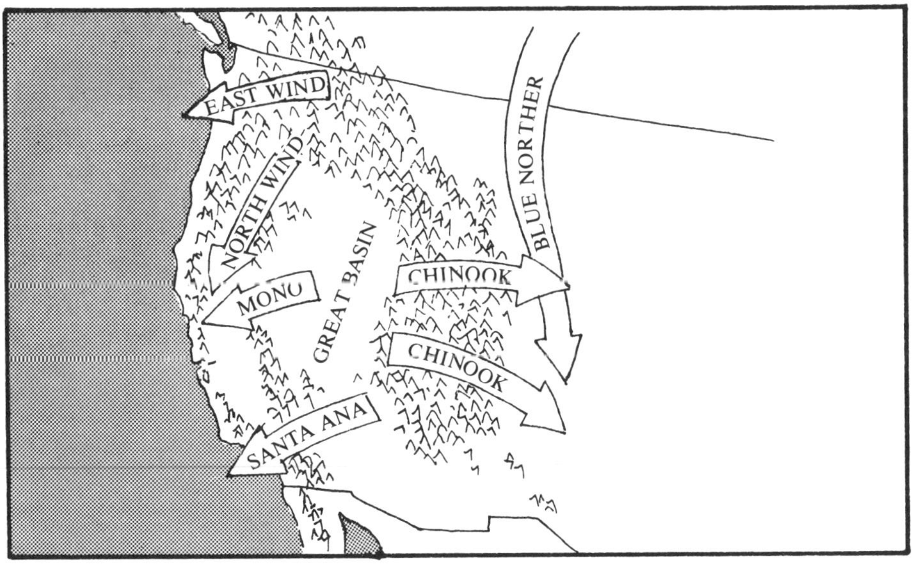

Figure 92 - North American Local Winds

A strong Santa Ana shows relatively little difference in day and night flow initially. However, as it weakens the sea breeze and upslope winds begin to reappear in the day while the Santa Ana winds remain aloft. At night the cooling of the land and subsequent stability of the air allows the Santa Ana to return in strong gusts. The flow may be wave related which causes the Santa

Ana to reach the surface in a spotty manner. Eventually the Santa Ana weakens and the normal daily local circulations resume.

The *blue norther* shown in the figure is not a foehn wind but is caused by stable air moving down from the north on the backside of a high being channeled by the Rockies to the south as it resists lifting. The blue norther is known as a cold, strong, often turbulent wind.

OLD WORLD WINDS

Perhaps most famous wind of Europe is the *mistral* which blows from the northern Alps south toward the Mediterranean. This wind increases in strength as it is channeled between the Central Massif and the southern Alps in France as shown in figure 93. The increase in strength is due to the venturi effect whereby a fluid moving through a constricted opening must move faster at the narrowest part. A stream of water can be seen to speed up when its banks narrow.

The mistral is caused by a high pressure system in the Azores region and a low pressure in the Mediterranean off the coast of Italy. It brings cold strong conditions to the southeast of France and may blow for several days.

An adjacent of the mistral is the *tramontane* which means "cross mountain." As it is applied in southern France though, the tramontane blows parallel to the Pyrenees in the lowlands between these mountains and the Central Massif. It is caused by the same pressure systems displaced slightly. Sometimes the mistral and tramontane blow simultaneously. With a high

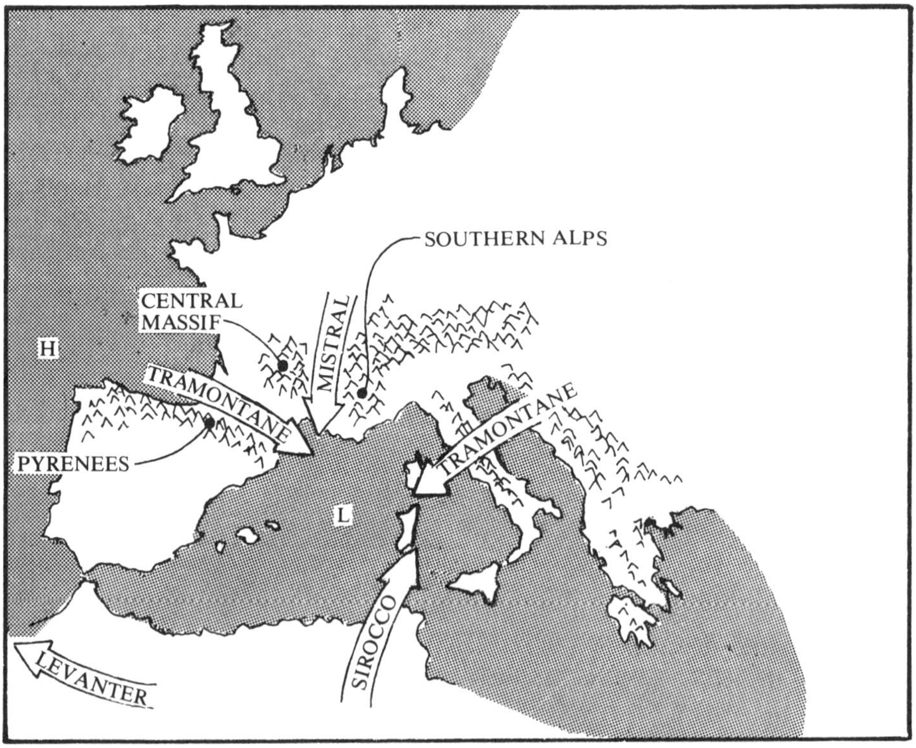

Figure 93 - European Local Winds

pressure system located in Europe, the true tramontane can blow across the spine of Italy creating dry foehn effects for the entire southeast coast.

The original foehn itself tends to be a name applied to any drying downslope wind blowing over the Alps in France, Switzerland, or Austria. Often these winds are accompanied by heavy clouds on the Italian side of the alps that create a wall of cloud as the initial descent takes place over the mountain known as a *foehn wall* as shown in figure 91. The foehn gaps also appear in this case.

Other winds of interest to traveling pilots are: The *harmattans* which are the often dust-laden winds blowing across the Atlas mountains to the atlantic. The *levantar* which blows from the Mid-east west to the Mediterranean and the *khamsin* that blows hot and dry from the south into Egypt. Finally we should mention the *sirocco* which originates in the Sahara Desert to bring dust-laden air across the Mediterranean to Sicily and Italy.

SUMMARY

The moving atmosphere brings us wind that affects our flights in many ways. Wind can provide soaring conditions or a free ride downstream, but it can also carry turbulence and sudden change. We pilots learn early to respect the wind. We also soon realize the need to understand it.

In this chapter we learned of some of the ways the wind interacts with the terrain then we saw how it changes as we moved away from the surface. The existence of the jet stream and its effect on flying weather was pointed out. We also learned that the air's structure is often layered which results in different wind flows at different altitudes. All this discussion is important to pilots wishing to know what to expect when they venture aloft.

Finally we took a look at special winds mainly originating in mountain areas. Knowledge of these winds is important for they can affect the flying conditions for days. Next we investigate the important matter of turbulence, then we look at more local conditions.

High winds in a building thunderstorm create ragged cumulus and turbulence.

CHAPTER VI

Turbulence—Unsteady Flow

Moving air often comes with its own little surprises: turbulence. Such surprises can be unpleasant, for a small craft suspended in mid-air feels every nuance and perturbation in the fluid. Of the many facets of the weather, turbulence is perhaps the most critical to pilot safety. We need to study it well—from afar.

Like many aspects of the weather, turbulence can not be readily seen but we can visualize its behavior. Knowing where turbulence lurks and when to expect it is a major step towards avoiding it. In this chapter, we will learn all about turbulence from the slightest texture to rodeo air.

THE MEANING OF TURBULENCE

If we could make the air visible by adding mist or dye in a large volume, we could actually see turbulence. What we would see would be swirls, whirls and roils of various sizes turning this way and that, interacting and breaking apart as they move along with the wind. You can readily witness a similar thing by watching the eddy action in a fast moving stream. The swirls or eddies are what we experience as gusts or turbulence when they pass by our body or wings as shown in figure 94. Here we see a passing eddy from a bird's eye view. A person standing at point A would first experience a gust from the left, then a lighter headwind, then a gust from the right and finally a return to the general wind direction as the eddy passes.

We thus can agree to a working definition of turbulence as *the random chaotic swirling of the air*. In truth, some forms of turbulence such as rotors and bumps caused by thermals can be somewhat organized, but randomness of the swirls is what characterizes most turbulence.

What constitutes significant turbulence to the pilot involved depends on his

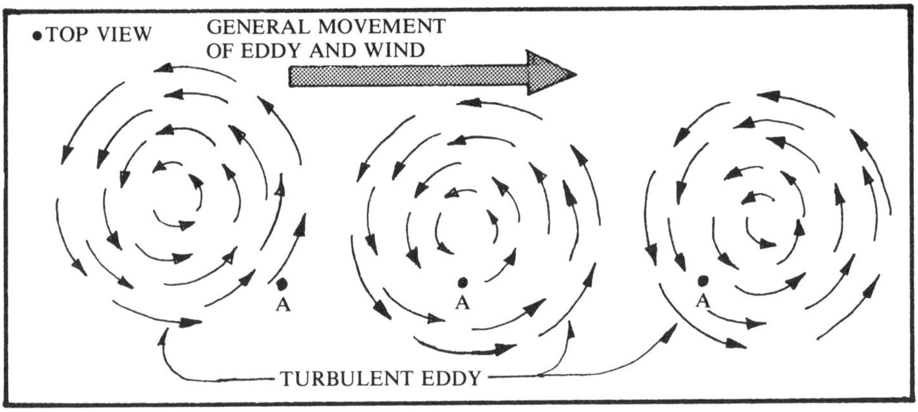

Figure 94 - The Meaning of Turbulence

perspective and choice of aircraft. To the man on the moon, the great swirls in our visible atmosphere are turbulence. On earth we experience these swirls merely as gradual weather changes. At the other extreme, a butterfly may experience extreme turbulence in a swirl that we feel as a gentle puff. As a rule of thumb, the smaller the aircraft and the lighter the wing loading, the more it will be affected by smaller eddies.

THE NATURE OF TURBULENCE

In the air turbulence affects our wings much like it affected our body in the illustration mentioned previously. The speed and direction of the relative wind will change as the eddy or swirl passes our wings. The effect depends on the intensity, size and orientation of the eddy.

Small eddies will feel like rapid bumps, much like when a speed boat skims a choppy lake. Eddies from several feet to about the span of your aircraft in diameter will be felt as larger bumps which may cause control problems or weightlessness if they are severe. Larger diameter eddies will be experienced as sudden lift, sink, rolling or yawing impetus. Finally, very large scale eddies will appear as wind speed or direction changes.

The dangers that turbulence poses to flying are several: A sudden gust may stall our craft which can have severe consequences close to the ground. Also loss of control can occur in turbulence when a wing gets lifted radically or pitch action occurs. Loss of control is also most dangerous close to the ground. In severe cases—most notably in rotors or strong thermals—rollovers or pitchovers can occur. Most of us prefer to do our flying right-side-up. Finally, severe turbulence can cause structural damage to our craft as it tweaks on the wings and rattles us around in our cage. Later in this chapter we'll look at escape procedures to follow when turbulence greets you with a slap on the back.

The life cycle of a turbulent eddy begins when it is formed by one of the three causes discussed in the next section. It then moves with the general wind flow and breaks down into smaller and smaller but more numerous eddies. This process continues downwind until the eddies are so small their energy of motion is converted directly to heat due to the viscosity of the air (about .01 inch or .25 mm in diameter at sea level). Essentially what takes place is an exchange of energy from large-scale motion to movement on a smaller and smaller dimension. A moving mass of air will lose much of its initial momentum through this mechanism.

Smaller eddies do not necessarily have less effect on our wings, for they may contain most of the energy of the larger eddy they developed from. You can readily see this in turbulent water when small, fast spinning swirls move inside slower, larger whirlpools. It is only after time and space that normal turbulent eddies lose their punch (rotors never give up as we shall see).

Turbulence with its snarls and gusts tends to spread out all properties of the air. For example, heat, moisture and pollution are dispersed in all directions by swirling air. Also, wind differences are evened out by turbulence. What this means to flying is that gradients are reduced by turbulence but the turbulence itself may pose more problems than the wind gradient.

THE CAUSES OF TURBULENCE

Here we are going to separate the development of turbulence into three sources. They are: mechanical, thermal and shearing actions. Each cause of turbulence is distinct so we will discuss them separately. However, we should be aware that they can appear in any combination. For example, mechanical turbulence and thermal turbulence are often both present close to the ground on hot, windy days.

MECHANICAL TURBULENCE

When a solid object obstructs the path of the wind—be it a mountain, forest, house or football lineman—the flow downwind from the object is disrupted. In very light flow the disruption may be gentle meanders, but as the flow velocity increases standing eddies may develop which give way to the random chaotic eddies of full-fledged turbulence. You can readily visualize this action by immersing your hand in water flowing at various velocities.

Figure 95 illustrates the turbulence caused by a solid object in the air flow. Note the difference at different wind velocities. Not only does the stronger wind produce more turbulence but it is more intense and travels further downstream. You will also see the presence of standing eddies in the figure. These are more or less stable swirls that stay in one place and are set up by the shape of the solid. In the atmosphere we call them *rotors.* Occasionally, these standing eddies or rotors may break away and move downstream to be replaced by a new eddy. Generally, though, they are stable and remain in place as long as the flow remains steady. Once the flow increases above a certain value, however, they are broken up and replaced with general turbulence.

The force that a solid object imparts to the air is equal and opposite to the force the air imparts to the object as Newton would have told us. This force can be felt by holding your hand out the window of a moving car. You can't see it but you can bet your hand is leaving a trail of turbulence. The force is caused by drag due to pressure differences on the front and back surfaces of your hand. Any solid object in the air flow works the

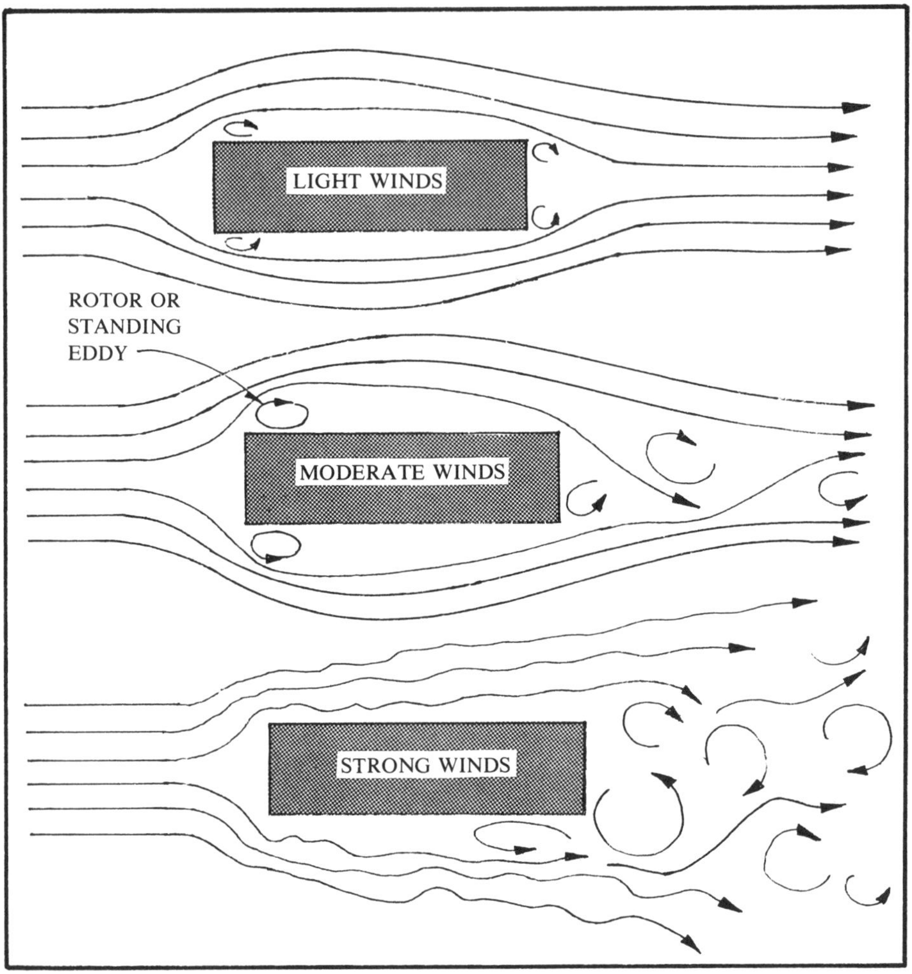

Figure 95 - Mechanical Turbulence

same way. Most of the energy acquired by the air from the drag forces on your moving hand goes directly into the creation of turbulence.

Besides the wind velocity, the shape of an object disrupting the flow is an important factor in determining whether or not the air is turbulent. If sharp edges or curves are present on the object, the air will have a difficult time moving uniformly over the entire surface due to the inertia of the air molecules. Figure 96 shows a variety of shapes and how they affect the flow of air. The first drawing has a cross section which presents the least resistance to the wind and therefore causes the least amount of turbulence. Airplane stabilizers and struts, boat hulls and even trees growing in a steady wind are often of this configuration. The other drawings illustrate how sharp edges or curves can initiate turbulence.

Now we have a fairly good idea how irregularities and obstructions from the size of mountains on down that lie in the path of the wind cause turbulence. Since solid objects other than birds and aircraft exist entirely on the earth's surface, turbulence from this source is usually limited to a layer below 1,500 feet (500m) above the highest object. We call this layer the friction layer as indicated in the chart in the previous chapter. In this layer some disruption of laminar or smooth flow is expected.

The size of the objects blocking the air's flow is called the *roughness* of the surface. The roughness determines the initial size of the turbulent eddies. Larger obstructions tend to create larger eddies, but these may quickly divide into smaller cells. Typically, an object creates an initial eddy from 1/10 to 1/7 its size. The chart below gives roughness values or eddy diameters for common terrain:

ROUGHNESS VALUES

City or Forest	6.6 ft —2 m
Surburban Homes	1.6 ft — 50 cm
Farm Crops Sage Brush	4.0 in — 10 cm
Mown Grass	.0.4 in— 1 cm
Ocean, Large Lakes	.012 in —0.3 mm

Once again, the actual effect roughness of a surface has on the wind is determined not only by the size of the obstructions but also the mean wind velocity. In light winds little or no turbulence may occur. In slightly higher winds turbulent eddies may form and the wind direction will become quite variable. In stronger winds (over 20 mph or 32 km/h) the turbulent eddies may become very intense and smaller and travel well downstream before they break up. In this case the variation in speed will be great but the meandering changes in direction is lessened.

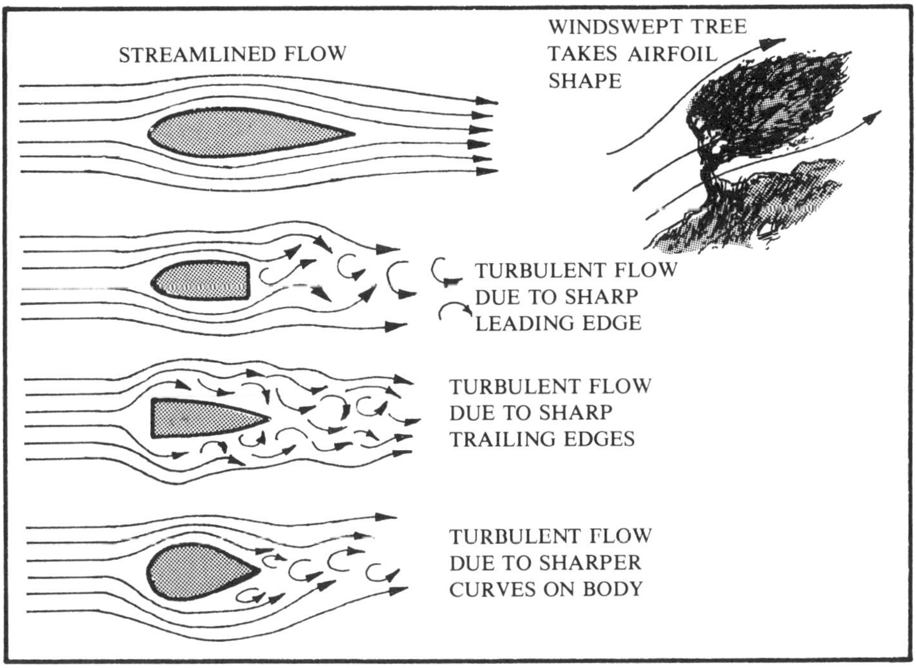

Figure 96 - The Effect of Shape on Turbulence

The force of the wind and the energy in turbulent eddies increase with the square of the velocity. Thus a wind blowing twice as hard as before will exert four times the force. A turbulent eddy in such an increased wind will be similarly more vigorous. This fact leads to the following guideline:

The strength of turbulence increases with the square of the wind velocity.

THERMAL TURBULENCE

The second source of turbulence in the atmosphere is convection currents or thermals. As shown in figure 97, when a thermal penetrates upward it disrupts the air it's passing through to form turbulent eddies and other velocity changes. As we shall see in Chapter IX and X, thermals themselves are usually organized with a mass of rolling lift in the center surrounded by sink around the perimeter. The act of flying through such a mass of air acting in concert often presents one with healthy sink followed abruptly by lift then another abrupt change to sink. Thus we have described the traditional "air pocket" by a bygone era.

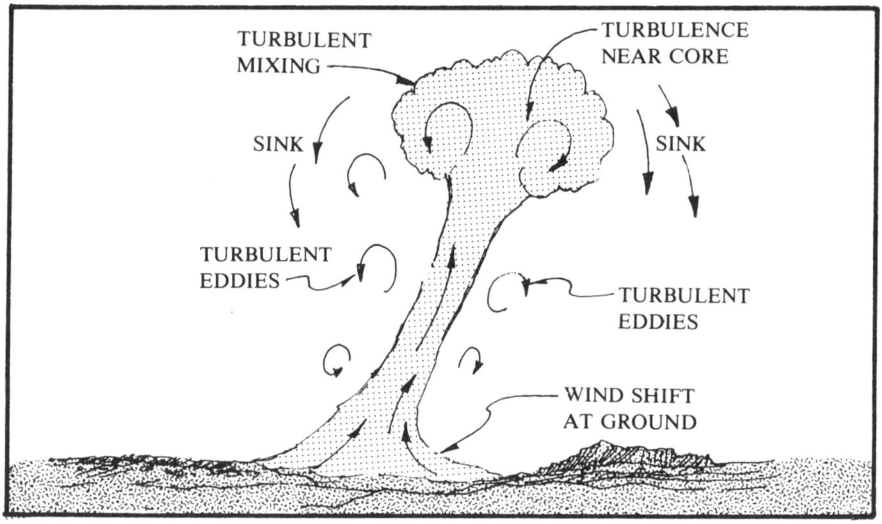

Figure 97 - Turbulence Caused by Thermals

The edges of some thermals are turbulent by anyone's standards. Extremely virulent thermals in hot desert areas can exert enough force to roll or pitch a small aircraft over on its back if it is caught with part of the wing in the up air and part in the down air. Fortunately thermals with such a bad attitude are quite rare and sport aviators generally fly safely in most thermal conditions.

Thermal turbulence is usually strongest in the lower 2 to 4 thousand feet (600 to 1300 m), but may reach to tens of thousands of feet in desert or in thunderstorm conditions. When thermal turbulence is added to mechanical turbulence in a flowing wind the results can be quite chaotic as shown in figure 98. Even when the general wind is still, thermals can create

Figure 98 - Thermal Disruption of the Airflow

ground turbulence as they suck in air from all directions when they lift off. Figure 99 illustrates the effect of thermals on the surface air movement.

When thermals rise they send air down from aloft to take their place. If a wind is blowing aloft, this downward moving air will be moving horizontally as well as vertically and will be felt as gusts on the surface. This is the source of "cat's paws" on water and the rush of air you can see in trees or fields of grass on windy thermaly days. In the air these gusts and packets of cool air are felt as moderate to strong turbulence.

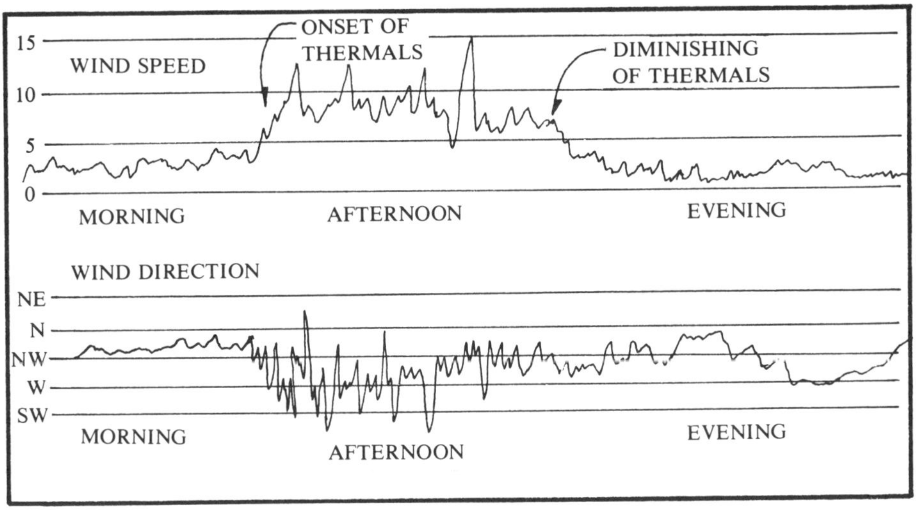

Figure 99 - Wind Velocity Variation Due To Thermals

SHEAR TURBULENCE

The third and final cause of turbulence in nature is through the mechanism of wind shear. The word shear means a cutting or tearing and when two layers of air lying next to each other move with different velocities (speed or direction), a shearing action takes place. In this case the boundary between the two layers becomes turbulent due to the friction of the opposing action as shown in figure 100.

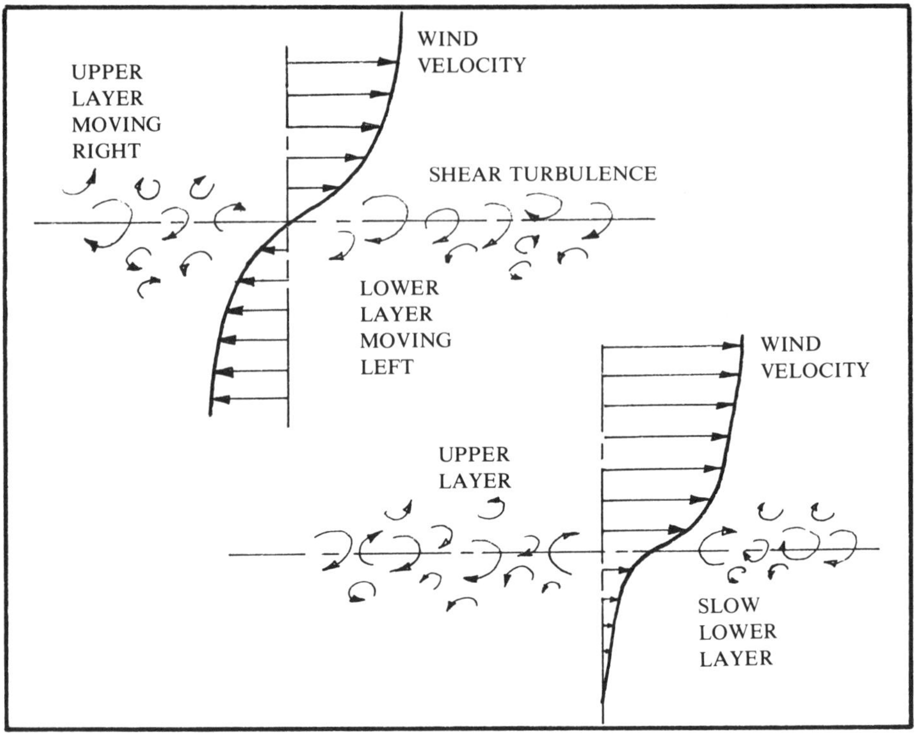

Figure 100 - Turbulence Caused by Shear

In truth, all turbulence is caused by shearing action, for mechanical turbulence is created by the roughness of the terrain creating a velocity gradient and thus a shearing action as depicted earlier in figure 83. A thermal penetrating aloft also creates a shearing action as it pushes its mass through the ambient air. However, we ignore these details and focus on shear turbulence as that produced between two layers or volumes of air rubbing each other the wrong way.

It has been said that it is impossible to separate the velocity distribution of the air from the temperature distribution. That is to say that when we have layers of air with different temperatures they also exhibit different velocities so that the presence of shear turbulence between the layers should be expected. We have seen in previous chapters that heating and cooling effects as well as high pressure systems create temperature inversions and jets at various levels. These are typical phenomena related to shear turbulence.

Indeed, the most likely place you will encounter shear turbulence, is near an inversion layer. This layer may be thousands of feet aloft as formed by the sinking air in a high pressure system, or it may be close to the ground at night when a low layer of air is rapidly cooled by the cooling ground. In the first case the inversion layer may stop the rise of thermals so that their general turbulence is added to the mixture. Figure 101 shows several situations where inversions and shear turbulence is common. In the last drawing we see how a valley can trap a pool of cool, dense air that slips off the slopes then stagnates

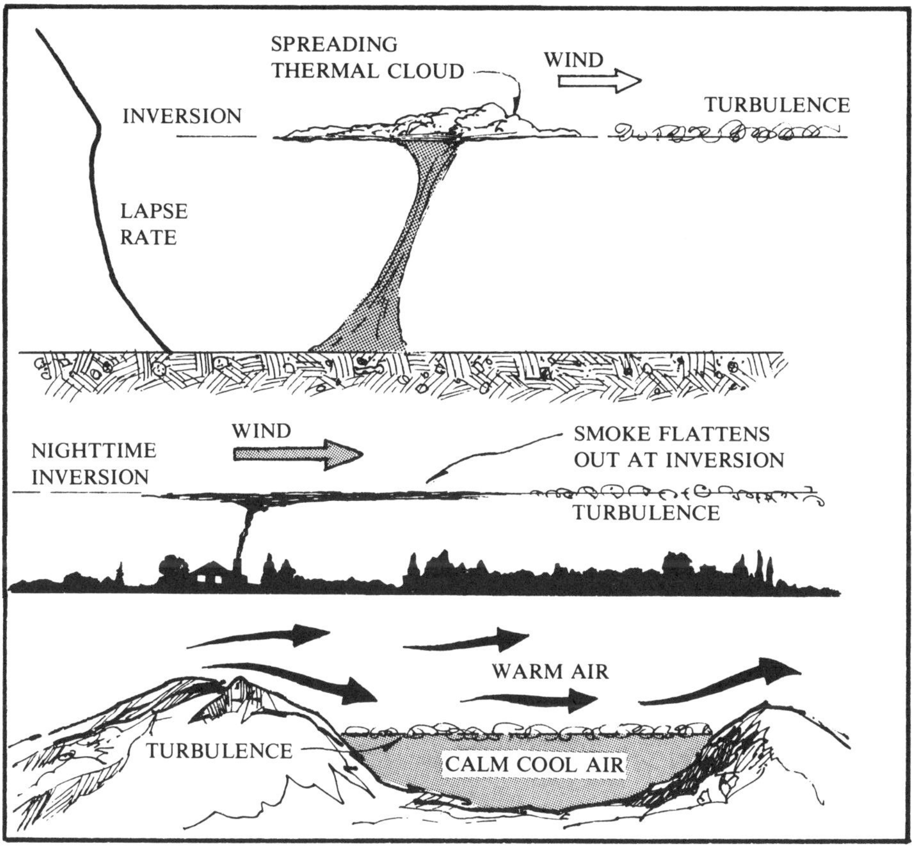

Figure 101 - Causes of Shear Layers

in the valley while the warmer winds aloft keep blowing to produce strong shear at the warm and cool air interface.

Also in high mountain areas the sudden cooling of the slopes as evening falls can result in a downslope breeze (see the next chapter for a more thorough discussion of this action) that thrusts out into the general valley air mass to create strong shear turbulence (see figure 102). This event occurs most frequently on an eastward facing slope with deep canyons at the end of a hot day when the sun suddenly slips below the crest to throw shadow across the entire east face.

Other places where shear turbulence frequently appears are fronts—cold, warm and seabreeze fronts (see the next chapter). This is pictured in figure 103. Note that the strength of the shear turbulence is determined by the relative velocities of the two air masses. Because seabreeze fronts can be quite vigorous, shear turbulence in their vicinity is often strong.

Shear turbulence tends to persist for a long time for the layers that produce it are often stable. Certain fronts can create shearing layers that last for days. Air layers of different temperatures and thus densities do not mix very readily. They are like oil and water. Thus they maintain their separate identity for an extended period as they rub against one another

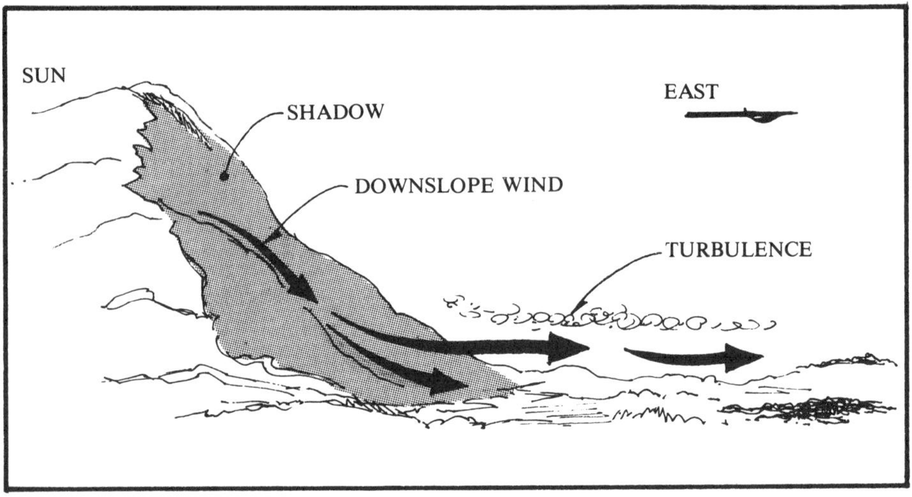

Figure 102 - Shear Caused by Downslope Winds

and mix in a narrow band.

Over the years shear turbulence has not been particularly hard on sport aviation. For the most part it is avoidable if one shuns inclement weather or uses prudent judgement when flying in the above described conditions. Shear turbulence can, however, be strong enough to pluck the feathers off an airplane but this generally only occurs near the upper level jet stream that frequents the higher latitudes.

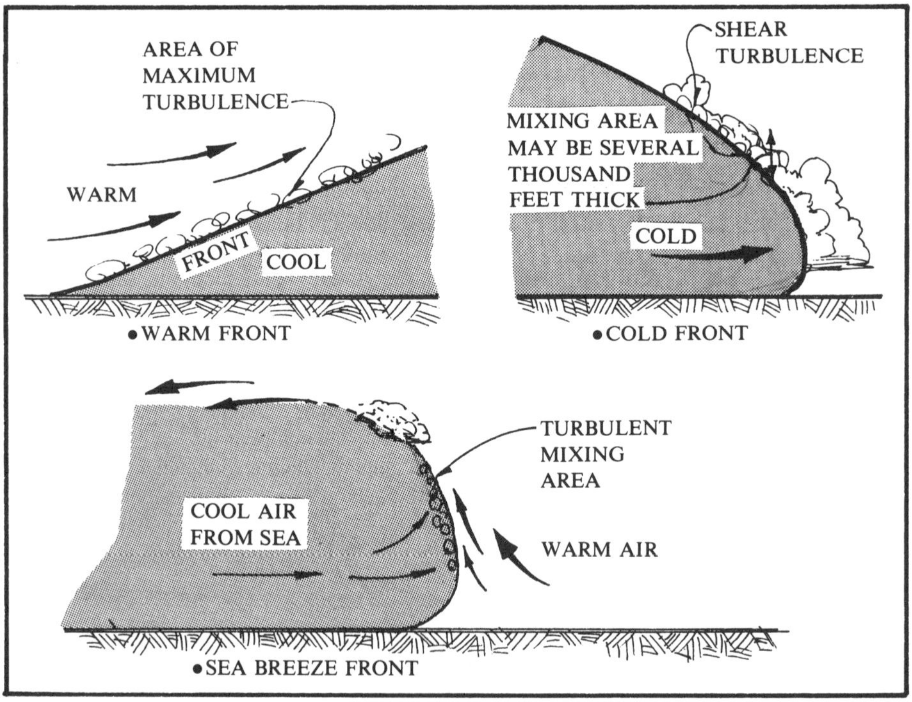

Figure 103 - Shear Turbulence at Fronts

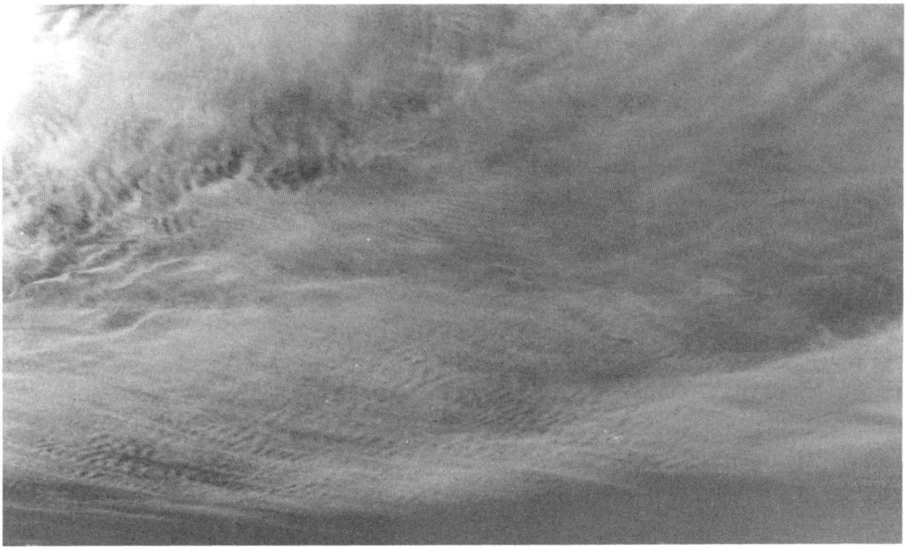

Jet stream related billow clouds resulting from shear between two layers.

VORTEX TURBULENCE

We have reviewed the three natural causes of turbulence above. There is a man-made cause which we should mention for the sake of completeness. This is wing tip vortices which are powerful swirls emanating from the tips of all wings. Because these swirls are so uniform they possess a lot of energy and can be readily felt in the air.

The vorticies from another aircraft about your same size will feel like a couple quick bumps or a force lifting your wing depending on how you hit them. This you can live with. Vorticies from aircraft larger than you can cause you more serious control problems or even structural damage. Avoid them to preserve your health. The heavier loaded, the less aerodynamically sleek and the higher an angle of attack an aircraft flies, the more violent are the vorticies and general wake turbulence.

ROTORS

In certain conditions around sharp terrain features standing eddies or rotors can exist and persist as long as the wind blows. We shall see examples of rotors in later illustrations.

Rotors appear most readily in stable conditions with light to moderate wind. In unstable conditions, thermals tend to break up the rotors and give them an erratic existence or eliminate them entirely. In stronger winds rotors usually get blown apart by the general turbulence that rages through their area of residence.

Rotors should be avoided in flight because of the strong sink and control problems they offer. Flight along the rotor axis could roll you over. Rotors that exist below waves (see Chapter VIII) have broken airplanes.

DETECTING TURBULENCE

Turbulence is not hard to discern on the surface. For certain, any rapid velocity change—speed or direction—is most likely caused by turbulent swirls or thermals with their attendant turbulence. Each pilot should have a personal guideline to judge when a certain amount of turbulence is too much for his or her aircraft of choice and individual skills. For example, you may choose to limit yourself to conditions that change no more than 5 mph and 45° within 3 seconds. If the magnitude of change is greater or the time of change is less, you would conclude that the turbulence is too great at the time and happily choose to wait for more benign air which is often forthcoming in the later part of the day.

Any flexible objects that readily show gusts in the wind such as trees, crops, long grass, bodies of water, flags and windsocks are useful for detecting turbulence. Rapidly changing flags and snapping windsocks are particularly reliable turbulence indicators. Rising smoke is also ideal for as figure 104 illustrates, in turbulent conditions it undergoes undulations and diffuses rapidly while the smoke stream is fairly uniform on non-turbulent days. These indicators most often foretell mechanical and thermal turbulence.

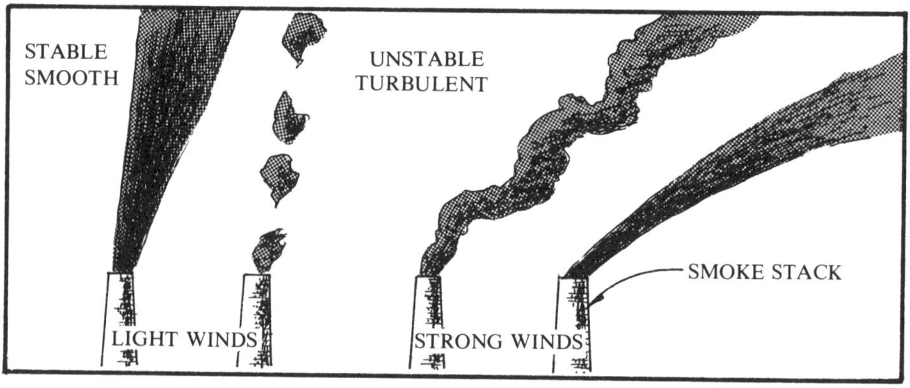

Figure 104 - Smoke as a Indication of Turbulence

Certain cloud types are related to turbulence as well, as noted in Chapter III. Cumulus clouds are most often associated with thermals and thus thermal turbulence. The strength of this turbulence is partially indicated by the vertical development and speed of build-up of the clouds which is related to thermal strength.

Shear turbulence is also related to certain types of characteristic clouds. At lower levels stratus layer clouds often exist at inversion layers since the cooler air near the boundary condenses the water vapor in the warmer air as they mix. Expect to find shear turbulence near the lower limit of stratus layers. At higher levels in the sky billow clouds (see figure 34) indicate the presence of shearing action. Often this shear is associated with the approach of a warm front and it is usually above the level of sport flying (15,000 feet—5,000 m or more) where such clouds appear.

A final cloud type that can signify the presence of turbulence is a wave cloud (see figure 33). Since strong rotors often exist in conjunction with waves, such clouds should serve as a warning to pilots of light aircraft. In Chapter VIII we cover waves in detail and see how to avoid rotors.

CONDITIONS AND TURBULENCE CYCLES

We should be able to see by now that hot, dry conditions are ideal for producing strong thermal turbulence. In addition, strong pressure gradients, whether due to local heating or general circulation, cause vigorous winds that can lead to virulent turbulence.

Changes in stability also relate to the type of turbulence likely to be present. Stable air surpresses thermals as well as other forms of vertical motion. Thus mechanical turbulence is somewhat surpressed and dies out sooner in stable conditions as does thermal turbulence. On the other hand, stable air is most readily associated with layered air and the resulting shear turbulence.

From the above we can form a general picture of what type of turbulence to expect at different times. Morning stable conditions often give way to afternoon instability followed by evening and nighttime stability. On the larger scale, winter stable conditions are replaced by spring instability followed by a mixed bag of summer (stability and instability in moist areas, instability in dry areas) then general instability in the fall as cold fronts move south. The chart below applies these generalities to the expected turbulence types.

	Turbulence		
	*MECHANICAL**	*THERMAL*	*SHEAR*
Stable Conditions	X		X
Unstable Conditions	X	X	
Morning			X
Afternoon	X	X	
Evening and Night	X (at end of windy day)	X (dying thermals)	X
Winter	X		X
Spring and Fall	X	X	
Summer	X	X (especially in dry areas)	X (only around fronts)

*Mechanical turbulence is expected only on windy days.

Of course there are many exceptions to the above chart. For example, thermals may occur in the winter after a cold front passage or in the desert on a sunny day. Shear turbulence can occur year around or at midday when fronts or pressure systems are in the vicinity. The cold, dense air of winter generally exhibits less thermal turbulence and mechanical turbulence doesn't spread so much. However, more energy is contained in denser air moving or spinning with the same velocity as thinner air.

SURFACE CONDITIONS

Because so much of sport aviation practices take place within a few hundred feet of the ground, we should pay special attention to surface effects. As we have seen, this lower level is the friction layer and we should always expect to encounter mechanical turbulence of some degree here when the wind is blowing.

SEASIDE TURBULENCE

Near the sea the airflow is often as smooth as whipped cream for several reasons. First, the wind moving over the water encounters very little roughness until it reaches the shore. Second, the air over the water is usually stable since it is cooled from below when the water is cooler than the air as is normal. Finally, the entire mass of air over the water is usually stable since it is generally descending during the day as we shall see in the next chapter. Flights along the coast of a major body of water rival those taken in a midnight calm for smoothness.

INLAND TURBULENCE

Inland the picture is much different. Mechanical and thermal turbulence may combine as shown previously in figure 98. Of course, mechanical turbulence exists downwind of all solid objects as shown in figure 105. Here we see how the turbulence spreads out and forms smaller eddies downwind. This spread depends on the wind velocity and the stability. A general rule for aviation is to stay as far away from the downwind side of an obstruction as its height times the wind velocity in miles per hour (half the velocity for km/h).

Safe Downwind Clearance = Object Height × Wind Velocity (mph)

As an example, a 20 foot house in a 15 mph wind requires 20x15 = 300 feet of downwind clearance while a 1000 foot mountain in 10 mph wind requires 10,000 feet or two miles of clearance.

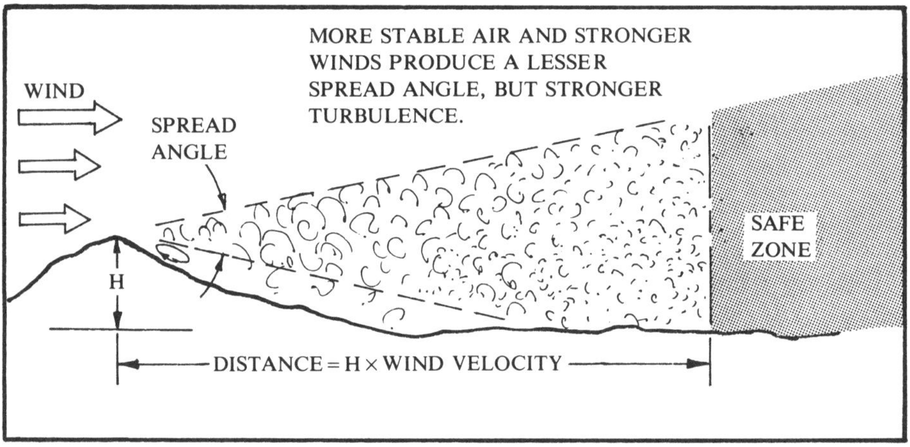

Figure 105 - Spreading and Dissipation of Turbulence

CLIFF FACES

Looking closer at the terrain, we can understand the effects of different shapes. Earlier in figure 95 we saw how sharp edges disrupt the air's flow. This is the action we should expect from a building. We can also apply this understanding to various shapes of hills as shown in figure 106. Here we see a rounded and gentle hill, sharper cliff-like faces in light and strong winds, then finally an undercut cliff. The gently rounded hill exhibits little or no turbulence even in moderate winds. Such a hill may be used for top

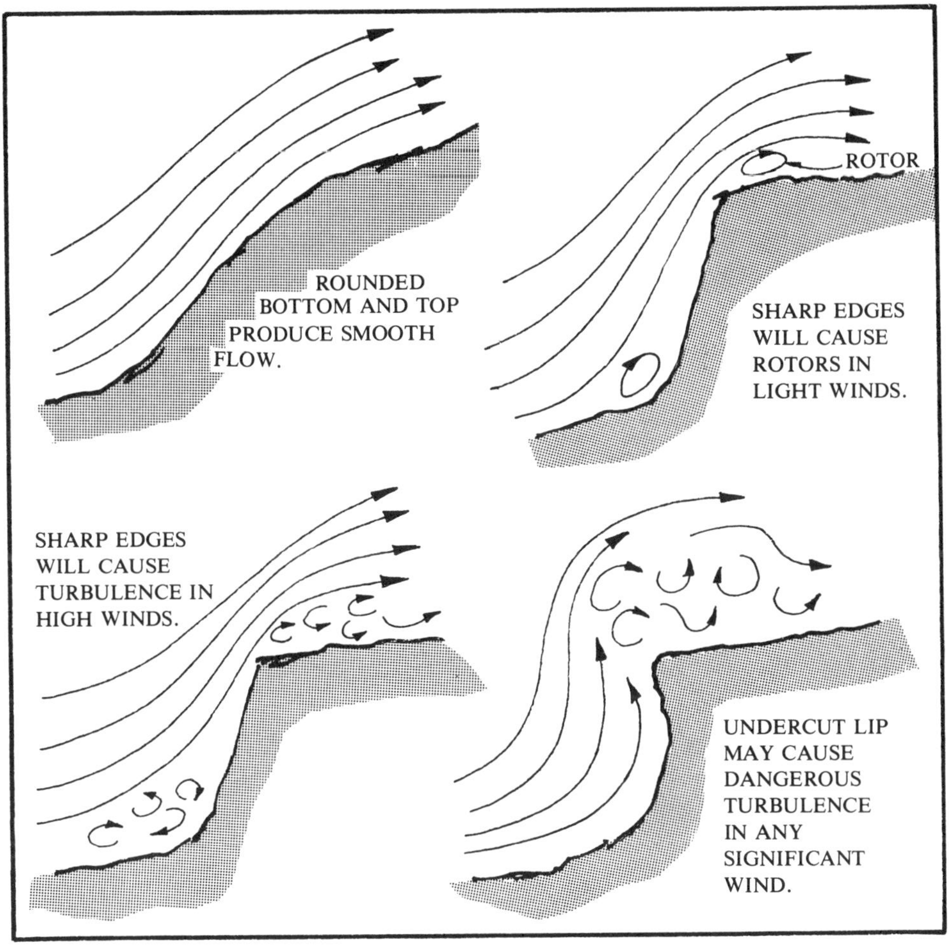

Figure 106 - Turbulence Near Hill Faces

landing of soaring aircraft as is the case with numerous grassy knobs in England and Point of the Mountain in Utah.

The sharp cliffs shown in the figure may always exhibit some form of turbulence in wind. Light wind may create standing eddies or rotors as shown. Stronger winds can produce more chaotic turbulence. The worse case is when the cliff face is actually undercut, for dangerous turbulence may form at the edge in all but the lightest of winds.

We examine these hill tops and cliff edges specifically because pilots in some forms of sport aviation launch from these points. In order to perform such an act successfully and safely the pilot must understand the nature of the turbulence likely to be present and how to deal with it. In general, the standard procedure is to use assistance and get the wing situated in the smooth airflow as much as possible. This requires moving to the edge of the cliff where the airflow just begins to break up. The presence of rotors may require several assistants and a quick release away from the cliff.

RIDGE TOPS

A condition closely related to turbulence at cliff faces is that which occurs at the top of ridges or mountains. Several common situations are shown in figure 107. Here we see hills with a variety of shallow downwind or leeside slopes. The hills with shallow downwind slopes only produce turbulence when the winds are quite strong. Hills with a steeper downwind side produce rotors in light winds and strong mixing turbulence in strong winds as shown. Note that a "bolster" eddy may exist above any irregularity on the slope as shown in the figure.

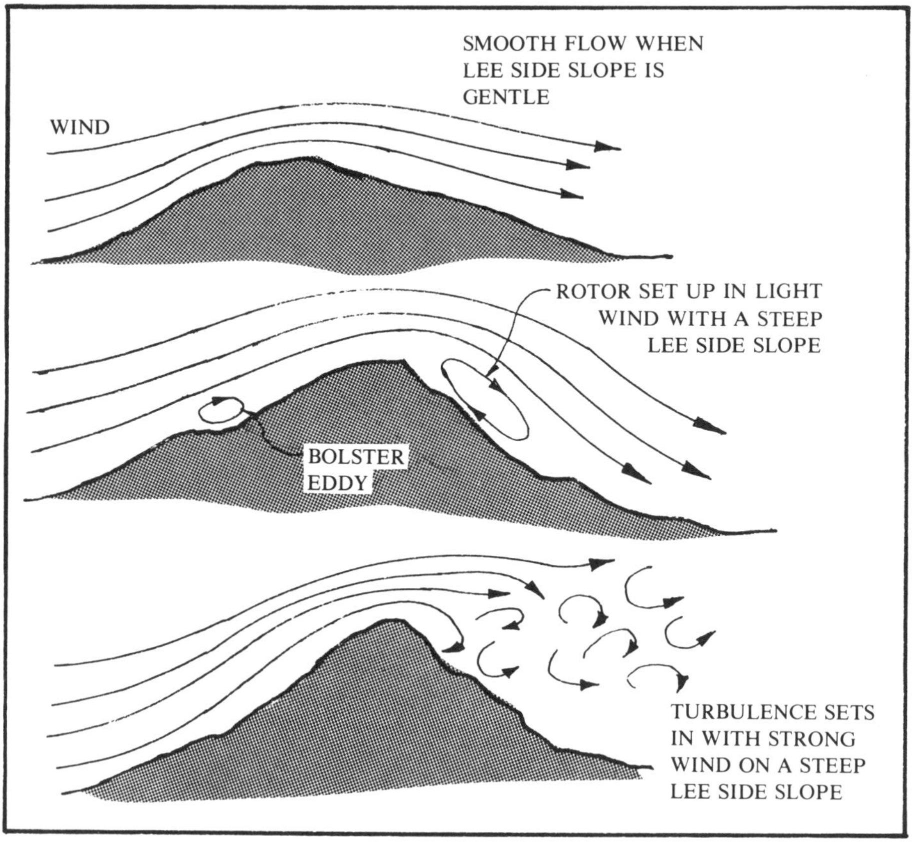

Figure 107 - Turbulence Behind Ridge Tops

A long ridge, tree line or a row of houses is more effective in producing turbulence than an isolated hill, clump of trees or house. As shown in figure 108, the wind can pass around a reasonably shaped hill with very little disturbance (in Chapter VIII we shall see that isolated hills produce less ridge lift than a long ridge for this reason).

It is possible to mistake the wind coming up the backside of a hill due to the lee side rotor as being the true wind as shown in figure 109. Taking off into this rotor would produce an erratic flight at best and slam you into the mountain in the worse case scenario. This author once observed a hang glider pilot make this mistake only to be knocked up on a wing, spun 180°

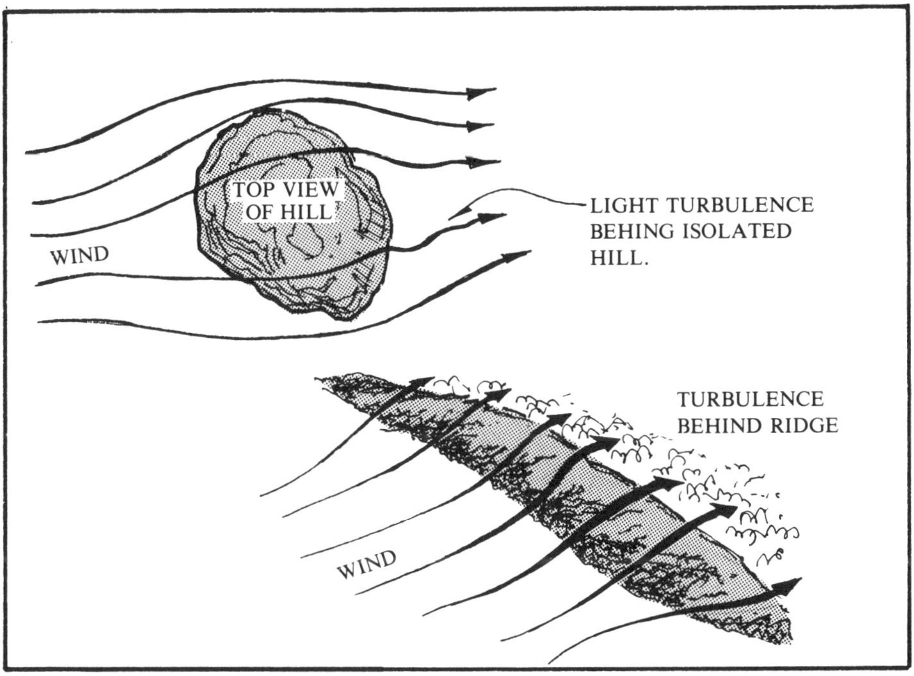

Figure 108 - Broader Hills Readily Disrupt The Wind Flow

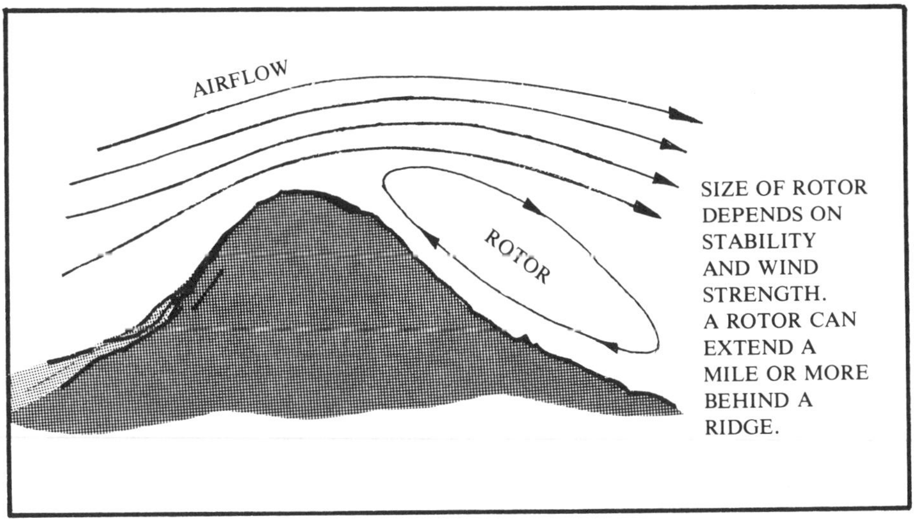

Figure 109 - Dangerous Lee Side Winds

and sent into a dive at the hill. He barely recovered.

On another occasion we were driving to a 1,000 foot west facing soarable ridge. The access road went up the valley on the east side of the mountain. Flags in this valley as far as a mile away from the mountain were indicating an *east* wind even though the true wind was west about 15 mph. A huge rotor existed behind the mountain that looked every bit like a steady, soarable east wind. This illustrates the importance of checking both sides of the mountain, the winds aloft and the forecast to be sure of the wind direction. (Note that during the approach of a cold front lower level winds may be southeast—northeast in the southern hemisphere—while the upper winds will often be westerly. This is a normal state of affairs as shown in figure 76 and you can assure yourself of the safety of the situation by checking the forecast.

The downwind side of a plateau, be it a cliff or slope also produces sink, rotors or turbulence as shown in figure 110. Launching from a cliff in a tailwind can be very dangerous in winds above a slight trickle.

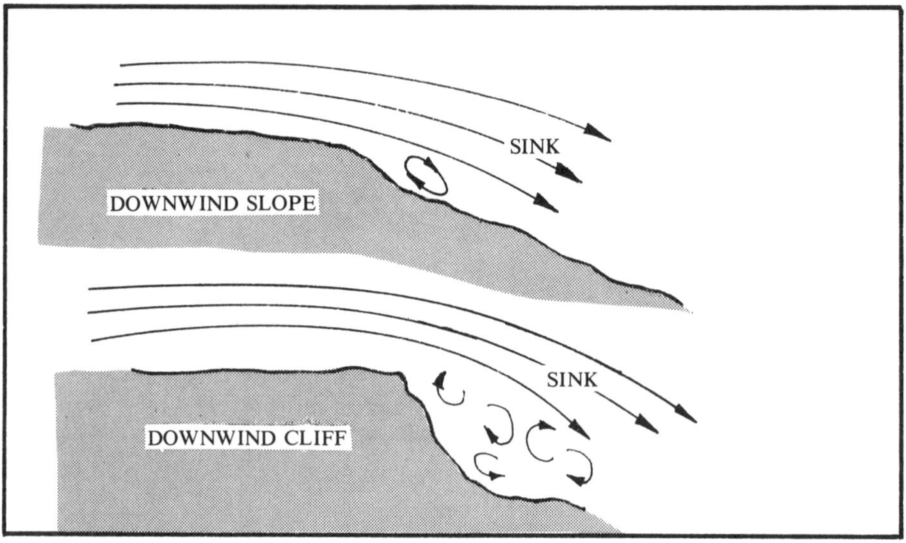

Figure 110 - Turbulence Downwind of a Plateau

GAPS, BOWLS, CANYONS AND GULLIES

The ridges and hills that soaring pilots utilize are often quite uniform so that they have their own little turbulent tricks. Gaps in a ridge line let the air flow through like water through a dam break. Figure 111 shows the flow and expected turbulence through an open gap with both a straight in and crossing wind. Winds higher than the general velocity should be expected in the gap as well due to convergence. In a long narrow defile with a constriction, the flow will appear as in figure 112 with higher winds expected at the constriction.

When the wind flows parallel to a gulley or narrow valley, the flow is generally smooth except for the disturbance of the sides and floor of the valley. However, when it is crossing such a long terrain feature, rotors or turbulence will exist in any significant wind. Figure 113 illustrates this matter. Lighter winds may produce a rotor filling the entire valley. However, thermals tend to break up rotors and create more random turbulence as in the strong wind case shown. When the winds are crossing these valleys the flow may be along the valley as it gets deflected by the opposite slope. In this case the turbulence may be limited to the proximity of the downwind

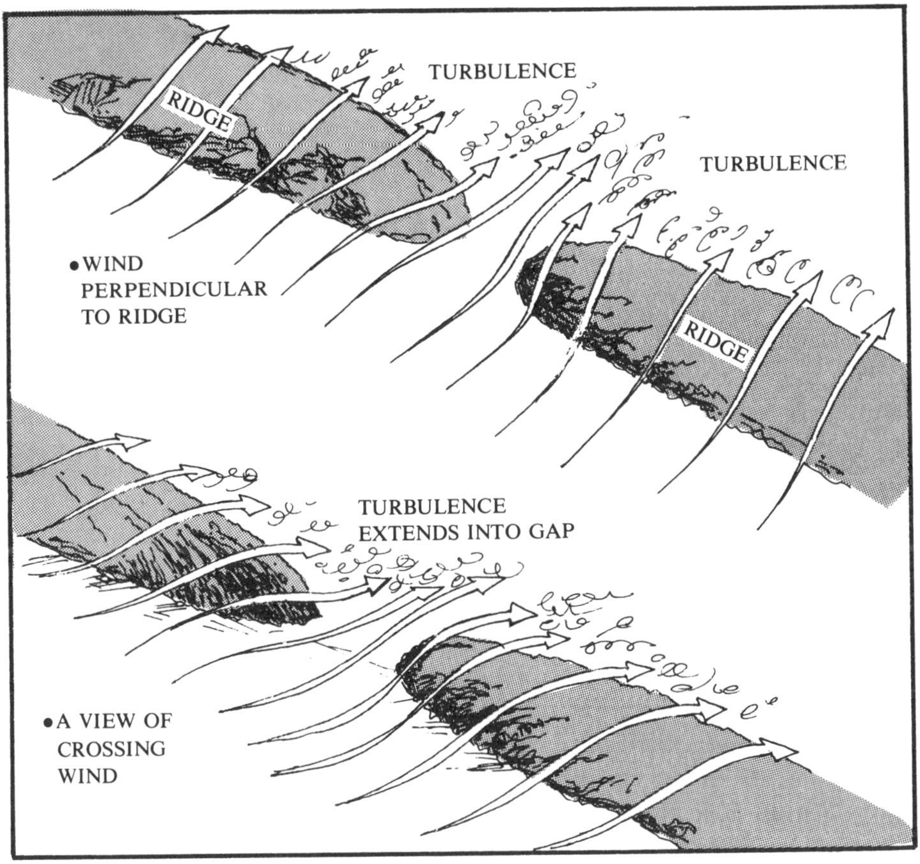

Figure 111 - Flow and Turbulence in a Gap

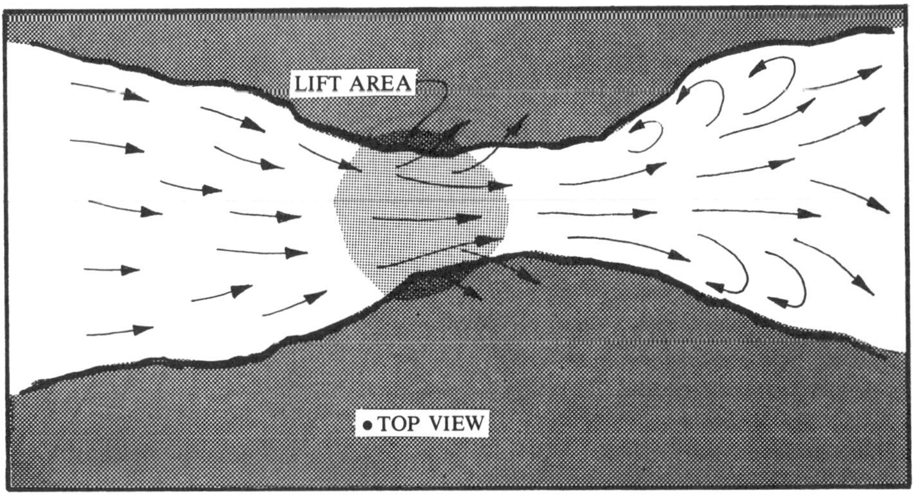

Figure 112 - Higher Winds in a Narrowing Channel

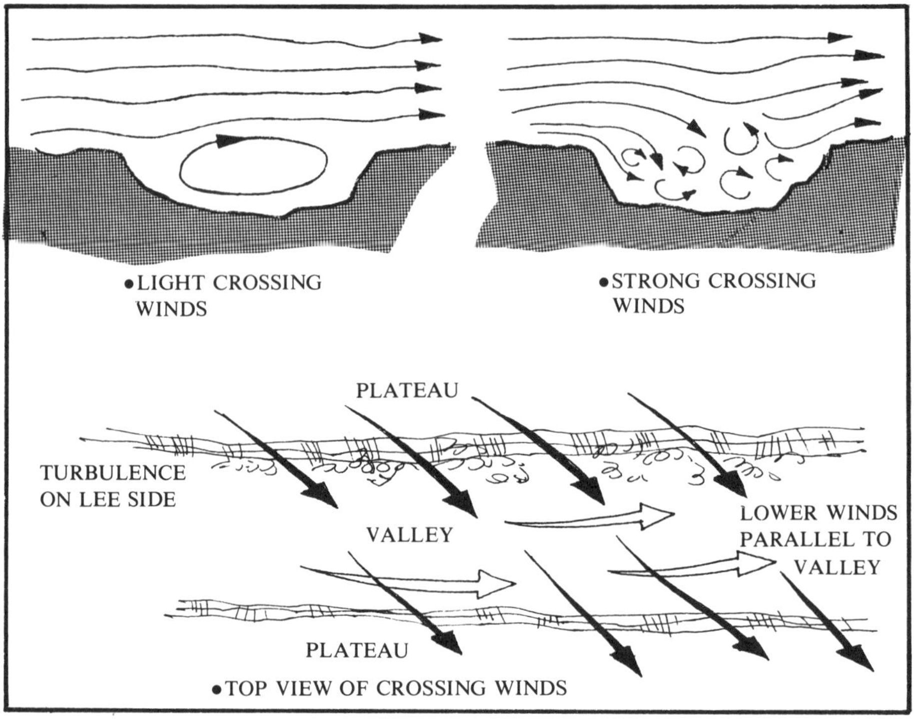

Figure 113 - Flow and Turbulence in a Valley

slope as shown in the figure.

Canyons in high mountain areas can produce their own brand of formidable turbulence. Heating effects and thermals often combine to create great sink and turbulence within the canyons while lift appears along the spines of ridges that border the canyons. The classic case of this type occurs in the Owens Valley in California as well as in the Alps and other rugged mountains. When the wind is crossing the general axis of these canyons, turbulence and sink can be even more severe and appears as in figure 114. The downwind side of the ridges and the depth of the gulley should be absolutely avoided in any conditions except a calm. Crossing such canyons requires ample altitude to reach from spine to spine. On smaller hills or mountains avoid protrusions and cuts in a crossing wind in a similar manner for the turbulence they produce as shown in figure 115. To cross such irregularities, simply loop around them upwind or above them with several hundred feet (at least 100 m) or more depending on their size and the wind velocity.

TREES

In many parts of the world trees are part of the everyday obstacles that pilots must dodge. They are also creators of turbulence. Soaring a tree-covered hill yields much bumpier flights in the absence of thermals when compared to a grassy or bare hill. The difference must be experienced to be believed.

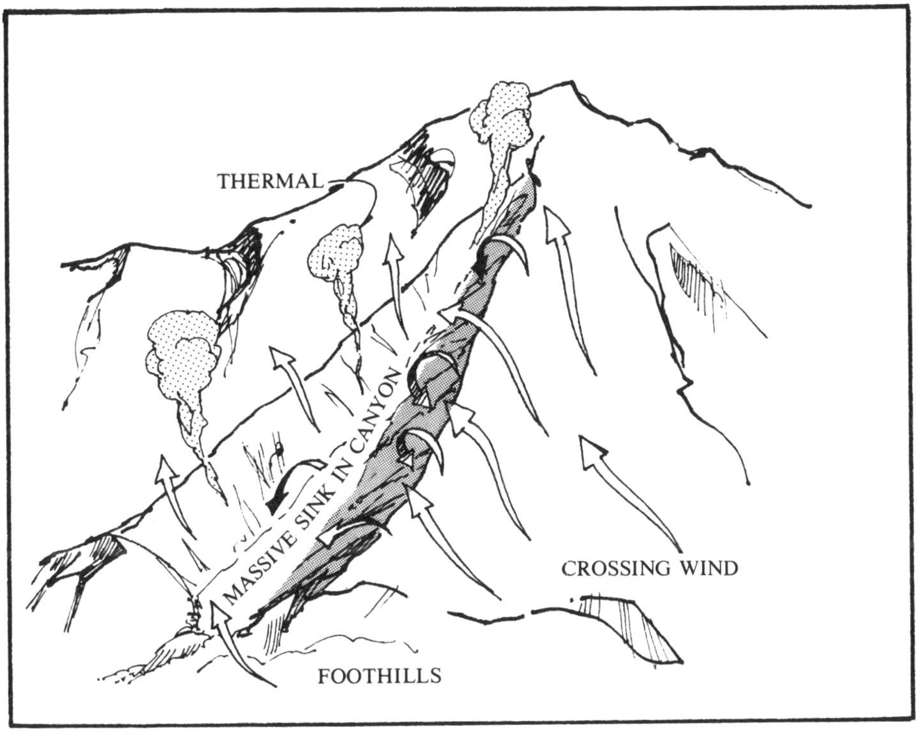

Figure 114 - High Mountain Canyon Flow

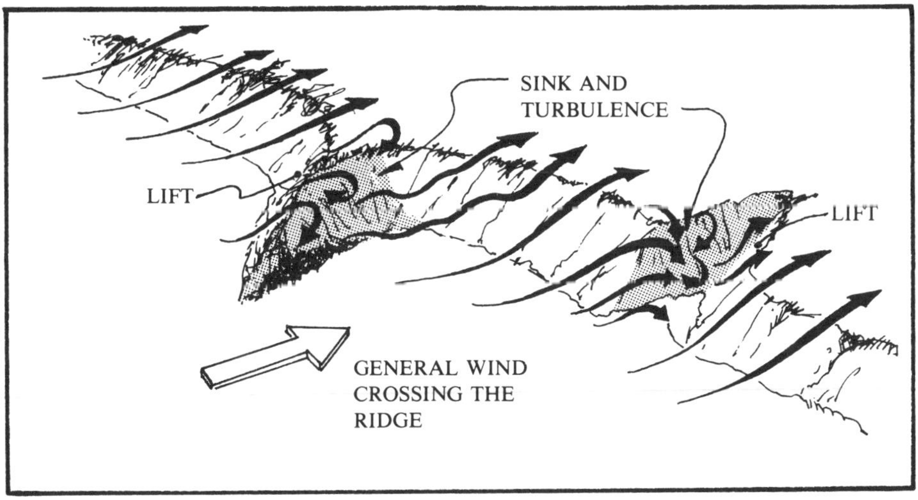

Figure 115 - Flow Irregularities on Ridges

A row of trees will naturally create many chaotic eddies and if the trees are sufficiently dense they can act like a solid wall. Often a pilot may feel a bit of lift on the upwind side of a tree line. Small aircraft have been known to soar "tree line lift." When the trees are leafless they produce less severe turbulence but they still chop up the air considerably. Figure 116 illustrates

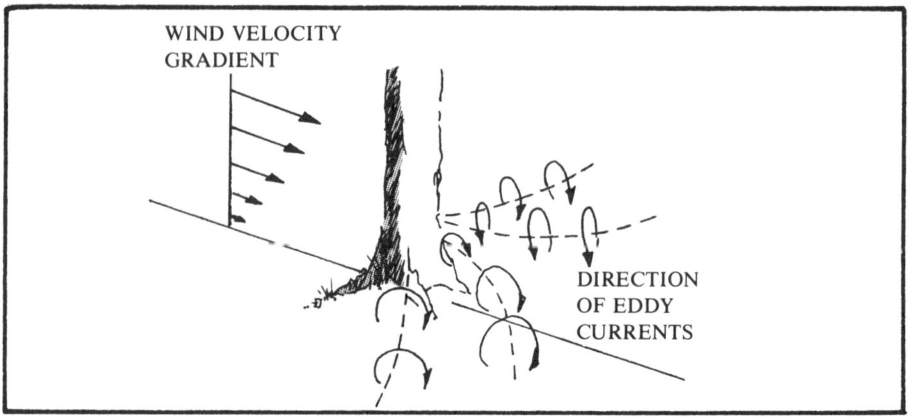

Figure 116 - Turbulence Downwind From a Tree

the turbulence created by a single tree trunk. Imagine a multitude of these trees all adding their contribution to the mixture.

In full leaf the trunk area will allow the wind to pass more readily than the crown space (area of leaves). Thus we have a wind profile as illustrated in figure 117. Note the strong change in velocity or gradient at the tops of the trees. This great change is called a *wind shadow*.

Attempting to land along a road or long slash in a forest of trees with a wind crossing the axis of the cut is dangerous, for the turbulence is just like that associated with a valley with a crossing wind. Also landing or taking off from a tree surrounded field must be attempted with caution for at some wind velocity turbulence will be too strong for safety.

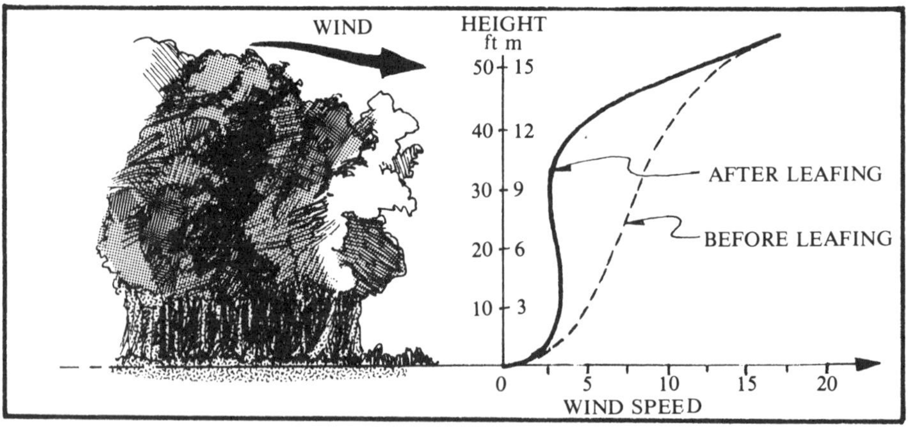

Figure 117 - Wind Shadow Due to Trees

WIND SHADOW

The blockage of the wind behind a tree line, building or hill is a wind shadow and can be associated with strong gusts (since the situation is not always permanent) and strong gradient. As we saw in Chapter V, a wind gradient always exists close to the ground. The more severe this gradient is, the more it affects our flying. Landing into the wind and encountering a wind shadow has the same effect on an aircraft as landing in any gradient—a stall can occur—except the gradient in a wind shadow is more severe. In a strong wind shadow it may be difficult to prevent a stall even when it is anticipated. Wise pilots avoid testing their skills in this matter by avoiding the downwind side of solids in any significant wind. If the encounter with a wind shadow is unavoidable it is best to pass through the extreme gradient area in a crosswind direction.

A special form of wind shadow occurs near the surface in super-heated conditions with a stable air mass. In this case a layer of hot air is formed that persists for some time before it releases, especially if some terrain feature blocks air movement close to the ground. This setup is especially common near the seas with the stable marine air moving in and the hot beach warming the air with dunes holding it in place.

At the top of the hot layer the wind speed may increase abruptly so a strong gradient exists as shown in figure 118. Notice how the wind profile follows the temperature profile (lapse rate) as is often the case. Landing in such conditions requires extra speed to compensate for the severe gradient.

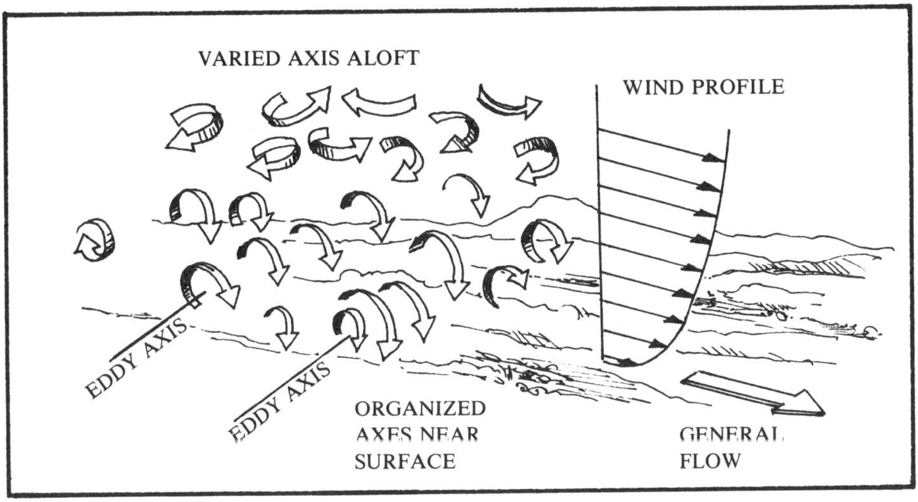

Figure 118 - Turbulence Orientation Near the Ground

FLYING IN TURBULENCE

Most of us would prefer *not* to fly in turbulence, but bumps are a fact of life in the aerial world. In fact, thermal pilots eagerly seek the bumps that herald a thermal and after time a pilot learns to enjoy moderate turbulence much like an experienced water skier looks for wakes to liven up the action.

Turbulence creates two problems for aviators: loss of control and stress on the aircraft. Strong turbulence can nose you over or lift a wing wildly or even stall you by rapidly changing your angle of attack. When this happens close to the ground it is disconcerting at best. Such turbulence can also produce gust loads that can break or fold certain aircraft. To combat the first problem we need to speed up for quicker control. To combat the second problem we need to slow down so as to reduce the suddenness of the gusts. It is obvious we need a compromise here. The tried and true rule in most aviation circles is to fly 1.5 times your stall speed in turbulence to help prevent an inadvertant stall and avoid overloading your wings.

Mechanical turbulence exists close to the terrain in its most virulent form. You can avoid it by remaining above it (several hundred feet up), waiting until the wind abates or landing in flat open terrain. Close to the ground mechanical turbulence has a preferred eddy orientation with the axis perpendicular to the wind as shown in figure 118 due to the rolling action caused by the ground. This is especially true in the lower several feet. Within 60 feet (20 m) of the surface the eddies become oriented in all directions with random energy. What this means is that a landing into the wind brings you face to face with turbulence that tends to cause pitch changes, unless it meets one wing only. Good control speed is in order during a turbulent landing.

Thermal turbulence can be anywhere from the ground up to the thermal-based clouds. However, it will be worse near inversion layers and in high winds. Sometimes these winds are only at certain levels or in specific areas so can be avoided. Despite the normal presence of higher winds aloft, thermal turbulence is often less severe the higher we climb for the thermal gets more organized and broadens out. In any case, the best way to avoid thermal turbulence is to wait until solar heating tapers off.

Shear turbulence generally can be escaped by descending below the shearing layer. If you are powering up and encounter shear, simply power back down. On the other hand, if you are powerless and descend into shear all you can do is hold on and drop below it. Shear rarely extends to the ground.

Avoiding turbulence from other aircraft is of utmost importance, especially if the aircraft is larger than you. The details are described in flying manuals, but the main idea is to avoid the downwind side of the other aircraft's path for several minutes. Helicopters create a terrific amount of turbulence that lingers below their flight path. Avoid this deadly air by heading the other way.

SUMMARY

Turbulence is with us on an intermittent basis at least until the sun burns out. We have to live with it, fly through it and avoid its most severe forms. We can do the latter by understanding how the various types of turbulence are created and what signs indicate their presence. The use of flowing water as a model and a little imagination help us visualize where the dragons lurk and where the flying is comfortable. Sport aviators should use skill and judgement to finesse themselves through the air rather than plow through the rough spots.

Soaring pilots choose to fly in thermals and a certain amount of wind which naturally introduce them to turbulence. Non-soaring pilots often pick and choose their conditions to minimize turbulence, but even so they occasionally run into textured air with a captial T. To feel at home in the air all aviators need to taste a bit of the rough stuff and swallow it with a smile. We'll leave the white-knuckle gnarly rides to the race car drivers, but accept a certain amount of bounce as being part of the aerial territory.

CHAPTER VII

Local Winds

The earth, its basins of water and its shroud of air undergo many overpowering processes from earthquakes and tides to tropical storms. We mere humans are but innocent bystanders in these great events. Even the passage of general weather systems render us as helpless opportunists taking advantage of any ride that comes our way like a hitchhiker on the highway.

But there are many smaller niches in the sky where matters flow to their own rhyme and rhythm. Fortunately for sport pilots these small-scale circulations incursions and currents are just about the right size to provide our eager wings with gentle lift.

We call this flow of air on the smaller scale local winds because they are produced by heating and pressure imbalances that occur over a score of miles (30 km) or less. This is true micrometeorology.

The important thing to note about local effects is that they are very accessible to pilots because they occur according to conditions that can be readily observed. In this chapter it is our goal to learn the causes of local winds, how to predict them, avoid their dangers and exploit their lift.

HEATING AND CIRCULATION

The main engine that drives most local effects is differential heating. This term simply means that an area is heated more than an adjacent area by the sun. We learned in Chapter I how different parts of the earth's surface are heated according to how much direct sunlight they receive and how much they absorb. We'll expand on that idea to see how local circulations work. These matters are very important to sport pilots for they greatly affect wind and soaring conditions.

Figure 119 shows what happens to air above both a warm and cool surface. To begin we see the entire area at the same temperature. The pressure lines or isobars shown at various heights above the surface are flat because the pressure of the air is uniform in the horizontal direction, as is the temperature. However, when solar heating begins, the area that absorbs

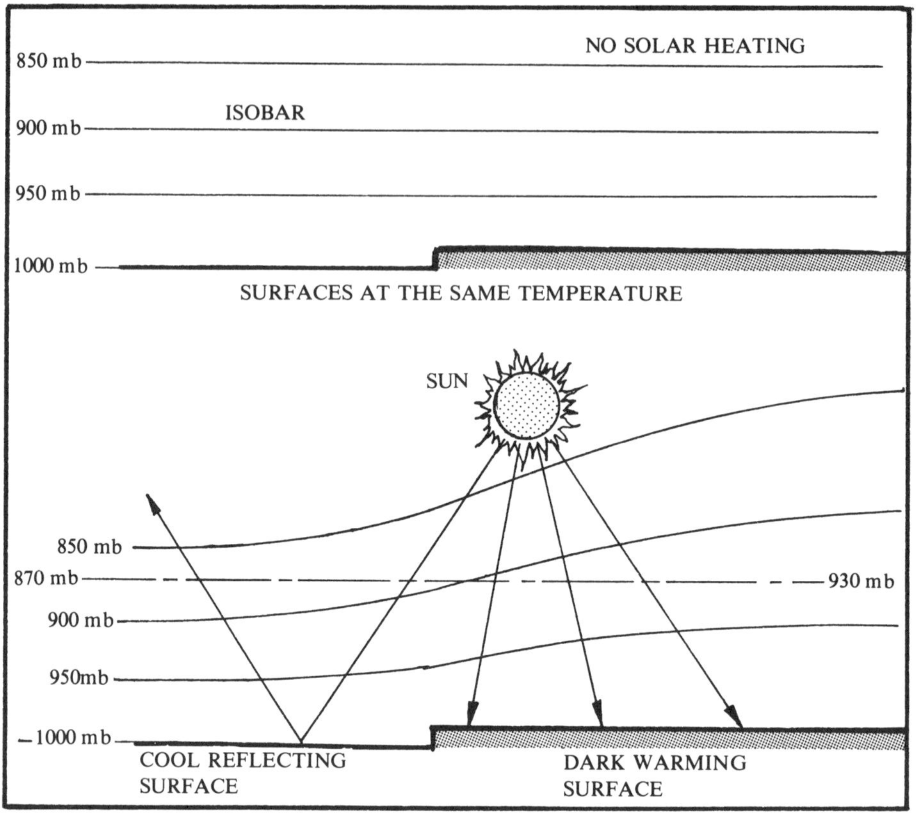

Figure 119 - Effects of Heating on Pressure

heat more readily rises in temperature and heats the air above it as shown in the lower drawing. The warmed air expands and the isobars are no longer flat, but sloped as shown. Remember that the pressure felt at any given height depends on how much air is bearing down above that height. Thus when the heated air expands in the vertical direction any given pressure except that at the surface is found at a higher level.

The situation shown at the bottom of figure 119 is not in equilibrum, for at any level—along the dashed line for instance—a higher pressure exists in the warm air than it does in the cooler air. Consequently a flow aloft begins from warm area to cool area as shown by the arrow.

As this process continues the pressure at the surface in the warm area is reduced since air aloft is flowing away, reducing the weight at the surface (see figure 120). At the same time, more air is being added above the cool surface so its pressure rises. Thus at the surface we have a return flow from the cool to the warm area. Above the cool area we find sinking air while the air continues to rise above the warmed surface. The result is a circulation as shown in the lower part of figure 120 that lasts as long as the solar heating continues. You can experience the same type of flow on a personal level. Go to the bathroom, close the door and take a hot shower. Once you are finished open the door while you're still wet. You will feel the cool air flowing inward on your legs while the warm air in the bathroom flows out aloft.

This circulation mechanism is responsible for sea breezes, upslope winds and other local winds as we shall see. At night when cooling occurs, a similar process takes place: the surface wind blows from the area that cools the fastest. We should also note that during the day strong thermals may break up the regular circulation created by differential heating if not stop it entirely.

We now summarize the important points:

> **Circulation Due to Heating**
> *Air flows from cool areas to warm areas; warm air rises, cool air sinks.*

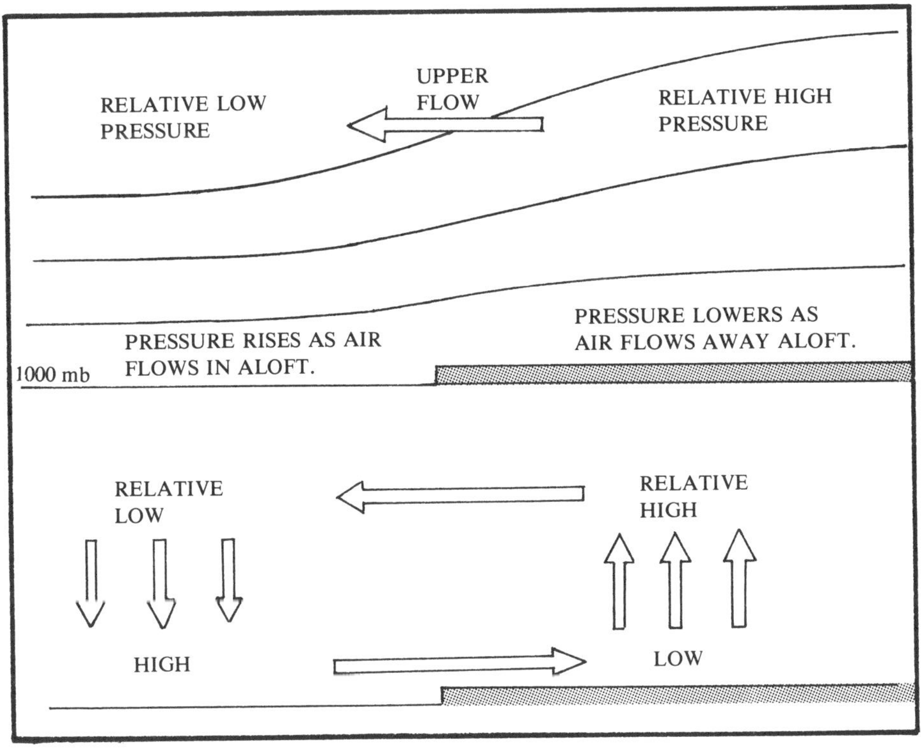

Figure 120 - Initiation of Circulation Due to Heating

THE SEA BREEZE

One of the most obvious places where warm and cool surfaces lie next to each other is near shorelines. Indeed, the sea breeze and land breeze familiar to all seaside dwellers is a classic example of the circulation due to local heating differences described above.

As we know, land heats up under the blazing sun an amount dependent on the ground cover. In any case it is always warmer than the sea during the day (except perhaps when the land is covered with snow). The sea heats very little on the surface for evaporation cools it, the sun's rays penetrate deep into the water instead of stopping at the surface to spread warmth and currents in the water distribute the heat downward. Consequently the daytime sea remains much cooler than the land.

A typical sea breeze situation is shown in figure 121. Also shown is the returning land breeze that sets in at night. The first inklings of sea breeze begin in the morning as soon as the sun begins to warm the land. It is usually well established by mid morning and reaches its peak by mid-afternoon when the sun's smile is the warmest.

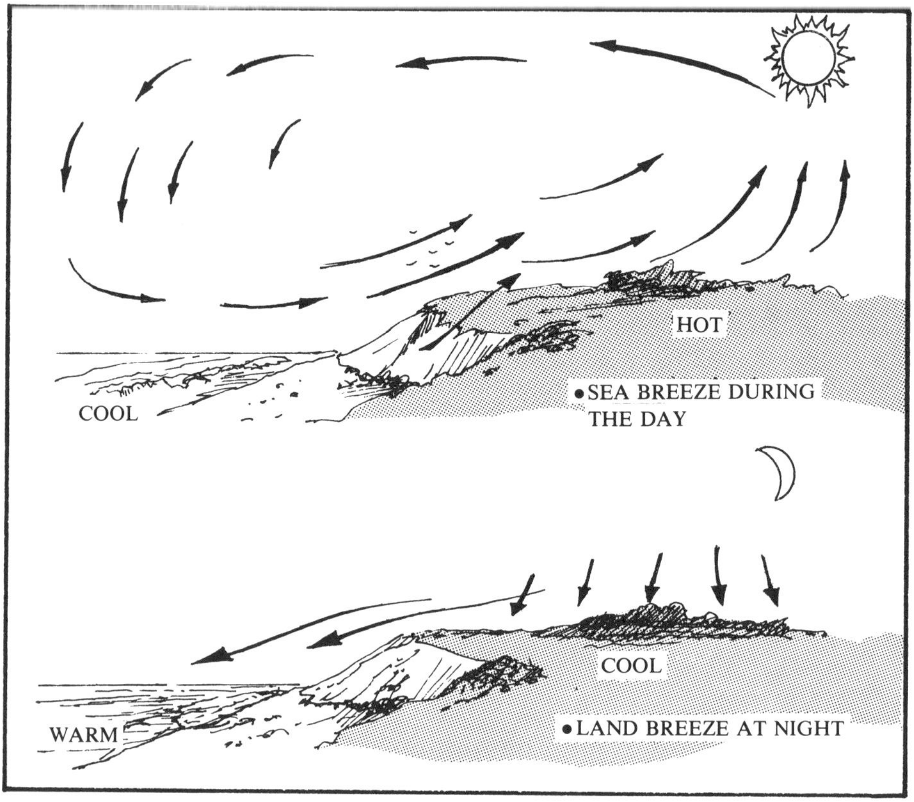

Figure 121 - Typical Sea Breeze and Land Breeze Situations

During the day-long flow of the sea breeze, Coriolis effect can serve to alter its direction to the right in the northern hemisphere and the left in the southern hemisphere so that it is nearly blowing parallel to the coastline, especially inland, by the time it diminishes in the evening. Figure 122 illustrates the changing sea breeze on a typical day.

SEA BREEZE AND GENERAL WIND

A sea breeze or lake breeze can set up near any body of water from the size of a small lake on up. The strength of the sea breeze will vary according to the temperature difference of the land and water and somewhat according to the size of the body of water. A very important factor in sea breeze behavior is the nature of the general wind.

The over-all wind caused by large-scale pressure systems affects the sea breeze in unexpected ways. We might think that a general wind blowing

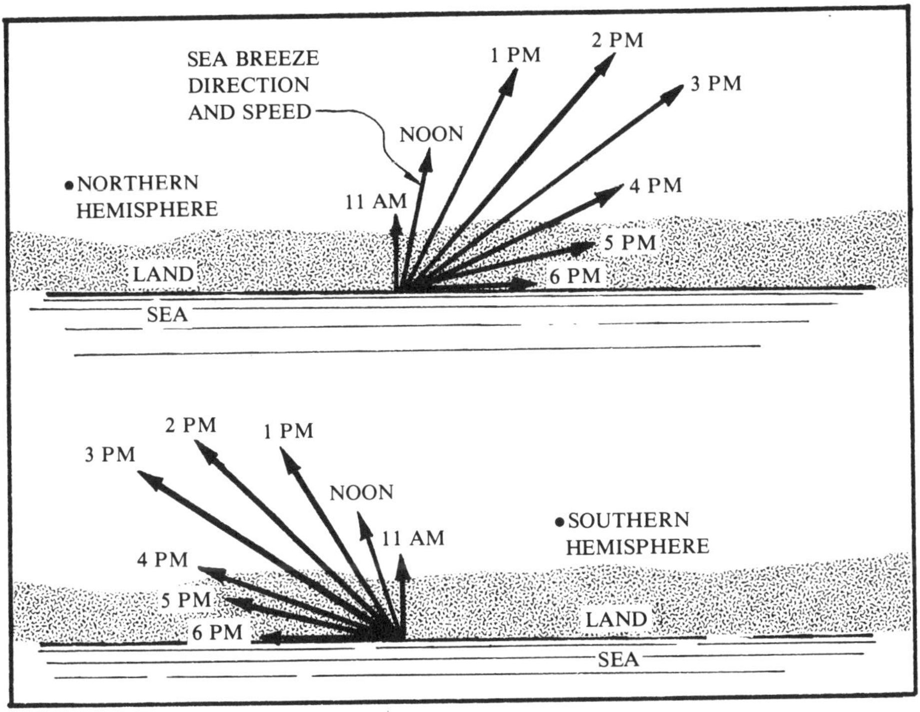

Figure 122 - Daily Sea Breeze Changes

from the land to the sea would prevent a sea breeze from forming, but this is not the case as long as the seaward wind is not too strong (over 15 mph or 24 km/h). If, for example, we have a slight breeze blowing out to sea it will be stopped at the surface by the heating effects and pressure differences thus created, while aloft it will assist in setting up the flow that begins the circulation. Remember, it is pressure differences and gravity that induce any wind flow and the strong heating differential near the sea creates its own pressure system that is able to overcome an opposing wind.

If we imagine a general wind blowing from sea to shore we may think that this will assist the sea breeze. In fact, such a wind prevents a sea breeze from forming for it carries cool, stable air inland to slow the morning heating and prevents the return flow aloft.

To fully understand how the general wind affects the sea breeze so we can make predictions, look at figure 123. Here we see a general wind blowing at an angle to a coastline. With your back to the wind if the land is on your right, a convergence zone is set up over the water offshore in the northern hemisphere. In the southern hemisphere this convergence happens when the land is to your left (back to the wind) as shown in the figure.

Convergence means "coming together". The reason the air comes together in these situations is because near the surface over land the air is slowed due to friction so it flows across the isobars as explained in Chapter IV. However, over water there is much less drag so the wind direction is close to that of the general wind aloft which follows the isobars. Consequently the wind on land at the surface is angled to the left of the upper wind direction and the surface flow over the water in the northern hemisphere. In the southern hemisphere this angle is to the right. Thus the wind over the water and land come together somewhat in a form of convergence as shown. In addition, the slowing of the wind over land creates convergence as the faster flowing wind over the sea runs into the slower wind.

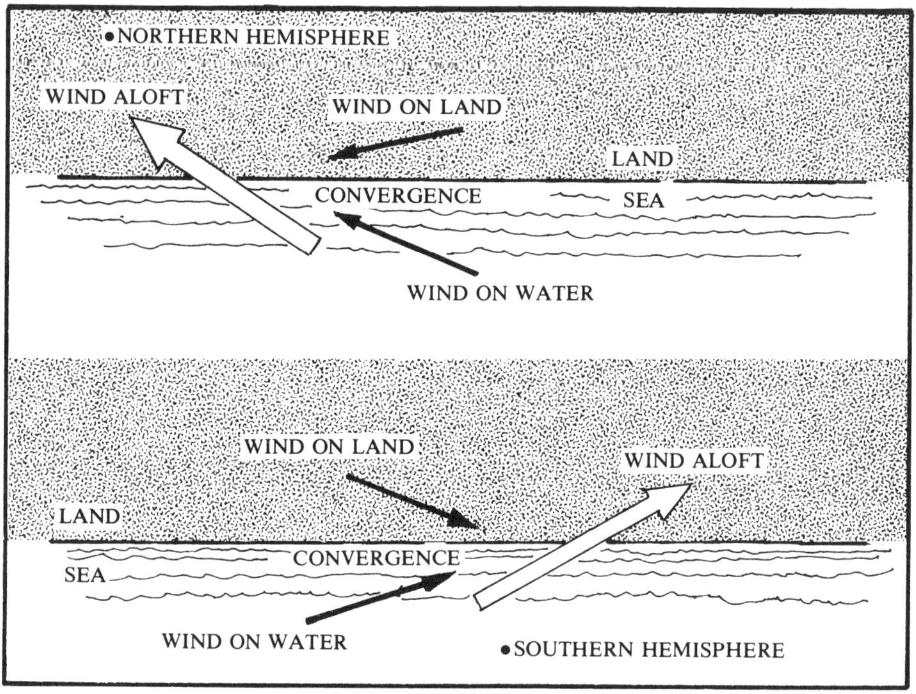

Figure 123 - Convergence Along a Shoreline

As a result of this convergence one of two things (or both) happens. The wind may increase velocity to get rid of the extra air if conditions are stable. Or, the air may rise up if it is neutrally stable or unstable in which case a band of clouds will form along the coast. In either case, the sea breeze is prevented from setting in because the convergence stops the sinking air at sea that is necessary for a sea breeze circulation.

The opposite situation is shown in figure 124. Here the breeze is from the land and crossing as shown for each hemisphere. The difference in direction and speed of the flow over land and over water creates a separation or divergence of air just offshore so a slowing or sinking effect occurs. This action tends to support the sea breeze formation and indeed the sea breeze is most pronounced when the general wind is from the land and crossing in the direction shown.

When the general wind is from the directions shown in quadrants 2 and 3 in figure 125 the effects of wind direction change and speed change over land tend to cancel each other out so neither convergence or divergence occurs.

SEA BREEZE PREDICTION

Because so many flying sites and airstrips are located near bodies of

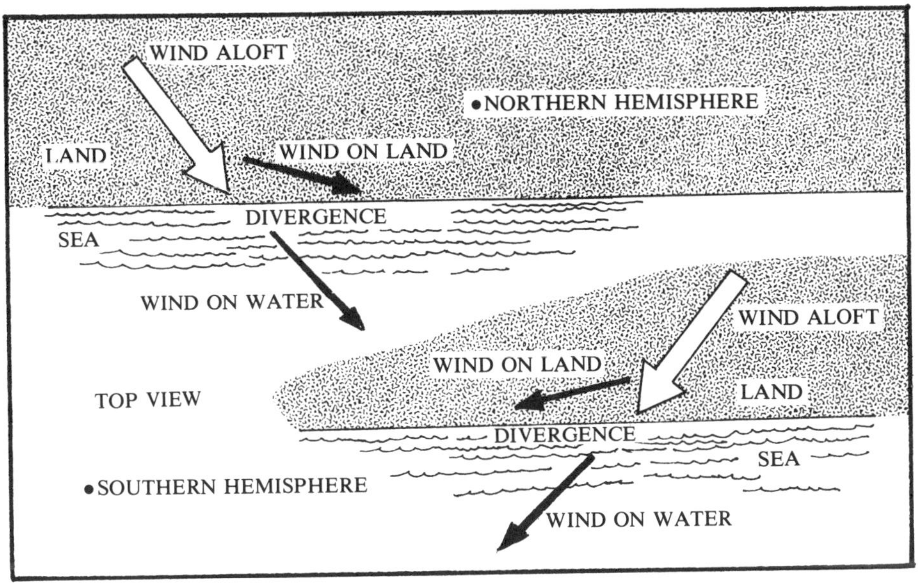

Figure 124 - Divergence Along a Shoreline

water, it is most useful for sport pilots to be able to predict sea breeze behavior. As shown in figure 125, the direction of the general (upper level) wind is the key to this prediction.

A summary of what happens when the general wind is from various quadrants in relation to the shoreline follow:

Quadrant I northern hemisphere; II southern hemisphere.

In this case the divergence feeds the subsidence and promotes formation of the sea breeze. This is the situation that brings the sea breeze farthest inland. Initially the offshore wind will calm by mid-morning then pick up in the onshore direction and progress inland.

Quadrant II northern hemisphere; I southern hemisphere.

The sea breeze under the influence of this general wind is opposed by any slight convergence set up offshore. However, the seaward flow aloft helps establish the sea breeze a couple miles out to sea whereby it eventually pushes to land when the heating increases by midday. The sea breeze in this case is not as strong as in the previous case.

Quadrant III northern hemisphere; IV southern hemisphere

When the general wind blows onshore we never have a true sea breeze for the wind aloft is prevented from returning to sea. However, when the wind is from this quadrant it means a low pressure system is over the land (remember the wind flows counter-clockwise around a low in the northern hemisphere). Thus the heating of the land that normally produces a relative low near the coast intensifies the pressure gradient (difference over distance) so the general wind is increased. This increase may be 10 or 20 knots in warmer areas such as the southwestern U.S. or Mediterranean coasts. Even though a true sea breeze is not in evidence here the cool, stable air arriving from over the sea can reach well inland.

Quadrant IV northern hemisphere; III southern hemisphere. Again no true sea breeze arises in this onshore wind as long as the general wind blows. However, because a high pressure system must exist on land for such a general wind direction to occur, the heating of the land lowers the pressure and opposes this general wind, slackening it. If the high pressure system and the general wind is weak, they may be totally eliminated by the heat of the day so that a sea breeze may set in all of a sudden later in the day. This late-starting sea breeze will never reach as far inland as those in other cases outlined.

When the wind is perpendicular or parallel to the shoreline the effects are a combination of the quadrants nearest to its direction. For example, a directly onshore wind would combine the effects of quadrant III and IV in the northern hemisphere.

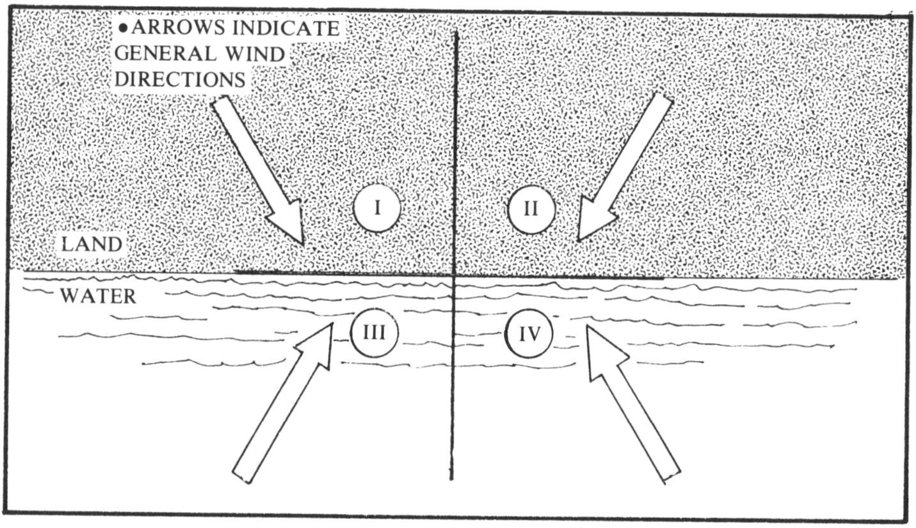

Figure 125 - Sea Breeze Quadrants

SEA BREEZE STRENGTH, REACH AND SEASONS

Besides the general wind factor outlined above, sea breeze strength is greatly influenced by the difference in temperature between the water and land. When desert areas lie near the sea as in Australia, the American southwest and around the Mediterranean, the sea breeze can easily reach 20 knots or more (1 knot = 1.15 mph or 1.84 km/h). In greener areas 10 to 15 knots is more common.

Regional differences in sea breeze are accounted for by the seasonal water temperatures. For example, in the eastern US and Europe water temperature varies from that on land most dramatically in the late spring and early summer which is when these regions experience the strongest sea breezes. These regions are also more dominated by changes in the general wind patterns and sea breezes are altered accordingly.

Desert areas receive their peak heating in mid-summer so sea breezes assailing such land areas tend to peak in summer to early fall. Australia and the southwestern US are cases of note. The Pacific coastline of North America is particularly prone to experience strong sea breezes for the cold Pacific upwelling currents adjacent to the nearby desert areas make for some remarkable temperature differences.

The orientation of the coastline has no real effect on the occurance of a sea breeze, but the presence of pressure systems does because of the general wind effects. A low in the southwestern US will afford the strongest sea breezes along the Pacific coast. In the eastern US low pressure systems bring clouds and rain so a Bermuda high bringing southwest winds along the Atlantic coast is the prescription for the best sea breeze. In Australia inland lows do the trick while a low over Europe or a high over North Africa creates the best Mediterranean sea breeze.

The inland reach of a sea breeze depends on how early it sets up and how strong it becomes. It is not unusual for the breeze to reach 20 to 30 miles (32 to 48 km) inland in moister areas. In desert areas, the sea breeze influence has been known to reach as far as 250 miles (400 km) but we must expect this is a situation where the general wind is blowing inland and a true sea breeze circulation is not occurring.

The main point of interest here is for pilots to know their local area and the general extent to which the sea breeze reaches. If mountains run parallel to the coast in your area then expect the sea breeze progress to be blocked except where passes and gaps provide an outlet valve to the push of the cool sea air.

SEA BREEZE EFFECTS

The sea breeze air tends to be stable. This is because in the process of sinking at sea its lapse rate becomes more uniform or stable (we described this stability process of sinking air in Chapter III) as shown in figure 126. Thus, even though the cool sea air moves over hot ground and is heated from below it does not readily produce thermals. Any convection that does form in this sea breeze mass is usually small, short-lived and rowdy. Satellite photos from space sometimes show this effect dramatically. We can readily look at the peninsula of Florida on a sunny day, for instance, and see a narrow 20-mile band of clear, stable air bordering the entire state while inland large inviting cumulus clouds are popping everywhere. The sea breeze surpresses all thermals along the coast on an otherwise unstable day.

Given the stable nature of the sea breeze we would think that soaring pilots would rather it went away. However, this is not the case. To begin, a steady sea breeze blowing on a coastal cliff or ridge can provide an abundance of relaxing lift.

Another important benefit of a sea breeze is the production of convergence lift. Because the sea breeze is stable it tends to flow around mountains and through passes rather than over the tops. Thus it often meets itself in back of the mountain and wells up as it comes together again. We will investigate convergence lift in the next chapter, but here we'll note that such lift is common enough in the coastal mountains of the Pacific northwest.

A well-known sea breeze effect occurs in the Los Angeles area basin when the sea breeze pushes around the Santa Monica and Santa Ana Mountains as shown later in figure 159. In this case the sea breeze is channeled through the plains areas to meet other arms of the breeze in the San Fernando Valley and the Lake Elsinore area. The convergence zones created are known as the San Fernando Convergence Zone (SFCZ) and the Elsinore Shear Line (ESL) respectively. Convergence zones are sometimes called shear lines since the two air masses do not always meet in harmony and some shearing may take place. However, the name shear line doesn't really convey the lifting process taking place.

A third source of lift a sea breeze presents is that in a sea breeze front. We cover this important matter in the next section.

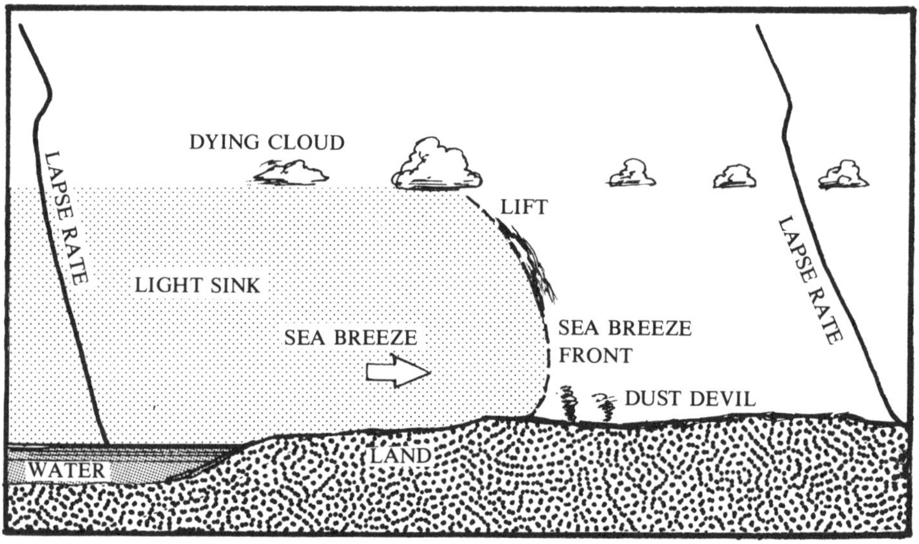

Figure 126 - The Sea Breeze Front

SEA BREEZE FRONTS

When a true sea breeze pushes inland it is constantly pushing against the land breeze. The cool marine air plows under the land flow which provides the return breeze aloft. As a result, the leading edge of the sea breeze acts like a miniature cold front and hence the name *sea breeze front*.

In figure 126 we can see some important features of the sea breeze front. First, note the area of lift in the warm air just in front of the front leading edge. This lift may be quite strong, but is generally in a very narrow band—a few hundred feet (100 meters) or less. Dust devils may accompany this lift and strong sink often exists in the warm air just beyond the lift as shown.

Cumulus thermal clouds are often evident in the warm mass overlying the land while the only clouds in the marine air mass are residual cumulus dying a slow death as they are left behind by the advancing front. There is usually little usable lift in the sea breeze air mass itself.

Right above the lift large cumulus may form if the land air mass possesses enough moisture. Such a cloud has been known to grow into a thunderstorm on occasion. If the air is relatively dry only a ragged wispy cloud may be evident or no cloud at all. At times some of the marine air may be entrained with the warm rising air to form a veil cloud as shown in the figure.

Locating a sea breeze front and finding the lift is mainly a process of noting the cloud positions and remaining in the warm air next to the front. This author was able to follow a sea breeze front for miles on the west coast of France by edging along the frontal clouds. Often the marine air will carry dust, smog and haze with it which disperses slowly due to the stability in this air mass. As a result the leading edge of the front is readily identifiable as the start of the dirty air. The sea breeze front that proceeds beyond Los Angeles carries the evidence of that city's excesses and is known as the *smog front*.

At other times the marine air may be fairly clear and the only way to identify the position of the front is by the sudden healthy turbulence as the marine air is entered. Generally the sea breeze front moves inland at a pace of less than 5 knots and it must be followed in order to exploit its lift. At times it moves faster—up to 15 knots or more—and pulses ahead in quick jumps. These pulses are likely due to perturbations in the general wind flow that starts the stable marine air oscillating like Jello.

The height of the sea breeze front and thus the height of usable lift varies greatly, but is typically around 3,000 feet (1,000 m). At the top of the sea breeze is the returning air from the land and it is often the height of the convection in this air that determines the height of the sea breeze effect. Indeed we can use this factor to predict the presence and strength of the sea breeze.

In general, with an inland convection level below 3,000 ft (1,000 m) the sea breeze will be weak at best. With a convection level from 3,000 to 6,000 feet (1,000 to 2,000 m) conditions are ideal for the production of the sea breeze front with plenty of lift and a deep inland penetration. A convection level above 6,000 feet (2,000 m) will also support a strong sea breeze front, but thunderstorms may develop due to the extensive lift located in one area.

Sea breeze fronts only develop when a true sea breeze pushes inland. Once again this occurs only when the wind is from quadrants I and II in figure 125 (both hemispheres).

COMPLICATED COASTLINES

Our earth's land masses rarely have the smooth contours we show in our idealized drawings. Undulations, peninsulas, bays, coves and nearby islands are the usual order of things along the coasts of the world. These varying shapes have their own effects on the sea breeze. Initially, the sea breeze will tend to set up perpendicular to the shore as shown in figure 127. However, after time the general sea breeze flow will override the small variances and produce a flow as shown.

Protrusions of land as well as long spits like those that line the Atlantic coast by the U.S. may have a sea breeze coming in from both sides that meets and wells upward with convergence lift over the land. Bays tend to produce very light sea breezes around their perimeter that is eventually overcome by the general flow with the greatest velocity normally felt on the right side of the bay.

The sea breeze flowing into a bay may force its way into a narrows and thus increase in velocity as well as produce some convergence lift. An example of this matter is the entrance to San Francisco Bay.

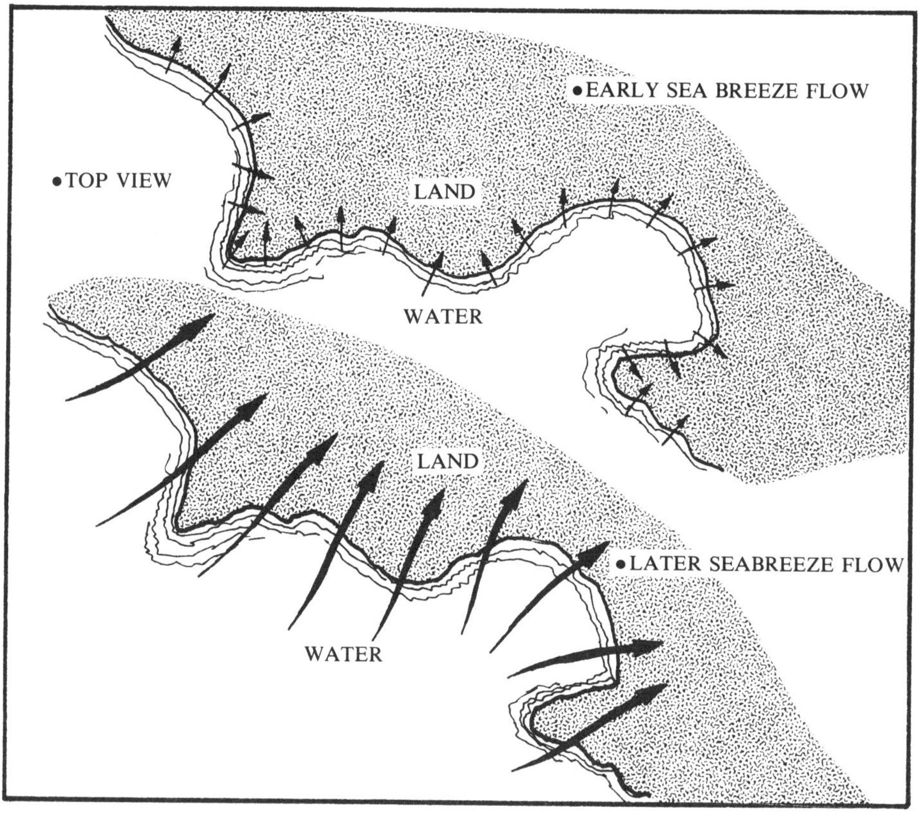

Figure 127 - Sea Breeze on a Complicated Coastline

ISLAND EFFECTS

Islands that stand just offshore of a large land mass such as Long Island, New York, Vancouver Island off British Columbia and the Isle of Wight south of England initiate a sea breeze blowing in from all directions. However, the overall sea breeze blowing in to the larger land mass soon overtakes the island weather and adjusts it to the general flow.

An isolated island will attempt to set up a sea breeze all around, but will be quickly cooled by the marine air if it is too small. In this case periods of sea breeze will be followed by periods of calm. A larger island, especially one with mountains will sustain a sea breeze combined with upslope winds that build the typical cumulus cap clouds that eventually slow the heating effects and thus the sea breeze.

Islands typically alter the wind patterns near their shore as shown in-figure 128. Here we see how the increased friction on the island compared to the sea slows the wind and causes it to flow across the isobars. This change in wind direction and speed meets the wind at sea along the left shore when facing with your back to the wind (in the northern hemisphere) in a convergence zone. This convergence causes stronger winds along this shore as shown.

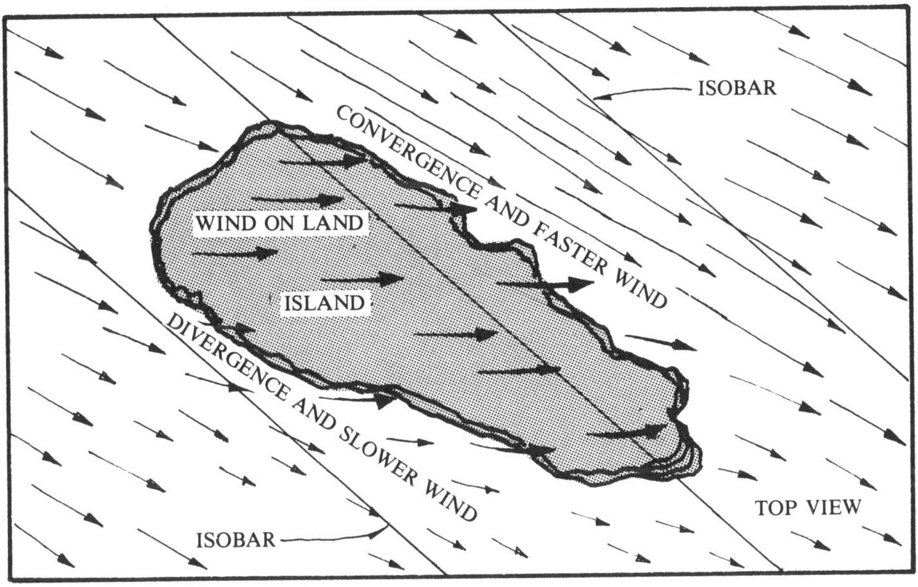

Figure 128 - Island Effects on a Wind

LAND BREEZES

Land breezes as mentioned earlier are the opposite of sea breezes. Usually the land has cooled to a temperature below that of the sea by a couple of hours after sunset at which time the land breeze begins drifting out to sea. The land breeze is gentle—usually 3 to 5 mph (5-8 km/h)—and shallow since there is no heating and thus no vertical movement.

In general, surfaces that heat most readily in the day are those that also cool most rapidly. Thus land breezes and other forms of return flow follow about the same path as the daytime circulation but in the opposite direction.

Pilots do not have much use for the land breeze since it happens at night, but we should be aware that a land breeze can set in suddenly earlier than normal when mountains exist near the sea for their late afternoon downslope winds assist the land breeze in getting started. Also, a land breeze tends to slide under a general onshore wind. Thus we should expect shear turbulence in the lower few hundred feet (100 m) when flying in the evening near a sea with a general on shore wind.

HEAT FRONTS

Sea breezes and sea breeze fronts are not the only important matters related to local heating. Anywhere that the surface is heated or cooled more than an adjacent area is a likely candidate for circulation. For example, a cloud bank or an area of smog that greatly cuts down the solar heating will make the underlying cool surface act just like a sea and produce a surface air flow from the cool to the hot area. If the general wind is light and opposes this flow we get a true heat front as shown in figure 129. The soaring conditions in this front are just like those in a sea breeze front due to the convergence present.

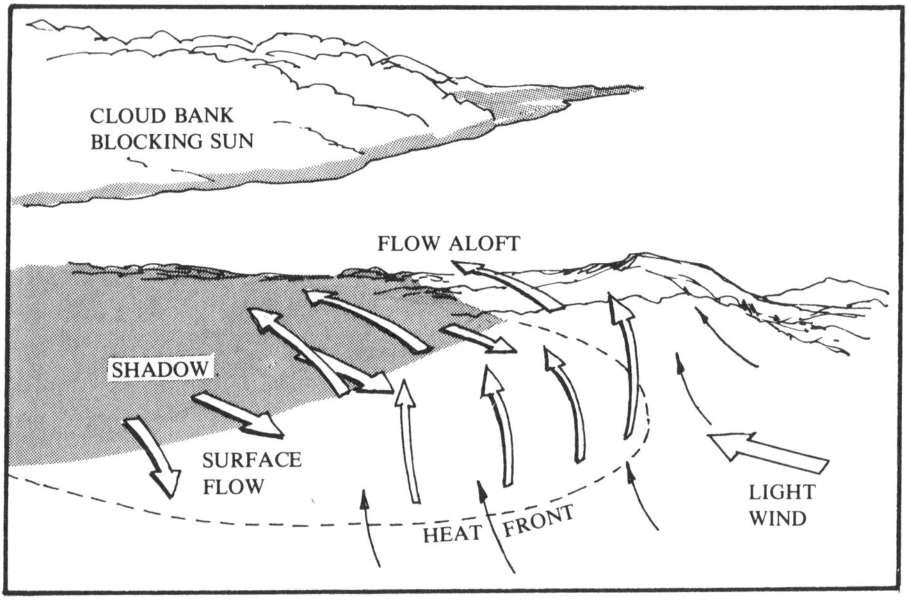

Figure 129 - Heat Front Due to Cloud Cover

The smog that inhabits the Log Angeles basin is carried inland along with the sea breeze and adds to the strength of the sea breeze front by cooling the area covered by the marine layer. The heat front in this case is known locally as the "smog front" although it is usually indentical to the sea breeze front. Other smog fronts can exist wherever industrial pollution blots out a large area of sunshine.

The fronts and winds in general associated with vast areas of clouds and smog will shift position as the cloud or smog moves with the upper winds. Often such cooling blankets move fast enough over new ground that they do not establish strong circulation and fronts but they do herald a shift in wind direction or strength as noted in Chapter V.

Water itself often has marked differences in its surface temperature due to upwelling currents or inflow from land sources. These temperature differences can create various local winds based on our circulation principle. A good example of this is the shears and convergence created near the coast of San Francisco by the outflow of the relatively warm bay water into the cold Pacific ocean.

Cloud bank blocking the sun over a large area.

A very important matter for soaring pilots is the heat front formed near plateaus. As shown in figure 130, the top of a plateau is heated readily during the day so the air overlying it becomes quite warm compared to the air over the lowland at the same altitude. The result is a small front with convergence and lift. This phenomenon is quite common in the plateaus of the American west and the central massif of France. For example, Blue Mountain near Dinosaur, Colorado, commonly exhibits such lift near its edge with a wind coming into the mountain despite a tailwind component to the general wind. Such a frontal type convergence zone has also been noted along the escarpments of the Sequachie Valley near Chattanooga, Tennessee.

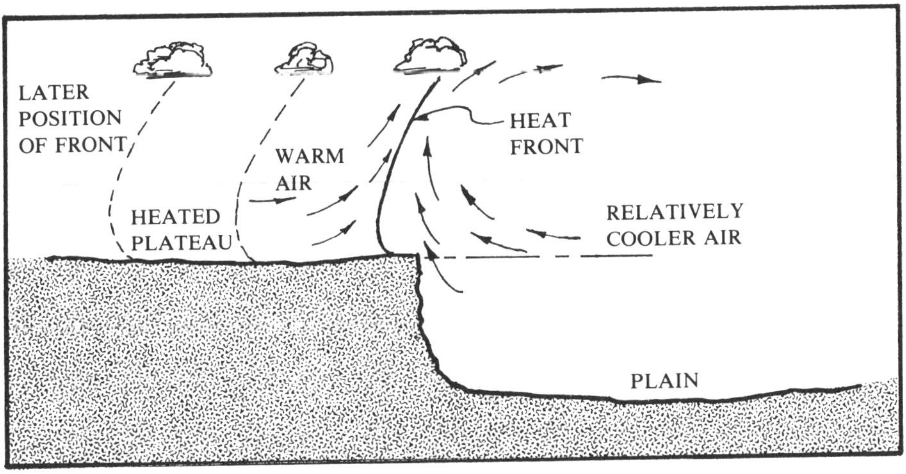

Figure 130 - Heat Front Near a Plateau

With time these frontal zones tend to move back from the edge of the plateau and may be more difficult to reach. However, a row of cumulus often indicates their presence and are enticing to the soaring pilot. We should note that general upslope breezes and thermals may be added to the effects of the heat front and may be hard to distinguish from the wind and lift of the front.

FROM FOREST TO FIELD

Large areas of forest remain considerably cooler than surrounding fields during the day as any raccoon with a fur coat can tell you. The reason for this is evaporation of moisture from the trees and the much greater surface area of leaves to absorb and distribute the sun's heating. As a result a daytime circulation is set up from cool forest to warm field in the manner shown in figure 131.

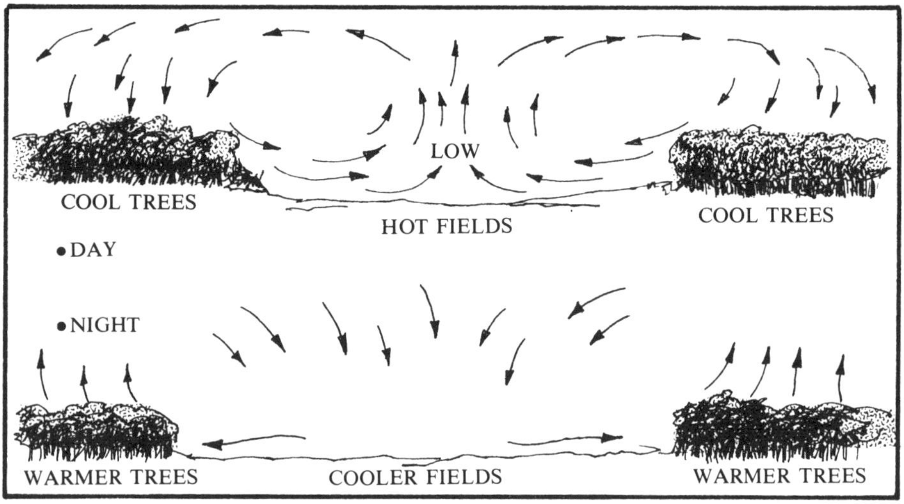

Figure 131 - Field to Forest Circulation

We should expect some lift over the fields, especially if they are surrounded by forests and convergence occurs as shown. This circulation situation normally doesn't produce fronts, however, for the circulation is weak—often too weak to sustain soaring flight.

In the evening the forest areas retain heat longer than the open fields so the flow is reversed as shown. At times forests can release enough rising hot air to keep a light aircraft up at the end of a hot day.

UPSLOPE WINDS

Every tourist who has stood on a mountain admiring the view has felt an upslope breeze. These breezes which are so familiar to soaring pilots blow up any and every slope during the day as long as they aren't overpowered by the general wind.

The creation of an upslope breeze depends on heat from the sun warming the slope and thus heating the overlying air. As the air expands a pressure difference is created much like with sea breeze production as shown in figure

132. The upslope wind generally starts at the top of the slope in early morning with a few puffs and works its way down. Slopes facing the morning sun naturally start their flow earlier. Westward and northward facing slopes may not exhibit an upslope breeze until late morning.

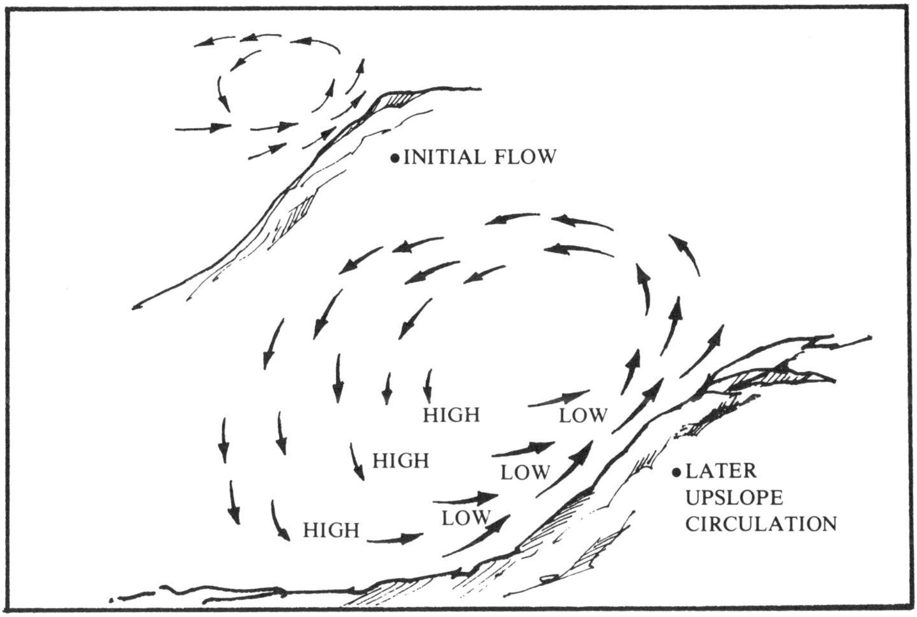

Figure 132 - Upslope Breezes

The strongest upslope breeze occurs not at the time of greatest solar insolation, but when the slope itself is collecting the most direct sunlight. This will be before noon for eastern slopes and later than peak sunshine for western slopes. Concave slopes will collect more heat than convex slopes and thus exhibit a stronger upslope breeze. Bowls are good producers of upslope flow.

The maximum strength of an upslope breeze is usually no more than 15 knots but can be much stronger on high, hot mountains. The upslope breeze is strongest and thickest at the top of the mountain. Typically the thickness of an upslope breeze is 300 feet (90 m) or less—sometimes much less. Even light upslope breezes are often soarable for they contain bouyant air that goes straight up at the top of the mountain (unless it is tilted by the push of the general wind) and often is mixed with thermals.

Indeed, the upslope wind may promote the production of thermals. A mountain that early on produces a good upslope flow usually produces the first thermals and cumulus clouds above it. This early establishment of lift can set the pattern for the day as such a mountain continues to pump lift while the areas around have their thermals suppressed by the sinking air around the lift of the early producing mountain. Black Mountain in the Owens Valley in California seems to operate in this manner.

There tends to be a return flow to the valley in front of the slopes as shown in the figure. Also, sinking air occurs in the valley as the surface air moves out and up. This sinking air becomes more stable as it drops and cools the surface, thereby suppressing thermal production for a distance in front of the slope.

It is not hard to predict the existence of an upslope breeze, for any day with good sunshine and not too much tail wind component will show some upslope flow. Instability is not necessary and the multiple release of thermals will disrupt the steady nature of an upslope breeze and may even reverse its direction temporarily. Other names for the upslope breeze are *valley winds* or *anabatic winds.*

DOWNSLOPE WINDS

What goes up usually comes down—eventually. The breezes in the mountains are no exception. In the evening when the slopes begin cooling they cool the air above them which then slides off the slope to the valley below. This downslope wind—also known as a *mountain wind* or *catabatic wind*—begins at the bottom of the slope and works its way to the top.

Figure 133 shows how a downslope flow can be initiated even when the upslope wind still blows further up. It is not uncommon for soarable upslope winds to blow at a mountain top while downslope winds exist at the base of the mountain in the evening. This is especially true if canyons, gullies, gorges or ravines open out at the base of the mountain for these cuts in the terrain drain the cooling air most effectively.

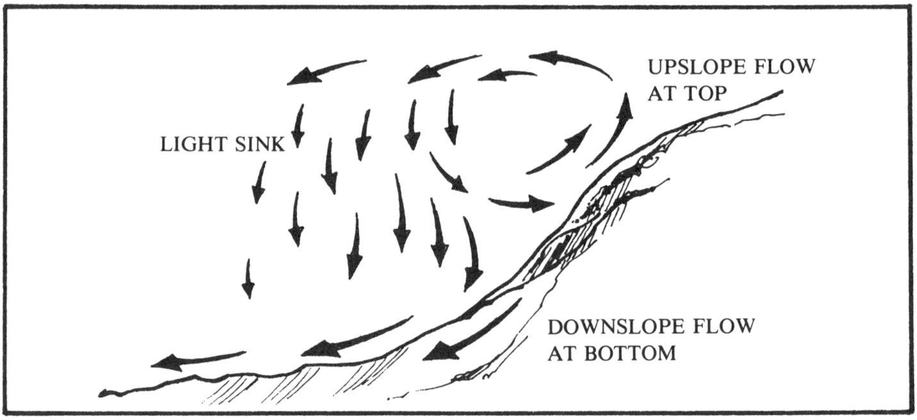

Figure 133 - Downslope Breezes

The principle mechanism that sustains the downslope flow is gravity, so we can imagine the flow as if it were water—the cool dense air will seek the lowest point and readily flow down the steepest slopes. The downslope flow is stable and confined to a shallow layer on the slope unless it gets dammed by an obstruction like a secondary ridge.

On clear days heating is greatest and cooling most rapid as the heat radiates into a clear sky. In this case the downslope flow can be sudden and strong. Occasionally pilots get surprised by a quick reversal of wind and shear turbulence in the valley in front of a mountain that just shed a layer of rapidly cooling air. This downslope air can come in puffs or cycles as it begins to establish itself and flows from various slopes join together. Eventually, however, the downslope wind becomes a steady cool breeze that lasts all night and only dies out in the morning sun when the upslope breeze takes over.

The cool air slipping off the mountains may suddenly suppress thermals and eliminate soaring prospects. The early and sudden occurrence of the downslope breeze is most common on eastern facing slopes since the sun disappears from these slopes very quickly while western facing slopes enjoy sunlight late into the evening and the horizon shadow gradually works its way up the mountain.

GRAVITY WINDS

A gravity wind is a special type of downslope breeze and is often known as a *Bora*. Gravity winds usually originate at high elevations over permanently cold areas such as glaciers or snowpacks. These winds are composed of very dense air and flow down and over the terrain much like the mist given off by dry ice settles to the lowest available place.

Gravity winds can reach high velocities when flowing through passes and should be considered turbulent and worthy of caution. These winds are found most often at the edges of snowfields, cave mouths and glaciers. Figure 134 shows how the upslope breeze is altered by a gravity wind. The mistral in the Alps can be considered in part a gravity wind.

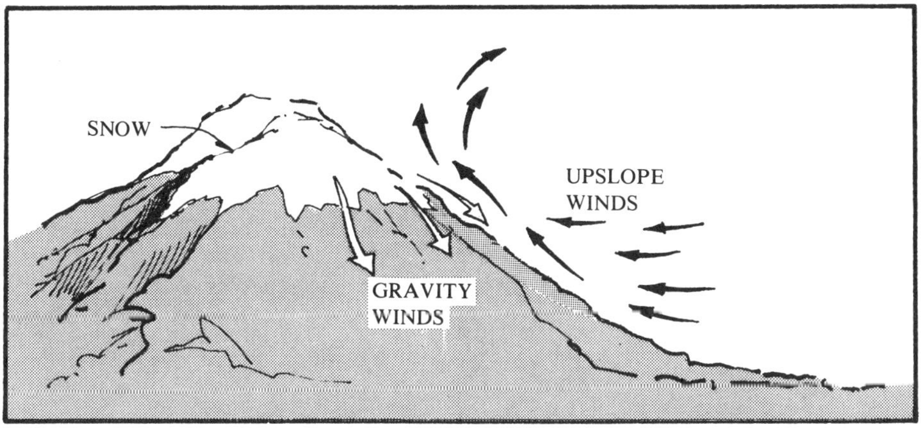

Figure 134 - Gravity Wind and Upslope Breeze

MAGIC AIR AND WONDER WINDS

At times in the evening some remarkably smooth and abundant lift sets up along a mountain. Soaring pilots have aptly named these conditions "wonder winds". The mechanism that seems to be at work is shown in figure 135. Here we see a cross section of a valley whose eastern side has just passed into shadow and begun to produce a downward flow. This cool air sliding into the valley pushes under the air that has been warmed throughout the day, lifting it and sending it up the other side of the valley to join the upslope breeze still in progress on the sun facing slope. The rising valley air is bouyant and becomes even less stable as it rises to afford excellent lift.

Wonder winds occur most often up long ridges (rather than isolated mountains) with narrow valleys, say 3 to 5 miles (5 to 8 km) across, or a valley width of 10 to 20 times the mountain height in higher mountains. The ridges that form the backbone of the eastern U.S. and the narrow Rhone valley through the Alps are two primary formations condusive to wonder winds.

These winds are most likely to set up at the end of a sunny day with relatively weak thermals and light winds. Strong thermals tend to empty the valley of heat and strong winds also provide more mixing and reduce the well of warm air in the valley. Often wave activity occurs in the evening of a convective day (as we shall see in the next chapter) and this phenomenon must be expected to augment wonder winds at times, but many windless or light days which cannot support waves will end up with wonder winds.

"Magic air" is the name applied to more widespread lift that extends well into the valley. These conditions may exist when wind is very light as well as in moderate wind. The explanation for this cornucopia of lift is shown in the bottom of figure 135. Here we see more lifting of the warm valley pool as downslope breezes on both sides of the valley undercut the warm air. As this air lifts it becomes less stable until it auto-convects and continues to lift.

Magic air affords lift over a wide area of the valley. Often it exhibits pockets of greater lift as it rises in large thermal cells not confined by ground sources. Sometimes it is smooth in a wide zone in which case we have to expect that wave activity is a factor in producing the lift.

Wonder winds and magic air are difficult to predict with any assurance; they must be exploited as they come. If lift at the mountain top is dwindl-

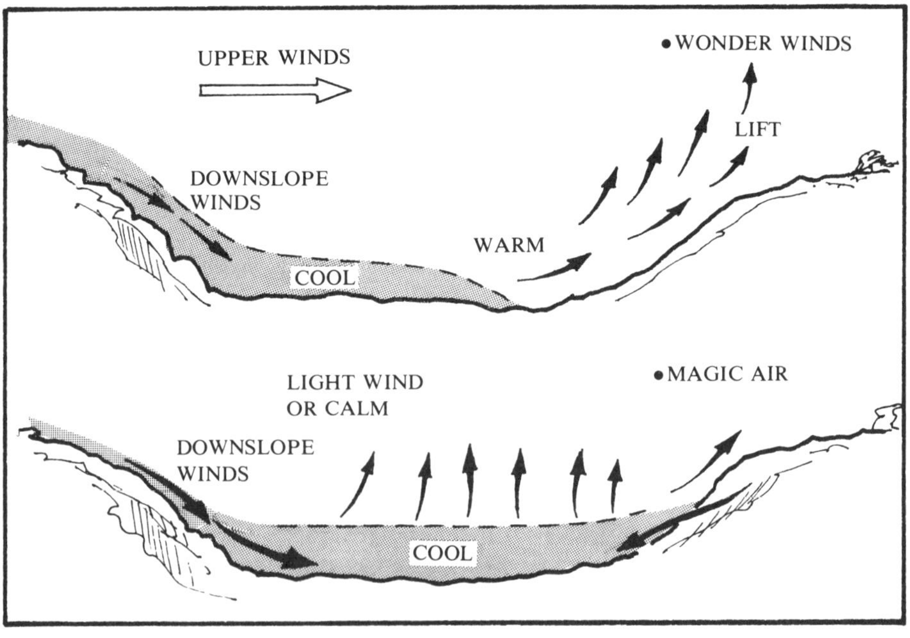

Figure 135 - Magic Air and Wonder Winds

ing in the evening, however, it is reasonable to try to find lift over the valley where the combination of late thermal releases from heated pavement and forests along with the production of magic air may prove to prolong a marvelous flight.

UP VALLEY AND DOWN VALLEY WINDS

Long, gradually sloping valleys will exhibit a daily cycle just like steeper slopes. The wind will tend to flow up these valleys during the day and down them at night. This valley flow is combined with the up and down slope flow in the manner shown in figure 136.

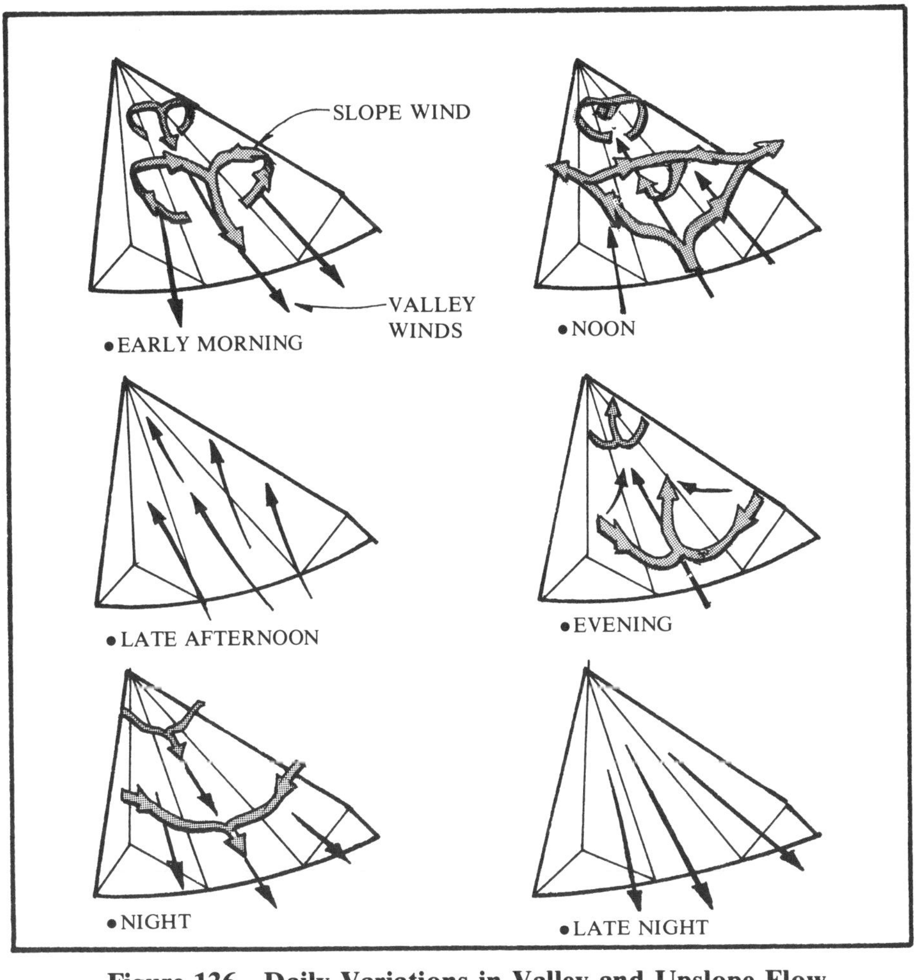

Figure 136 - Daily Variations in Valley and Upslope Flow

To begin, the early morning sun initiates the upslope breezes while the down valley wind continues to drain the canyons and shadowed slopes. Soon the down valley wind stops and begins reversing by noon. By the late afternoon the up valley wind is powerful and even alters the upslope wind on the mountain sides. Towards evening the downslope winds begin while the up valley wind continues. Note the converging air in the middle of the valley. In a couple of hours the up valley wind has stopped and turned to become a down valley wind. This wind builds in strength into the night to alter even the downslope wind. In the morning the cycle begins anew.

The presence of uneven terrain, valley constrictions and general wind alters the neat plan outlined above. An overall wind can often prevent the flow up one side slope or even channel the wind for days in one direction along the valley. However, understanding how the daily cycle works helps us predict wind behavior for the local wind will be added or subtracted from the general wind. An overall wind up the valley will be increased during the day and decreased at night.

There are a few additional points to remember about valley winds. First, they are affected by Coriolis effect so that they want to turn to the right in the northern hemisphere and left south of the equator. This makes them want to climb the right side of the valley as they head up it (in the northern hemisphere) and more lift should be expected on this side if all other matters are equal. A transit time along the valley of only a few hours is all it takes to turn the valley wind.

In addition we note that valley winds can be quite strong as they are continually fed by falling air. In fact, a reverse gradient can occur whereby the wind in the valley can be stronger than the wind a few hundred feet up (100m) and higher. The valley wind in the evening may exist as a river of cool air under a warm inversion layer and thus exhibit considerable shear turbulence. Pilots landing or operating in long valleys of this type should be wary of such turbulence and vigilant for a sudden wind direction and speed change in the evening.

COMPLICATED SYSTEMS

We will close this chapter on local conditions by noting that some areas possess weather altered by all of the above mentioned conditions. For example, Mt. St. Pierre in Quebec and St. Andre in the southern Alps as well as many other areas with high mountains, seas or lakes and plateaus combine the effects of sea breeze, heat fronts, slope and valley winds, convergence and deflection of the general wind to provide very complicated wind, lift and sink patterns. Many times in these areas wind indicators will change direction frequently or several streamers in one field will point to entirely different wind directions for long periods of time.

The key to understanding such difficult micro-systems is to understand all the elements we have covered and put them together with observation. Knowing that circulation patterns on the local level set up according to the current heating conditions and that these conditions change thereby moving the patterns is a major key to the mystery of the multitude of effects we can encounter. To indicate how dramatic the combination of local effects and large pressure systems can be, we cite the example that occurred in Mohave, California on November 18, 1991. At 6:00 am the wind was 100 mph (160 km/h) due to a strong gravity wind aided by a moving pressure system. One hour later, after some valley heating, the wind was calm.

SUMMARY

In this chapter we narrowed our sights to focus on smaller-scale effects. These are conditions that happen on the local level and are driven by daily sunshine. They have a cyclic nature.

We are not content to merely name the various circulations sea breeze, heat fronts or upslope breeze, for we can greatly enhance our enjoyment of flight if we learn how to use these local effects. The way we understand the atmosphere's short-term mood swings is to understand the importance of differential or unbalanced heating. We have witnessed the effects of such heating all our lives: the slight draft of air down a cold winter window, the shimmer of warm air rising up a telephone pole, the indraft of cool air around a campfire (notice that your body blocks this air so the smoke always seems to blow towards you!).

We now have a solid background on the whys and wheres of wind and weather in our atmosphere. We have looked at the big picture down to the small screen. To complete our integration with our chosen lofty environment we now turn our attention to the creation and utilization of vertical wind.

Stacked lenticular clouds indicate wave lift.

CHAPTER VIII

Soaring Conditions Lifting Air

To defy gravity by using favorable upward air currents is the quest in much of sport aviation. Even in motorized flight disciplines, pilots who know the secrets of lifted air can often extend their performance or enhance their safety. Being at home in the air means flying in harmony with the ebb and flow of the sky's tides and currents. To do this we must understand the air's behavior in the vertical as well as the horizontal dimension.

In this chapter we look at the sky in light of not so much what to avoid, but what to find. What we want to find is lifting air. There are a number of mechanisms that cause such lift and we'll investigate them in turn. The result will hopefully be a greater understanding of the azure world we choose to visit and a chance to rival the hawks on high if only for brief but wonderous spells of time.

TO SOAR

The word soar comes from the Latin *ex aura* meaning "to air." To pilots, soaring doesn't just mean taking to the air, as in getting airborne, but as in being at home in the air enough to use lift and prolong a flight. Soaring is staying up with the use of control, skill and the knowledge of where to place oneself in the evolving sky.

The requirement for soaring can be simply stated: we must find air rising with a vertical component equal to or greater than our aircraft's minimum rate of sink. There are various causes of rising air to be found in nature. They are wind deflection, waves, convergence, frontal movement and thermals. The last one is so important that we cover it separately in the next two chapters.

RIDGE LIFT

Wind hitting a mountain, ridge or hill gets deflected the same way water bulges over submered rocks and logs. If the terrain obstruction is broad enough in the direction perpendicular to the wind, it will flow over the obstruction, thus creating lift as shown in figure 137. We call such lift *ridge lift, slope lift, dynamic lift* or *orographic lift.* The terms mean the same thing and are used interchangeably.

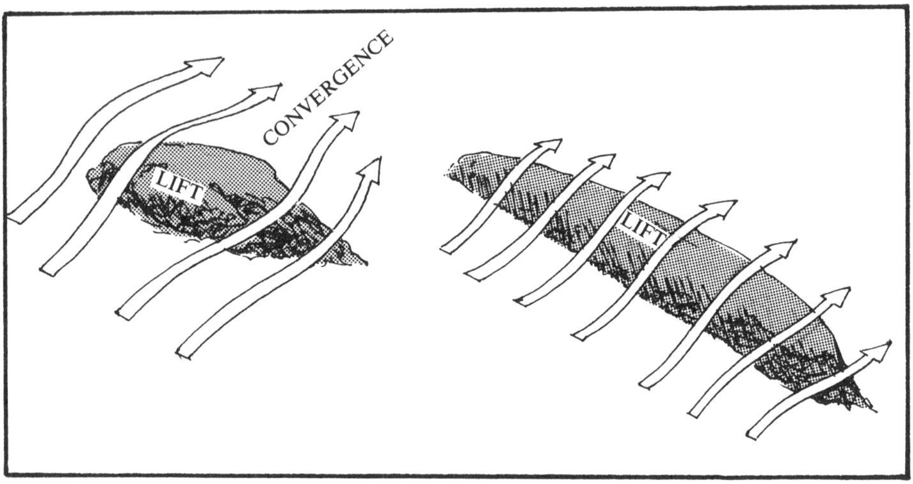

Figure 137 - The Creation of Ridge Lift

Various shapes of terrain deflect the air different amounts. In general, the steeper the slope, the greater the vertical component of the air for the same wind velocity. This is shown by the wind streamlines in figure 138. The graph in figure 139 shows the maximum lift on different slopes in different winds. For example, on a 40° slope a 15 mph wind has a maximum vertical component of 780 feet per minute (13 ft/s or 4 m/s). On a ridge or mountain with various slopes we expect to find the best lift over the steepest terrain assuming a perpendicular wind into the faces of all slopes.

Since all motorless aircraft (except balloons), birds and butterflies sink with respect to the air they can only stay up in part of the lifting air over a mountain. The area of sustained flight is known as the soarable envelope and appears in profile for different hill shapes and wind velocities as shown in figure 140. Note that the steeper slopes have a higher soarable envelope. Also the maximum lift line A-B is angled slightly forward. In more stable conditions the soarable envelope moves forward but is not as high. Finally notice the moving back of the envelope and maximum lift as wind is increased. The ideal shape for ridge lift seems to be the concave shape that gradually gets steeper.

COMPLEX SHAPES

In the real world the wind isn't always blowing straight into idealized hill shapes as shown in the above figures. Contours and crossing winds serve to complicate matters. The best shape for collecting and augmenting

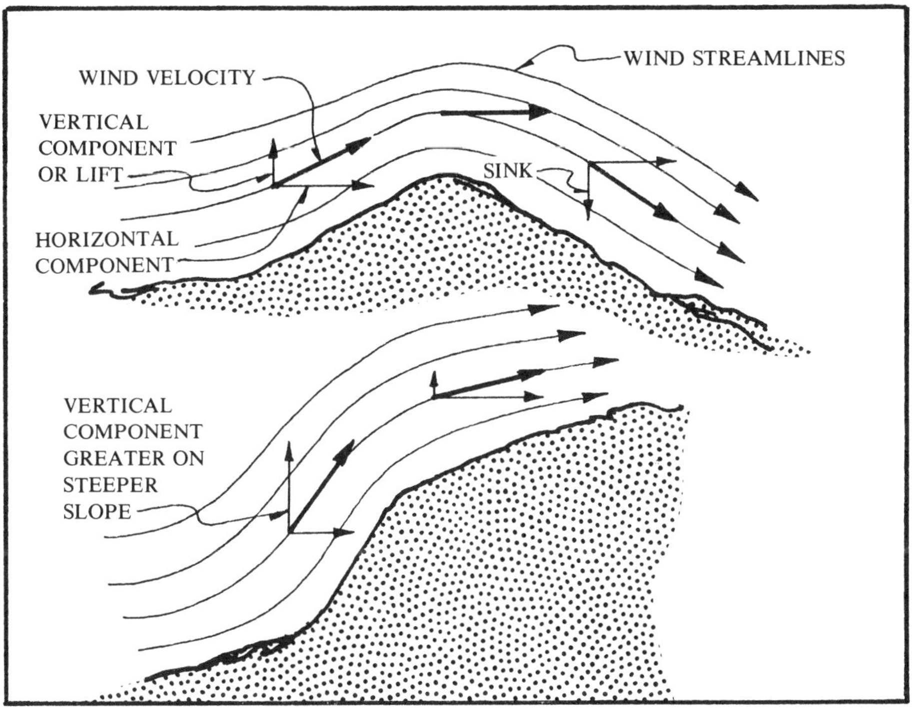

Figure 138 - Variations in Ridge Lift

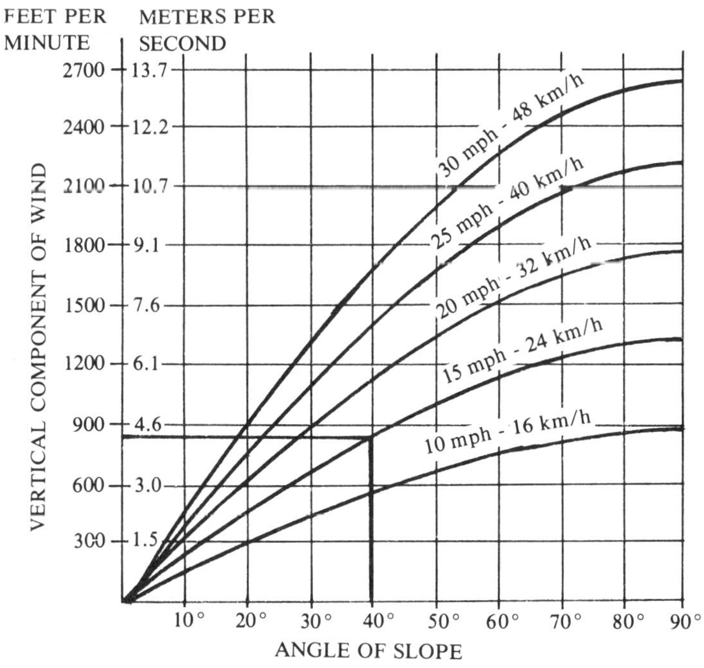

Figure 139 - Maximum Lift as Slope Angle and Wind Speed Changes

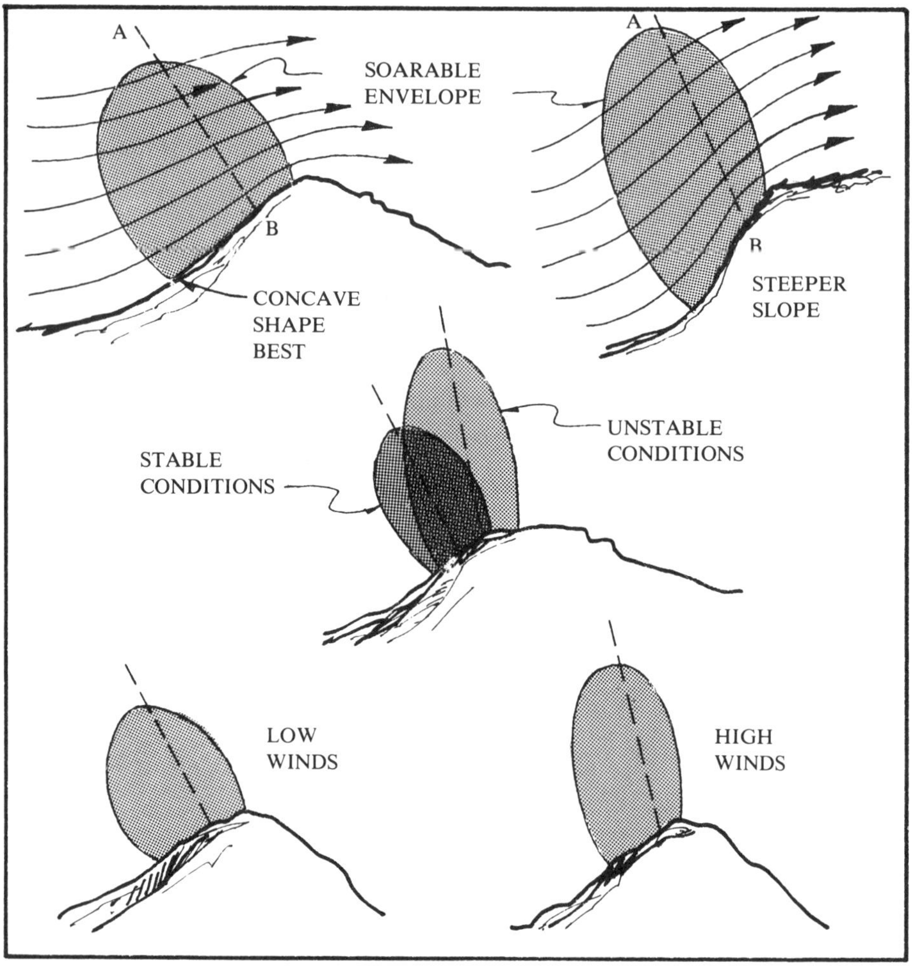

Figure 140 - Variations in the Soarable Envelope

ridge lift is a bowl as shown in figure 141. Here the air wells up all over. The opposite shape also shown is a protuberance. If the wind is straight toward this bulge it may be deflected and produce reduced lift.

When the wind is crossing on the face of a mountain less lift is produced because it is partially deflected sideways as well as up. In fact, the steeper the slope the more the deflection. The lift on a vertical cliff face is more sensitive to direction of the wind than a more gentle slope. The graph in figure 142 indicates how the maximum lift changes on different angles of slopes as the wind angle to the slope changes. For example, a vertical cliff (angle of slope = 90°) will lose half its lift if the angle of the wind changes only 30° off the perpendicular. On the other hand, a 15° slope won't lose half its lift until the angle of the wind is 60° off. However, even at the half value, the lift over a cliff is still greater than that over a 15° slope at full value in the same wind.

The above illustration points out the benefit in locating oneself on faces, outcroppings or sides of a bowl that face into the wind as much as

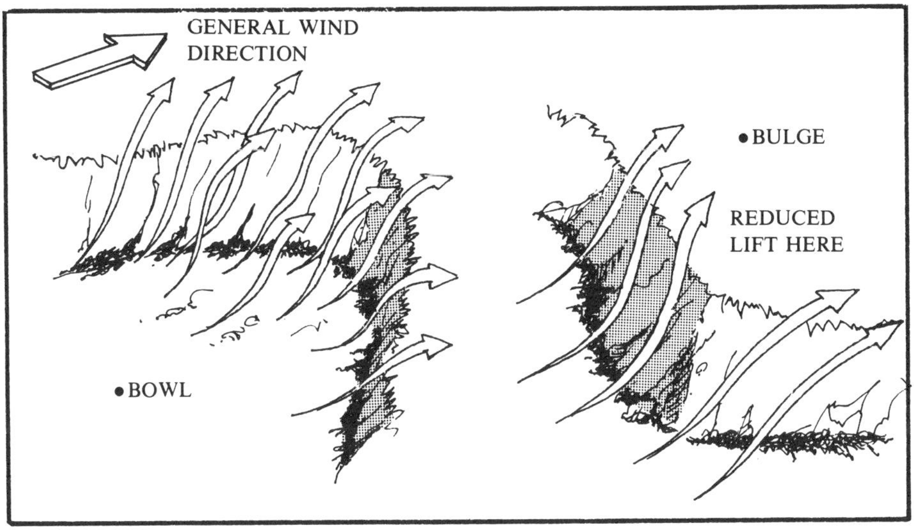

Figure 141 - Effects of Ridge Shape Variations

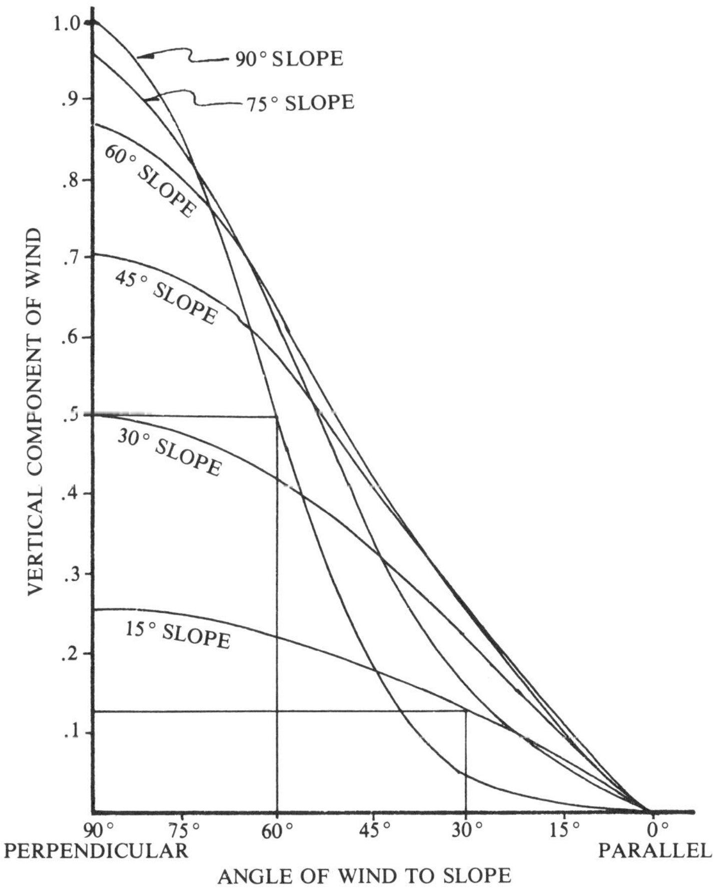

Figure 142 - Variations in Lift on Different Slope in Different Crosswinds

possible. Figure 143 illustrates a complex mountain and how an outcropping produces good lift while the rest of the mountain either exhibits sink and turbulence or poor lift. This drawing is an approximation of what was encountered on a real flight at Mont Revard in the French Alps. The crossing wind eliminated most lift, but it was possible to soar on the outcropping and in the gorge where lift is indicated.

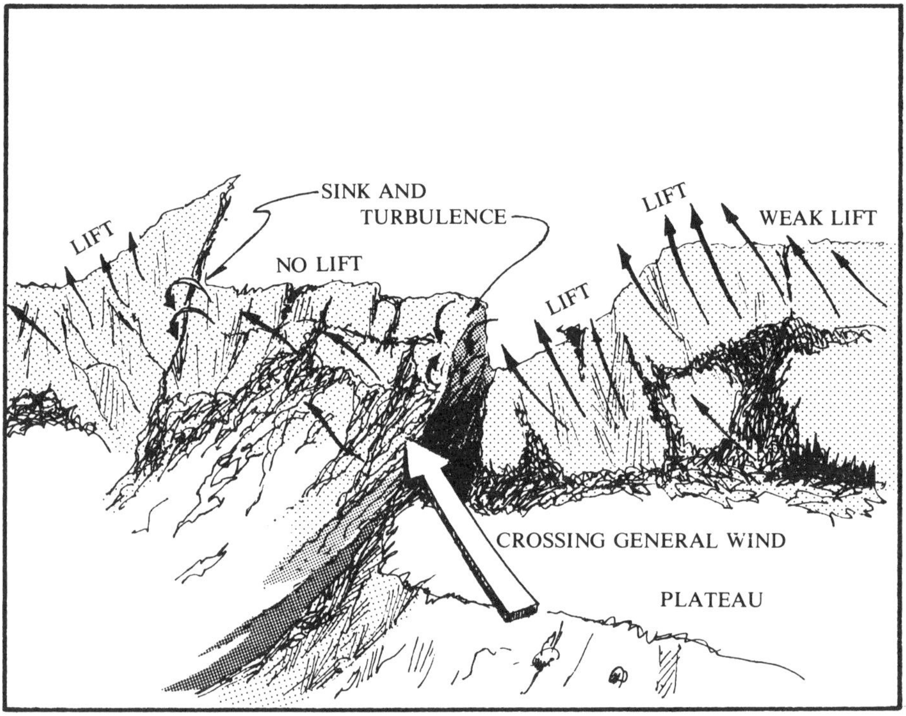

Figure 143 - Localization of Lift in a Crossing Wind

CANYONS AND GAPS

We have dealt previously with canyons and their lift and sink patterns (see Chapter VI). Here we simply note that they often produce good lift at their tops if the wind is straight in, as well as lift on their side that faces a crossing wind. Beware of their downwind or lee side for turbulence and note that massive sink can lurk at a surprising height above this downwind face.

Gaps are typically openings in a ridge that lets the air pass directly through. With a straight-in wind the situation appears as in figure 144. Here we see the wind escaping through the gap, altering the flow near the gap, producing convergence and an acceleration of the flow in the gap. Turbulence downwind of the gap is also expected as illustrated previously in figure 111. Due to the accelerated flow in the gap it may be unsafe to attempt to exploit the convergence lift. Do not risk getting flushed through the gap in high winds.

Gaps that don't go through the mountain may serve as bowls and actually collect the lift if they are not deep. Deeper gaps create lift further

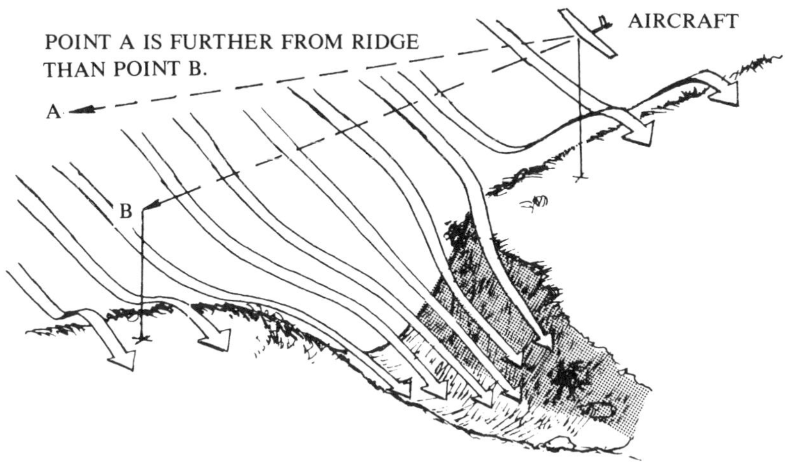

Figure 144 - Wind Flow in a Gap

back which may be out of reach to slower craft.

Both type of gaps require you to avoid the lee side and aim for the upwind side to maximize altitude as shown in the figure. In general, the lower you are the further out front you should arc your path across a gap in order to avoid the strong wind in the mouth of the gap. Crossing with a slight tailwind component is easier than with a straight-in wind. A headwind component is most difficult especially as the wind velocity increases.

RIDGE SOARING PROBLEMS

One of the realities of ridge soaring is the fact that turbulence from the drag of the slope exists lower down. Thus more control speed is required when soaring low, and this increases your sink rate. There is a noticeable difference in the amount of turbulence produced on a tree-covered slope and a grass-covered slope. Less powerful eddies linger over the smoother ground cover.

There is a gradient to the wind speed as it moves up a slope as shown in figure 145. This can have a tendency to lift the outside wing and turn you

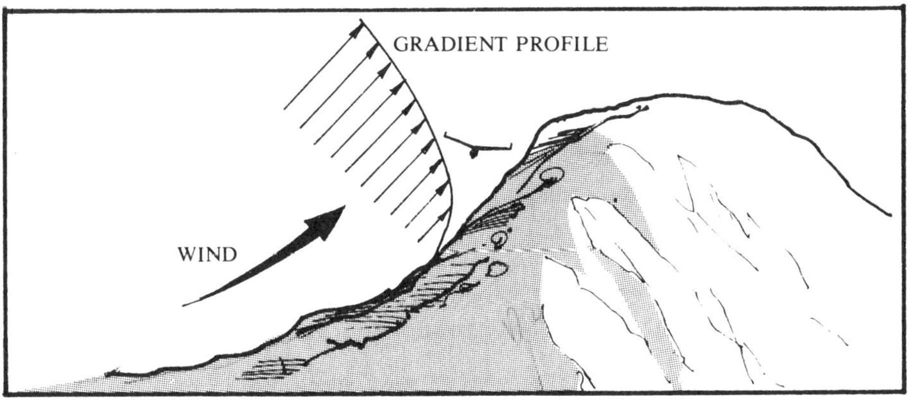

Figure 145 - Wind Gradient up a Ridge Face

into the slope. You must combat this tendency by maintaining some maneuvering speed when scratching on a ridge face.

Over the top of a ridge there can be an accelerated layer of wind known as a *venturi*. The process is just like that in your car carburetor whereby a constriction of the opening makes the air rush through faster (the same thing happens in a gap). As we can see in figure 146 the faster flow is limited to less than the height of the mountain. It is very real down low, however, and many inexperienced pilots have been caught unaware in the venturi over a ridge. Generally, it requires winds of at least 12 mph (20 km/h) and a ridge to create a significant venturi. Isolated mountains do not normally produce venturis.

Because the air seeks the path of least resistance, often stronger wind is found over the low part of a long ridge if the low part is of small extent. Thus, in lighter conditions the best lift may actually be on the lower portion of the ridge provided thermals and upslope winds aren't a factor.

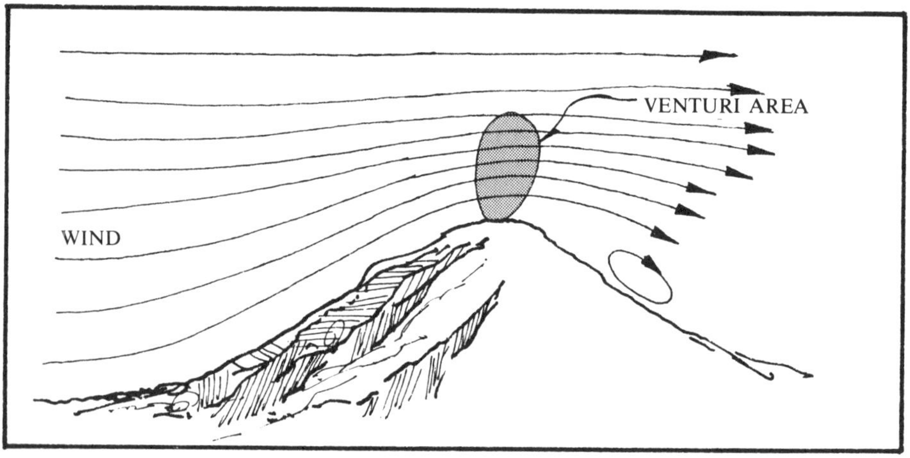

Figure 146 - The Venturi Above a Ridge

VARIABLE LIFT

Ridge lift would seem to be a fairly dependable source of up air, as long as the wind blows. Indeed it is, on some days, but at other times the same wind speed and direction can provide little or no lift. This state of affairs can be very confusing not to mention frustrating.

The answer to this mystery is the realization that other forms of lift combine with ridge lift to produce the overall lift patterns. Thermals especially have a profound effect on ridge lift. Typically thermals are enhanced by ridge lift (except when the wind is so strong it blows thermals apart). However, the sink between thermals can greatly reduce ridge lift. It has been this author's experience many times when soaring along the ridges in the eastern US that there are big holes in the lift. Even though the trees are showing plenty of wind activity, the lift isn't there.

Thermals often line up in streets (see Chapter X). Between these streets is vigorous sink. When streets are crossing a ridge the sink in between them is enough to kill all ridge lift and put you on the ground. Even on cloudless days streets can occur, so such wide stretches of sink should be expected whenever healthy thermals are found on a ridge. The best plan when traveling along a ridge in these conditions is to stop and get high in the thermals in order to glide across the sinking area.

Of course, such streets do not always occur and are rarely encountered in late afternoon and evening ridge soaring. Also such matters are only important on long ridges, for more isolated hills are thermal generators. You should be aware that stopping to work thermals to the max on a ridge run will slow you down considerably, so you must be observant to decide which conditions exist. Your first tentative ventures along the ridge on a given day should give you an idea whether or not the lift is continuous or localized.

As we mentioned, upslope breezes add to the effect of ridge lift. Consequently clouds over a certain area can reduce the ridge lift in that area if upslope breezes shut down. Evening downslope flow can also have a deleterious affect on ridge lift, but usually it begins low on the slope and simply gradually reduces the incoming wind velocity as evening falls.

Upslope breezes in general feel like ridge lift (deflected flow) except they have less horizontal component. Also, true ridge lift generally requires more velocity to afford soarability because it is not as bouyant since it is not necessarily heated. In any case on a sunny day we cannot separate the upslope flow from the ridge lift—they work together. However, on soarable days with no wind aloft we are riding upslope air (often studded with thermals) and when the sun is hidden by a layer of clouds we can be sure that our magic carpet is ridge lift.

WAVE LIFT

The air is a light fluid and like all fluids it can experience waves. In fact, if you wish to see an object lesson on atmospheric waves, go to your nearest friendly stream and watch what happens downstream of a submerged rock or log. You'll see lift in front of the submerged object which corresponds to ridge lift, while behind it you'll see a series of ripples or waves. These waves can be quite large in a fast moving, deep stream.

Waves in the atmosphere are produced by a similar disturbance. Simply replace the rock or log with a mountain or ridge and you have the required setup. However, only certain atmospheric conditions produce waves. If we look at figure 147 we see the effect of wind blowing over a ridge in unstable, neutral and stable conditions. Note that only the stable situation tends to create an undulating pattern. This is because a lifted stable layer tends to return to its original level once it passes the mountain. However its downward momentum causes it to overshoot its preferred level so its stability brings it back up. Again it overshoots and continues this process downwind to oscillate up and down as if it were on a big soft spring. Thus we have our first requirement: a stable layer.

The next thing we need is ample wind. We find that generally waves require an average wind speed of at least 15 knots (26 km/h) at the mountain top. In addition, the wind must be fairly perpendicular to the ridge, not change direction with altitude and should show a general increase from the surface to the tropopause. These requirements are summarized in figure

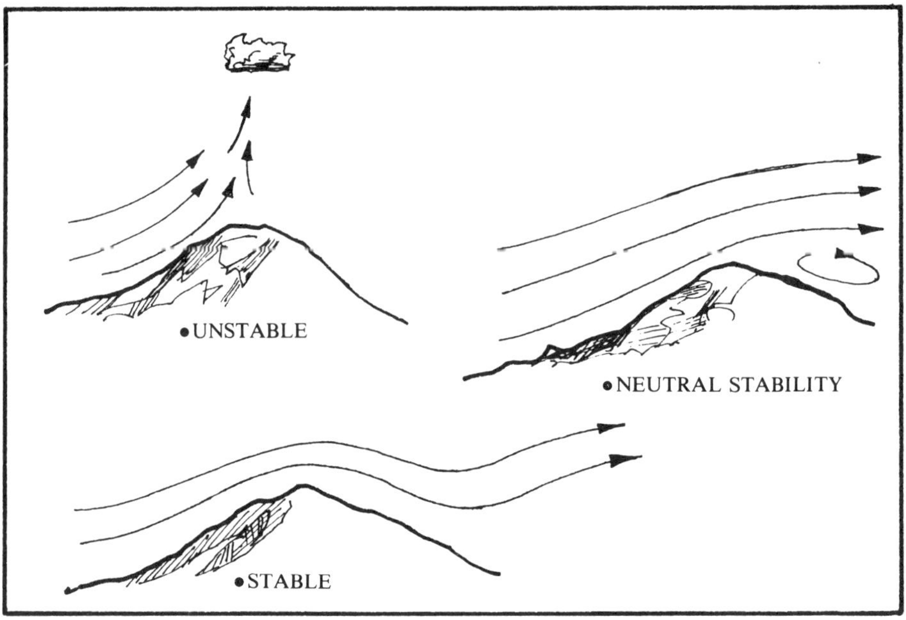

Figure 147 - Airflow Over a Mountain as Stability Varies

148. Note that the lapse rate indicates a layer of stable air lying above the mountain. This is the ideal case, for an unstable layer below and above the stable layer create what can be described as a springboard for the stable layer to bounce on once the mountain begins the oscillation.

The shape of the wave-producing mountain is a factor in wave strength. The ideal mountain is shown in the figure. Basically the upwind side is concave, the back is steep and the mountain size is about that of the first wave.

A long ridge or mountain is the best wave producer. Short ridges and hills allow the air to flow around their sides and interfere with wave formation. The length of a ridge for optimum formation should be a

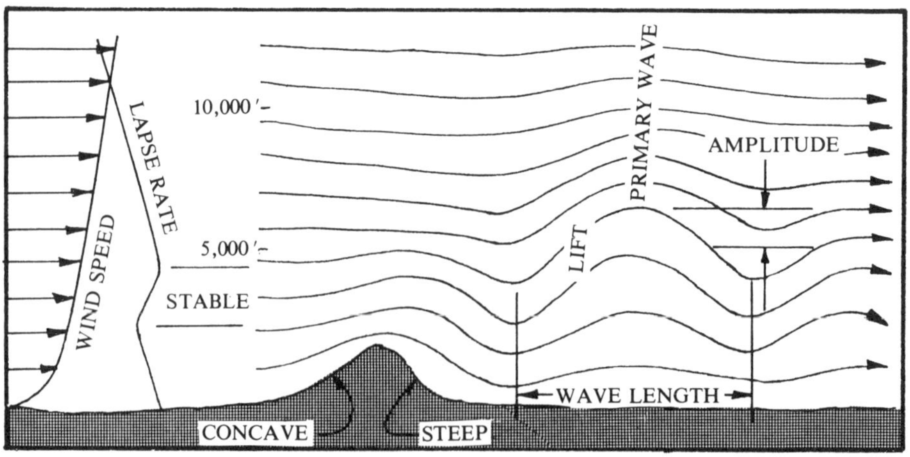

Figure 148 - Wave Requirements

minimum of one wavelength. Wave can be produced behind isolated hills as shown in figure 149. However, the wave will be small and die out quickly downwind. A perfect wave generator can produce a series of waves that extend for hundreds of miles downwind.

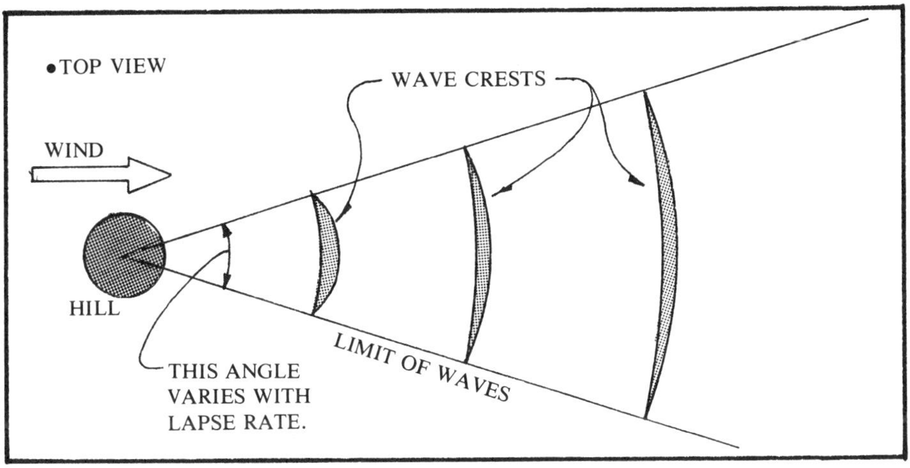

Figure 149 - Wave Action Behind an Isolated Hill

The sudden drop off of a plateau can produce a wave as shown in figure 150. Waves formed in this manner often appear downwind of the Allegheny plateau as it drops into the Appalachian ridges along the entire eastern United States. Due to cooling effects such a wave is prone to shift upwind because the wavelength shortens as the air gets denser.

Waves can actually be produced by any object in the wind's path.

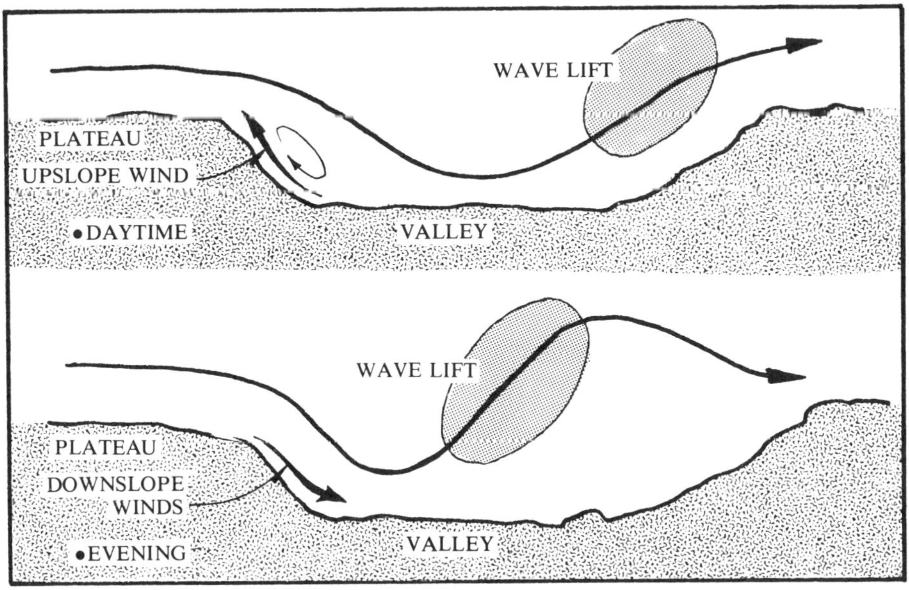

Figure 150 - Waves Downwind from a Plateau

Waves in the rockies.

Model sailplane pilots have been known to soar their small craft in waves produced by buildings, fences, small bumps and drop offs in the terrain. The process on such a small scale appears to take place in lighter winds than required for larger waves.

WAVE PROPERTIES

Two important properties of waves are amplitude and wavelength. These are labeled in figure 148. Amplitude can best be understood as how far up or down a particle of air travels as it passes through a wave. The wavelength is the distance from crest to crest.

The amplitude of a wave depends very much on the lapse rate profile of the air and the wind profile as mentioned previously. In addition, moist air is condusive to high amplitude waves. The lower the mountain is, the more important the type of ground cover. Low mountains (below 1,000 ft-300m) covered with trees or rocks outcroppings are less likely to produce waves than mountains with smooth slopes (grass, dirt or snow covered).

The wavelength of atmospheric waves depends on the lapse rate and the wind velocity. The spacing between crests can be around a mile to twenty miles (2 to 32 km). A spacing of six miles (10 km) is common. As a rule of thumb, the wavelength in miles is 1/5 the wind velocity (in mph).

The wavelength is of great importance when multiple mountains or ridges exist. The reason for this can be seen in figure 151. Here we have a mountain downwind from the wave-producing mountain that is in phase with the wave. That is, the upward thrust from the second mountain is right where it boosts the wave. This is called constructive interference.

The lower drawing shows the same terrain with a slightly longer wavelength. Here we see the downwind mountain lifts the air at the wrong time so that the wave is eliminated. This action is called destructive interference. When a multitude of hills exist in an area partial constructive and destructive interference can greatly complicate the wave pattern.

Here is a summary of requirements for good usable wave production:

Wave Requirements

WIND—At least 15 mph (24 km/h), perpendicular to the ridge, unchanging in direction aloft and increasing in speed with altitude.

STABILITY—The lapse rate should show instability below a stable layer and instability above. The more stable and narrow this layer is, the higher the wave amplitude.

MOUNTAIN—The shape that best produces a wave has a cross section identical to the wave. The height should be 500 feet (170m) or more for human carrying aircraft. Multiple bumps downwind should be in phase with the wave.

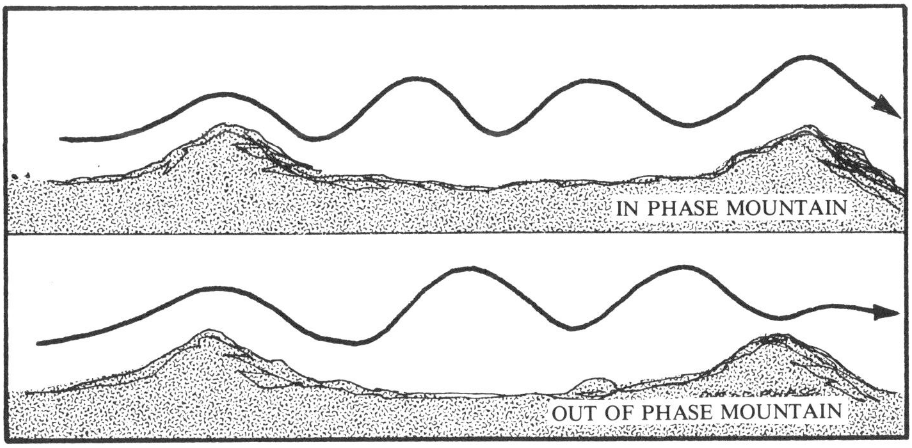

Figure 151 - Constructive and Destructive Interference in Waves

WAVE CLOUDS

Waves can exist in perfectly clear air, but ofen the lifting can produce a form of cloud specific to waves. These are lenticulars as described in Chapter III. Lenticulars or "lennies" for short form on the crests of the waves as shown in cross section in figure 152. They are stationary over the ground because they form at their leading edge in the upflow and dissolve at their downwind edge in the sinking air. In three dimensions lennies look like long flat pale bands or flying saucers in narrower waves.

Occasionally wave clouds can be stacked on top of one another which merely reflects vertical humidity layers in the atmosphere. The presence or absence of clouds doesn't seem to affect the waves, but note that moist air is wave-prone. In especially humid conditions layer-type clouds can be present in the stable layer. Any wave action will show up as a hole or slot in this layer. A gap in the clouds behind the mountain is known as the foehn gap since it occurs often during foehn conditions in the Alps. If the cloudbase is low, the layer may have a uniform base and only the foehn gap indicates the presence of a wave (see figure 153).

Below the crests of the wave a roll cloud may appear as shown in figure

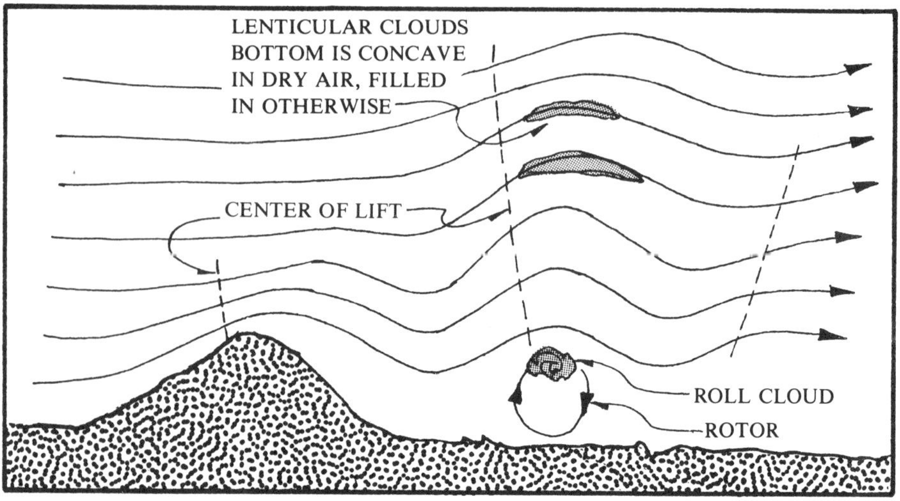

Figure 152 - Lenticular and Roll Clouds Associated with Waves

152. The nature of this roll cloud tends to be ragged, often dark and rolling as its name implies. It is formed by the rising air of the rotor that often exists below the crests of a wave. Rotor clouds are not necessarily present when such a rotor is active.

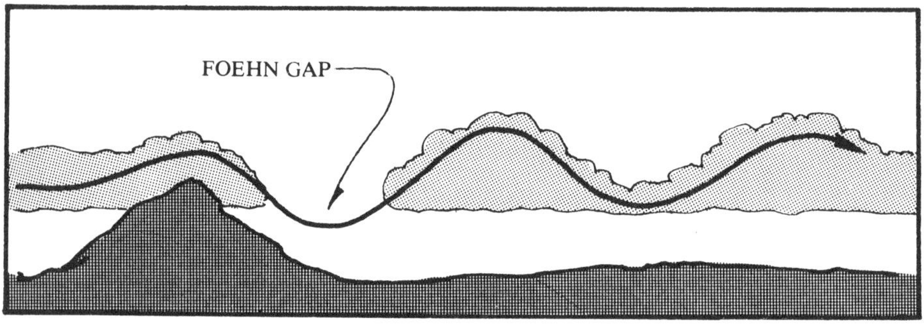

Figure 153 - Foehn Gap as Evidence of Wave Production

WAVE DANGERS

Before we see how to utilize waves, we should form an appreciation for their dangers lest we rush out and go aerial surfing unawares. By far the greatest danger in a wave is the rotor and severe turbulence it represents. Not only is it often an overpowering eddy, but it can be composed of strong shears and random gusts. The effects can be quite violent.

Rotors can reach to the ground and may be blundered into by pilots not thinking about wave prospects. Observation is the key to avoiding rotors. When rotors and waves in general are present, bands of alternating calm and strong winds may appear on the surface below the crests and troughs, respectively. A rotor will show up as an erratic and violent wind on the ground.

If you are in a wave the best way to avoid a rotor on descent is to drop down in front of the wave-producing or reinforcing mountain. If you are too far behind this mountain and have enough height, fly as far downwind

Multilayered waves in New Mexico. Note the cumulus nature of some of these waves.

as possible to descend since rotors are less likely and less powerful several waves behind the mountain. If neither of these strategies are possible, descend in the rising air and move forward as you descend to avoid the rotor near the ground. If you can't descend—a distinct possibility in a good wave—simply fly upwind or downwind to the next area of sink and move progressively downwind as you descend to avoid the rotor.

The next most serious danger related to wave flying is the possibility of overstrong lift carrying you to excessive heights and into excessive winds. The lift in a wave can carry you above the capabilities of your oxygen if you are so equipped and into air that will freeze your bones. Sailplanes have ascended to over 49,000 feet (15 km) in a wave and it was still going up. If your aircraft is slow you will not be able to penetrate upwind as you rise into faster moving air. You'll have to escape the wave downwind. Hopefully you have landing options in this direction.

Other problems with waves include massive leeside sink behind the primary mountain and the sudden filling in or shift of clouds as conditions are slightly altered. Watching wave clouds carefully reveals that they sometimes change their shape perhaps as disturbances or a variable air mass moves through.

The most powerful waves exist in the strongest winds and the highest mountains. The Sierra Nevada mountains in California are particularly noted for magnum sized waves that set up over the Owens Valley. The "Sierra wave" has provided record-breaking flights and aircraft-breaking rotors.

FLYING IN WAVES

Waves tend to appear most frequently in the morning, late afternoon and evening at the end of a fine soaring day. The lower unstable layer may produce thermals all day then give way to wave action as the thermals die. Perhaps the reason that waves don't exist during the day in these conditions is that the strong thermals break up the steady laminar flow necessary for waves.

When the thermal height is reduced the wave sets in. This is precisely the situation experienced by a dozen pilots thermaling along Bald Eagle Mountain near Lock Haven, Pennsylvania on August 31, 1991. Pilots were typically reaching several thousand feet over the mountain in thermals until the wave set in around 5:00 PM. By flying a mile or more in front of the mountain they climbed in the wave to over 7,000 ft (2,300 m) to reach the bottom of the wave cloud.

On another occasion this author and two other pilots were in a competition in the Sequachie Valley of Tennessee. We were soaring the ridge at 2:00 PM trying to gain height to cross the valley. At about three thousand feet over the mountain top a wave set in and we began climbing without circling. The lift was smooth, steady and widespread. We climbed to over 9,000 ft (3,000 m) at which time the stronger wind aloft made us speed up so much our climb rate diminished. The air was still going up when we pulled out.

We can learn a couple things from the above incidents. First we must acknowledge that waves are somewhat elusive and often arrive as a surprise. The important matter is to be aware of their possibility at all times and take advantage of them when they are detected. Secondly, it is clear that waves can arise anytime during the day and our rule of thumb about when they appear most often is just a guideline, not an absolute. Waves are actually more common than most pilots realize—it's just that they are often out of reach or setting up at night or in winter when flying activity is limited. One researcher has estimated that waves exist two thirds of the time in mountainous areas.

The vertical velocity you can reach in a wave depends on the steepness of the wave and the velocity of the wind through the wave. Shorter wavelengths tend to be associated with steeper waves. The maximum vertical velocity is found in a layer between 5,000 and 10,000 feet (1,800 to 3,000 m) in most places and considerably higher in the Rockies, Sierras and Alps. This velocity can be over 2,000 feet per minute (10 m/sec) but is usually below 400 feet per minute (2 m/sec). Waves have been recorded up to 100,000 feet (30 km) but typically run out of impetus much lower than that.

Using the lift in waves requires that you fly in long paths parallel to the wave itself while remaining in the uprising air. This is a bit like ridge soaring since the wave is generally stationary. Figure 154 shows the areas of lift and sink in cross section. Notice that the wave action at the instigating mountain does not extend very high compared to the lift in the succeeding downwind crests. Of course, if you are soaring a mountain in phase with the wave, downwind from the primary mountain, you may be in the higher lift bands.

Flying from one crest to another is also possible to provide cross-country flights, but slow craft will have to proceed in the downwind directions to pick up successive waves. Soaring along the wave is usually more productive.

A common paranoia expressed by pilots in a wave is that they "can't get down." In fact they all have, but often it required some effort if not extreme maneuvers. The general rule for exiting the sky in a wave is simple:

To Leave Wave Lift
Fly upwind or downwind until sinking air is found. Remain in this sinking air by flying crosswind. As you get lower move a little downwind to avoid any possible rotor.

Since waves often occur towards evening the real problem is being caught aloft after dark. Start your escape procedure with plenty of time allowed for the slow descent.

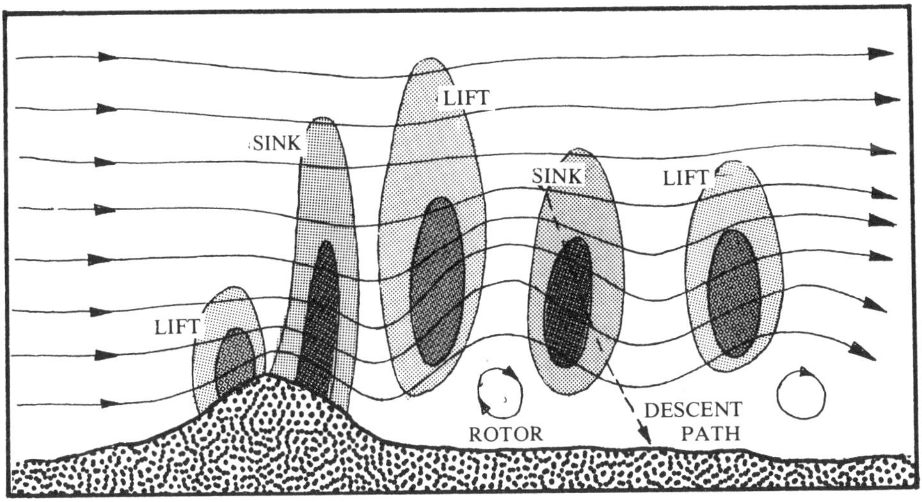

Figure 154 - Lift and Sink Areas in a Wave

FINDING WAVES

Ofen a wave announces itself by presenting glassy widespread lift after you have been struggling in thermal lift. If you find yourself climbing higher and higher well away from the terrain without the size limit or variability of a thermal, you are most likely in a wave. Other times you may blunder into a rotor area with its sudden turbulence. Suspect a wave when bands of wind or turbulence exist near the surface. It can generally be distinguished from thermal turbulence by its larger extent and more random nature out of the actual rotor eddy.

Lenticular clouds are the best way to identify waves other than discovering one in flight. Widespread layer clouds with holes are other good indicators. Also be aware of the prime times for wave production and the ideal conditions in order to best find them.

The overall weather situations most condusive to wave formation should be understood by soaring pilots. The approach of a warm front often produces waves for the warm air aloft represents a stable layer over the normally unstable lower air mass. Indeed we can often see high, flattened wave clouds during the approach of a warm front. However, these waves are generally out of reach of sport aviators. Within 10 to 20 hours before the frontal passage, however, the warm layer is much lower and wave soaring may be achieved. Unfortunately, in moist, temperate climates such a front usually is accompanied by thick stratus clouds which may mask the presence of the waves. Also ridge and thermal soaring prospects aren't very good in these conditions so many waves are likely missed as pilots catch up with their neglected chores on cloudy days.

The most frequent condition to encounter a wave is when a high pressure system moves across the area. Such a high has gently subsiding air which creates a temperature inversion ideal for wave production. Unfortunately the surface wind in the center of a high is very light so we must have the good fortune to be at the periphery of the high pressure system in order to find waves. When you find yourself under the benign influence of a high with 10 knot surface winds and more wind aloft keep an eye peeled for waves.

The prediction of waves is possible once you are familiar with their cause. The two most common times to find waves are immediately before a warm front arrival as mentioned or immediately after the passage of a cold front. The more vigorous these fronts are the more they are likely to produce waves due to stronger winds. Each flying area has a preferred wind direction and speed for maximizing wave production. It is very useful to observe wave clouds and determine what conditions produced them in order to be able to predict waves in the future. Noting the weather report on the day of their appearance can help you become a wave forecaster.

OTHER WAVE SOURCES

We have already mentioned shear waves and billow clouds in Chapter VI. We will expand the notion here to indicate that shear waves sometimes exist at the transition zones of fronts. Cold fronts in particular occasionally produce waves that pilots have exploited. Sea breezes and downslope flow also produce such waves at times as they plow under warmer air but these billows are normally too small to sustain soaring.

Thermal clouds with their huge mass can act like a convective barrier. The reason for this is that thermals originate from the ground with a slower airspeed than the faster moving air into which they are rising. Although we think of clouds as airy puffs, the air contained in them weighs tons and their inertia prevents them from moving right along with the wind as they rise aloft. Consequently the air flows around and over them. A line of such clouds can act like a barrier to produce waves just like a mountain or ridge. As shown in figure 155, lines of cumulus clouds known as "streets" can produce waves above their tops when the wind turns to cross the street at cloud level. These *thermal waves* as they are known can be reached on occasion by soaring up the upwind side of the cloud street. Such a penetrating cloud often produces ridge lift like a mountain when it rises into increasing winds. Using such "ridge" or wave lift requires soaring along the line of clouds to stay in the lift.

CONVERGENCE LIFT

As we learned earlier, convergence means coming together. When the

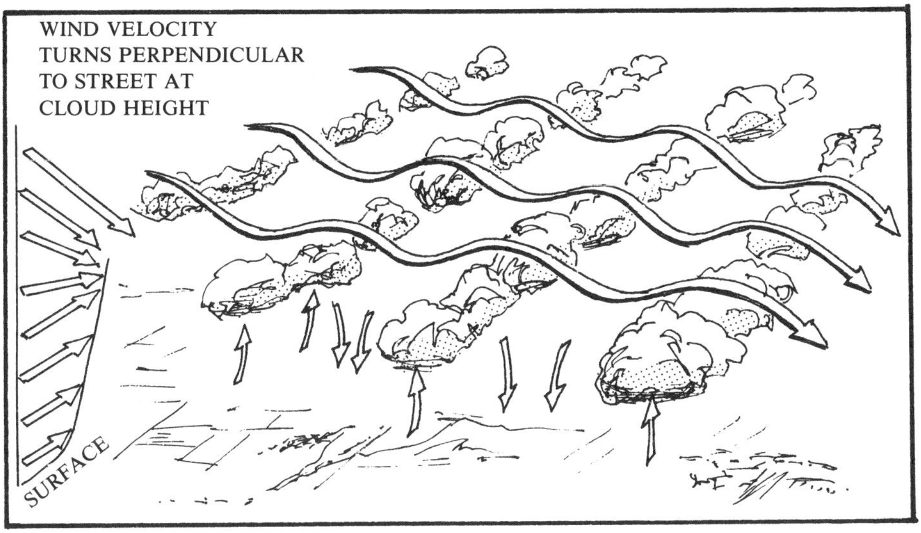

Figure 155 - Thermal Wave Production

air converges, if it can't go sideways, it goes up. That's what we want. Let us take a look at the many ways this can happen in nature.

We have previously seen how convergence occurs in large low pressure systems. The rising air that results can promote thermal production, but is too weak to provided soaring in itself. We have also seen the convergence produced when the wind passes through a narrows (figure 112) or a gap (figure 144). These forms of convergence lift are only usable when the winds are fairly light because of turbulence and penetration problems. Bowls, the ends of long box canyons or valleys whose end is blocked not only produce characteristic ridge lift but also some convergence lift that often occurs well away from the bowl crest as shown in figure 156.

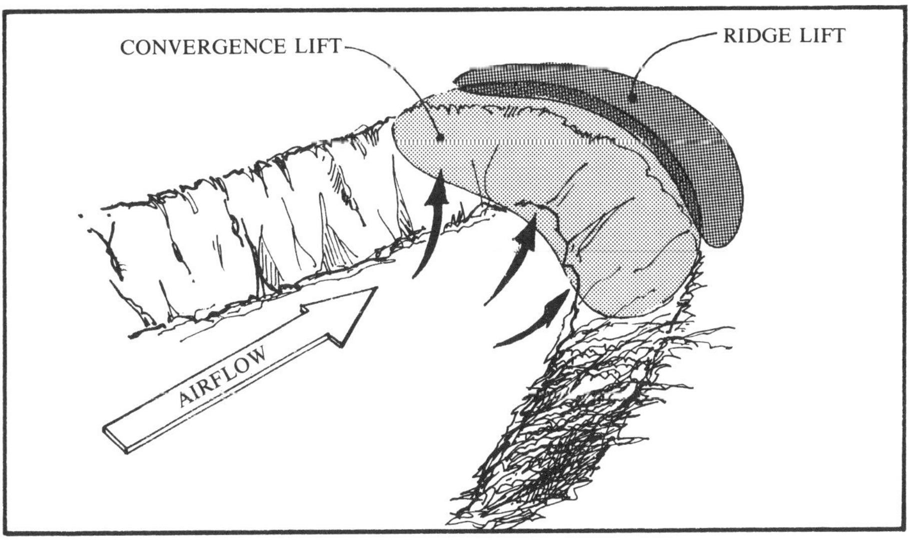

Figure 156 - Convergence Lift in a Bowl

At times an abrupt wind direction change can create a convergence zone, especially if the newly arriving wind is of greater speed. A situation like this happened in the spring of 1991 in a contest in Tennessee. The wind was light southeast into the mountain and pilots were soaring low on weak thermals. A widespread stratus cloud layer approached from the south bringing stronger south winds. The change from southeast to south winds progressed up the valley toward the north and was accompanied by a convergence zone that carried a few fortunate pilots to 5,000 feet (1,600 m) over the mountain and 25 miles (40 km) up the ridge.

This particular convergence zone was announced by the approaching layer cloud and was abrupt which accounted for its production of lift. Any time the wind direction is predicted to change as this was, we should be on the lookout for a convergence zone.

Light cumulus clouds marked the convergence in the above case, and this may be the only clue we have to such elusive convergence lift. In this situation the lift moves with the direction of the newly established wind. In reality this situation described here has all the characteristics of a heat front and was probably exactly that. However, the weather information services do not mention such things so the clues to look for are a predicted wind shift and cloud effects.

Another place a wind shift occurs with convergence is at the interface from water to land. This action isn't necessarily connected to the sea breeze. To see how this works look at figure 157. Here we see a wind blowing across the land then across a large lake then returning to land. Because of the friction on the land the wind is slowed near the surface and thus crosses the pressure isobars

Areas of clouds showing convergence zones in Rio de Janeiro.

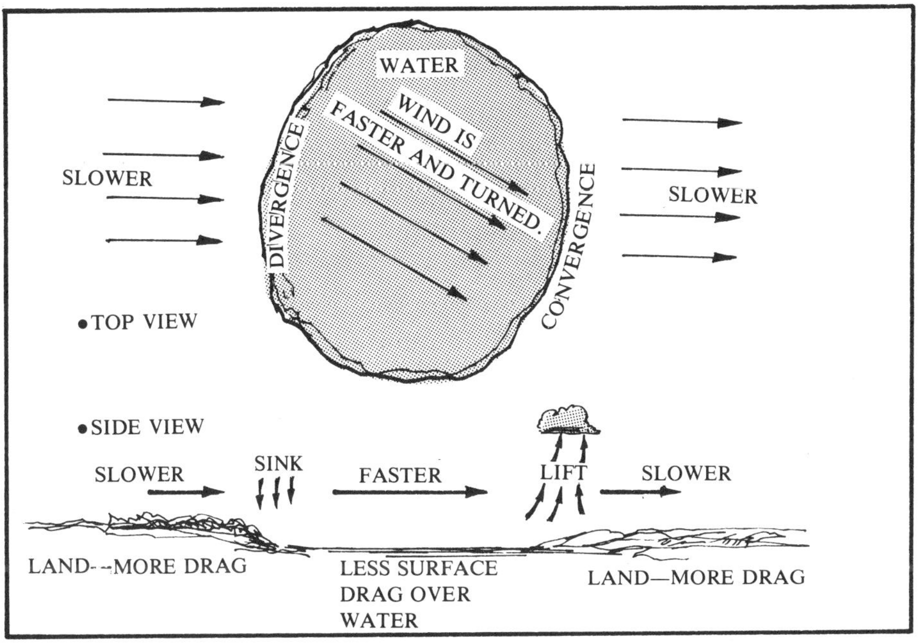

Figure 157 - Convergence Lift Near Water

as we learned in Chapter IV. Over the water less friction means the wind speeds up and follows the isobars more closely. As a result, the wind separates or diverges at the upwind side of the water and comes together or converges at the downwind side as shown. The convergence is stronger the more unstable the air. The shift in wind direction at the shores is 20 to 40 degrees and counterclockwise at the downwind shore in the northern hemisphere (clockwise in the southern hemisphere).

We have seen previously that waves can be produced behind a hill. Even when a trail of waves does not arise, convergence can occur to produce lift behind an isolated hill as seen in figure 158. Here the air filling in behind the hill creates convergence lift.

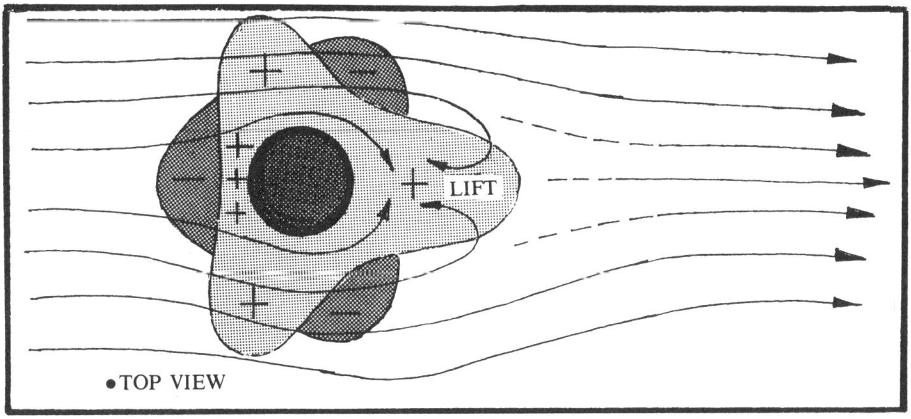

Figure 158 - Lift and Sink Around a Isolated Hill

A common situation where convergence occurs is when a sea breeze blows inland in mountainous areas. If the stable sea air meets an isolated hill it separates and flows around the hill more readily than going over it. When the breeze meets itself behind the hill it wells up in convergence as shown in figure 159. Complex mountain systems as shown can create convergence patterns in many areas even away from the sea breeze. When exploring for convergence lift of this type it is important to remember the dangers of leeside rotors, downdrafts and turbulence (see figure 110).

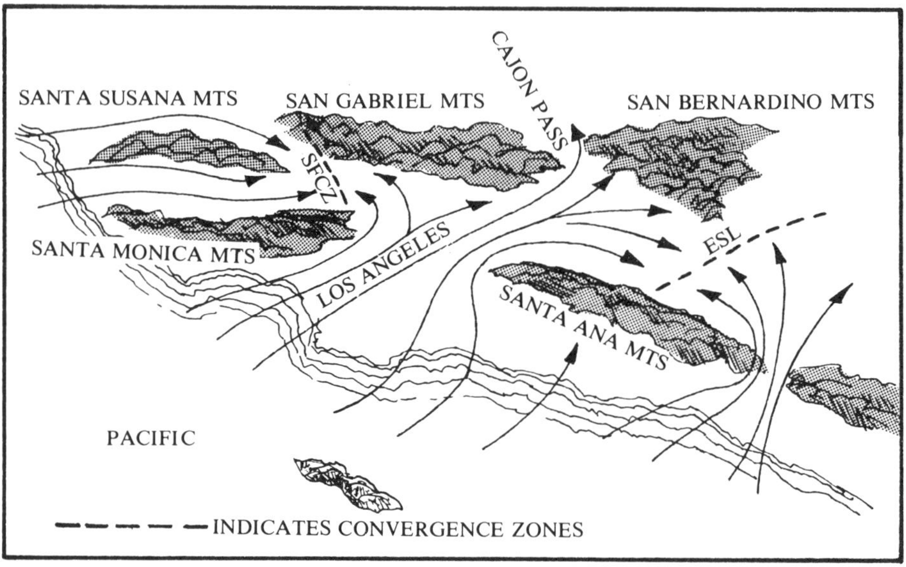

Figure 159 - Convergence on Complex Terrain Near the Sea

We have learned about local circulations in Chapter VII and the convergence that can occur when a heat source such as a field is surrounded by trees (see figure 131). Also we investigated the convergence that happens in the middle of a valley when both sides of the valley produce downslope winds in the evening. This is shown in figure 160 and in the

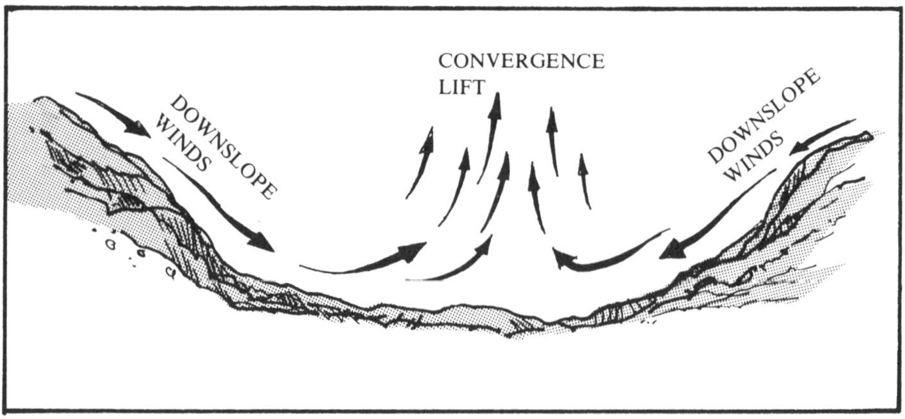

Figure 160 - Evening Convergence in a Valley

evening condition of figure 136. This lift is usually light and widespread and deserves its name: magic air.

A special convergence situation occurs when an up valley wind meets a down valley wind. This can occur when a general wind oriented down the valley does combat with the up valley wind that tries to establish itself due to heating differences. The up valley wind may flow in the morning until mixing of the lower atmosphere brings the general winds down to oppose the up valley flow. Another way a down valley flow can occur during the day is if particular cooling takes place at the head of the valley. An example of these principles occur in the upper Rhone Valley in Switzerland. In this case the up valley wind heading northeast is met in the afternoon by a strong flow dropping from the cool glacier and forests that adorn the head of the valley. This down valley wind is the well-known "Grimsel snake" that converges and shears with the up valley wind. This convergence can march up and down the valley and change the surface wind direction abruptly.

Perhaps the most useful form of convergence occurs at the top of a hill or mountain when upslope winds from both sides meet at the top as shown in figure 161. The flow up one side alone may not be strong enough to sustain flight while the almost vertical currents above the hill provide abundant lift. In this case a little thermal or ridge lift will be needed to get over the mountain. If the flow up one side of the hill is stronger (perhaps it faces the sun more directly) a shearing action may take place above the mountain as shown. If a general wind is blowing the rising air above the mountain the soarable envelope will be tilted as shown in the figure. Too much wind will change a convergence situation to a situation where ridge lift occurs on one side of the mountain with a leeside rotor on the other.

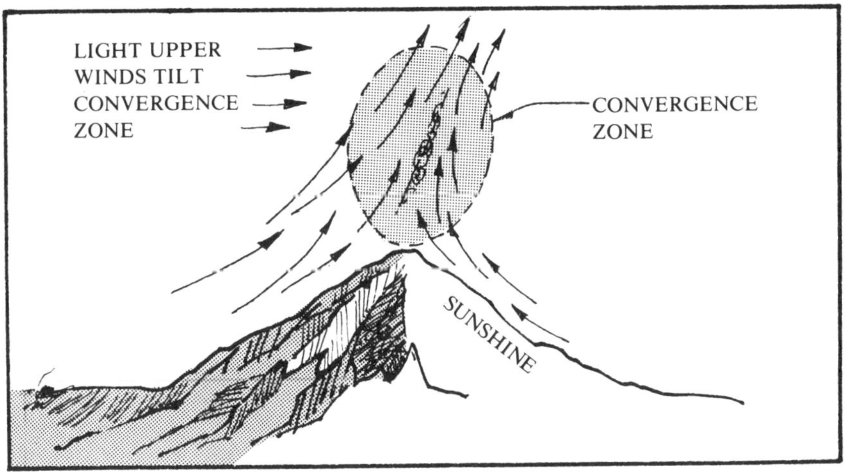

Figure 161 - Convergence at the top of a Hill

FLYING IN CONVERGENCE

The lift in a convergence zone is often like a carpet of smooth, bouyant air, spread over a wide area. This is because convergence flow is often moving straight up and less than 3 mph (5 km/h) of vertical flow is needed to sustain a modern soaring craft be it paraglider, hang glider or sailplane.

Even less is required for a radio controlled model. With such a light flow turbulence can be nonexistent. Also, after spending time in the confines of thermal lift or being limited to a narrow zone of ridge lift, the broader area of convergence lift is liberating indeed.

This benign state of affairs is not always the case, however. Thermals are ofen aided and abetted by convergence and move up through the flow with their roughhouse nature. Also any shearing action will add its share of rolling air. If the vertical shear is too great due to one side of the vertical flow rising faster than the other, it is best to stay on one side—that with the strongest lift, which is to the left in figure 161.

Often clouds accompany convergence lift. These clouds may be rows of thick cumulus over a ridge or a cap cloud over an isolated mountain. Other times small cumulus may form along the line of convergence. If conditions are dry only the slightest wisps of clouds may appear to identify the convergence area. Occasionally these wisps are oriented vertically as a narrow band of converging air rises. Finally, layer clouds of limited extent can occur over a slowly rising uniform convergence zone. These various clouds are shown in figure 162. Of course, convergence that doesn't rise to great heights, such as that in a valley center in the evening, will not form clouds at all since the air doesn't cool to the point of condensation unless it is very moist.

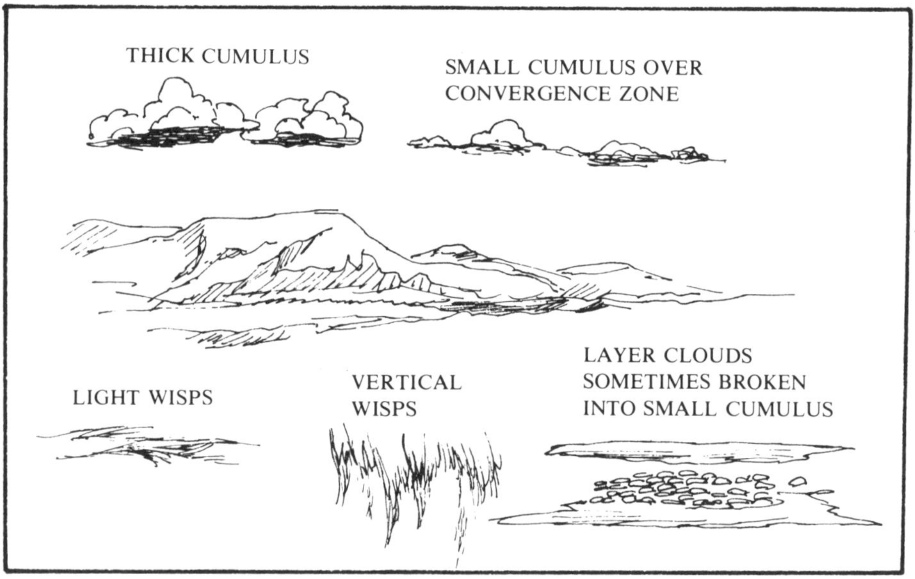

Figure 162 - Convergence Related Clouds

Over mountain peaks and chains it is very common to see clouds with two different base altitudes. This is a sure sign of convergence above the peaks because the different level clouds are formed in air that comes up opposite sides of the mountain as shown in figure 163. The air on the side with the lower cloud base contains more moisture and reaches saturation at a lower level. This situation is most frequent when one side of the mountain faces a nearby sea or large lake.

Figure 163 - Clouds with Different Bases Indicating Convergence

Flying in convergence lift often times requires you to recognize what is taking place so that you can stay in the convergence area. Thermals may be marked by clouds and drift with the general wind; ridge lift is a semi-permanent resident above the ridge top, but convergence can be short-lived and elusive. The zone of lift between two converging flows can move back and forth or disappear all together only to crop up a bit later in a new location as surges move through the flows. You may have to make long passes to stay in a narrow zone of convergence or circle in one place if the lifting area is small.

Usually convergence lift is light and requires patience in order to achieve much altitude. However, strong convergence can occur over mountain peaks. The most important point concerning flying in convergence is to expect it at any time—especially in the latter part of the day, above hills and in changeable conditions—recognize it quickly and figure out its extent to put it to maximum use.

FRONTAL LIFT

In reality the lift in fronts is simply a special form of convergence for there is a coming together of the air as one air mass pushes towards another. We separate it here because frontal lift has its own unique problems and behavior. Generally we are concerned with cold air moving towards warmer air and lifting the warm air mass. The opposite case (warm air advancing) doesn't provide much lift due to the slight slope involved.

We have previously discussed the use of sea breeze fronts in detail (Chapter VII). This is a type of cold front and is perhaps the most useful for sport aviators. Other heat fronts covered in the same chapter are also very beneficial to altitude fans, but these may be harder to diagnose since they are not usually marked by a front of humid air.

Synoptic (large-scale) cold fronts sometimes have their moments as producers of exploitable lift. That a passing cold front represents abundant lift can be told by the massive cloud build-up that often announces these fronts. Indeed, the real problem with utilizing such lift is the presence of thunderstorms (see Chapter XI) that frequently precede the front. However, there are plenty of dry and weak cold fronts that pass through, some with just the right amount of lift to provide good soaring and mild cloud development.

The technique for using frontal lift is to stay in the warm air sector close to the front boundary as shown in figure 164. To do this you should fly parallel to the front. Long distances may be covered in this manner, but if the front is dry it may be difficult to tell exactly where it lies. Remember, a front moves along the ground (unless it stalls) so the lift conditions must be monitored carefully in order to stay with the front. Weak fronts often move less than 15 mph (24 km/h) so remaining in the front as it travels down wind is a slow way to make progress.

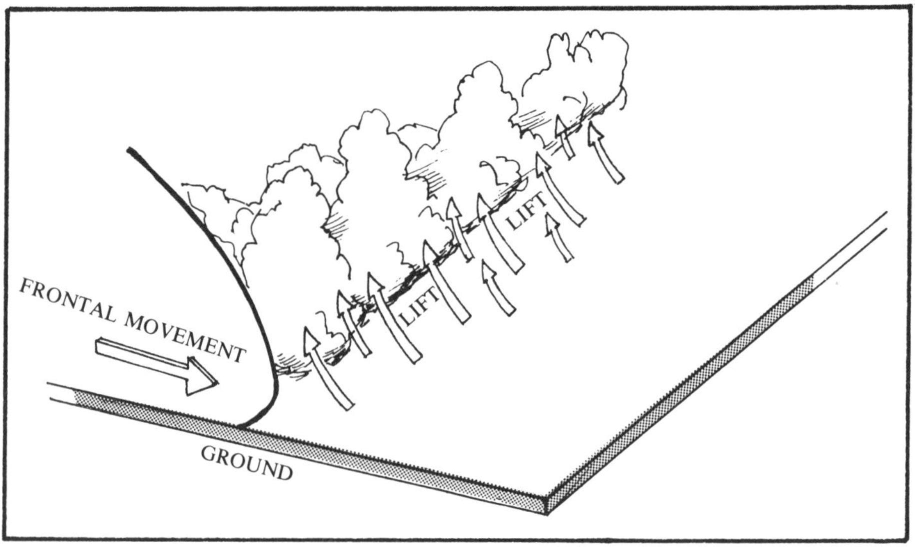

Figure 164 - Frontal Lift

Fronts with more moisture will be easier to determine and exploit. In this case you should remain at the leading edge of the cloud as that is often the area of the best lift and is the best place to engineer an escape if lift gets too strong or the cloud threatens to develop to storm proportions. Beware of the unpredictable nature of moisture-containing fronts. When they encounter air of greater humidity cloud base can drop rapidly. Sometimes pressure waves proceed a front that develops squall lines with deadly thunderstorms. Serious cold fronts are no place for sport aviators for such fronts can produce the most terrible storms that rage in our atmosphere.

LIFT INDICATORS

To use lift we must know where it is and when it is occurring. We can only look at the secondary signs of lift, for the neat little arrows in our drawings are unfortunately not evident in the sky. Clouds are obvious signs of lifting air. We have discussed the different types that accompany the various forms of lift we encounter in our aerial quest. Other signs are surface wind variations, tree movement, smoke and birds.

Changing wind on the ground accompanies thermal production as we shall see in the next chapter, but winds of different directions blowing steadily in two nearby locations is a good sign of convergence. Trees lining a ridge are great indicators of soaring winds. They tell speed and direction by how vigorously and how they bend. In light vertical air they rustle in front of a ridge or mountain even though a wind cannot be felt at the top of the mountain.

Smoke shows the characteristics of the air it is polluting. Besides wind velocity and turbulence (see figure 104) smoke can indicate stability and the presence of lift. Figure 165 shows how smoke will drift along in ground winds until it meets an area of lift. Puffing smoke is usually an indicator of good lifting air. We should mention that smoke from fires or large stacks is an artificial thermal that certainly generates lift. If we consider how much this air can be heated above the surrounding air we can imagine some fairly strong lift. In fact, besides the breathing problem that smoke presents, the possibility of excessively strong turbulence rules out smoke from forest fires, burning debris or industrial stacks as of no interest to all but the most desperate pilots.

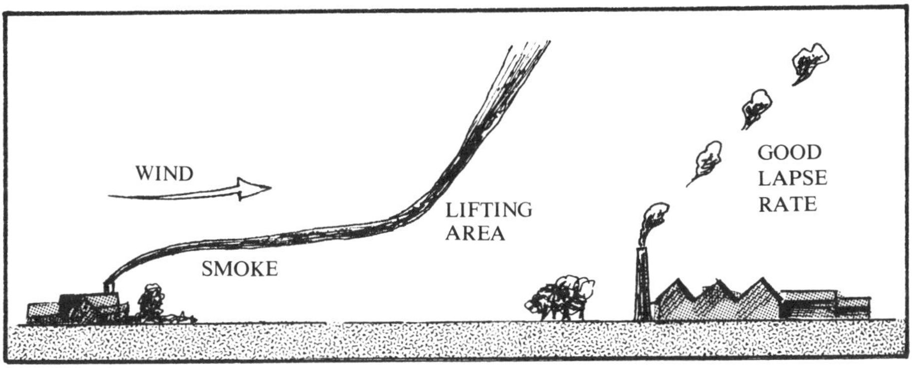

Figure 165 - Smoke Indicating Lift

Birds of a soaring bent are most useful to pilots as lift indicators. Hawks generally mark thermals down low and are very helpful because their performance is similar to that of sport aircraft. Black vultures are also ideal for this reason. Turkey vultures are found nearly everywhere and soar almost exclusively when they are on the wing. They are more often found at altitude covering more ground than hawks since vultures travel long distance in search of an odoriferous meal while hawks are more territorial. The problem with turkey vultures is that they generally can soar in lighter lift and in closer than is possible for a sport aircraft. Also they can handle very high wind with strong turbulence and they have a nasty habit of flying in leeside rotor. However, taking these qualities into consideration we can readily learn to read the lift conditions as well as the wind's behavior by carefully watching turkey vultures. Another bird species we should mention is swallows. While these birds don't soar technically they do feast on insects that are carried aloft on updrafts. Usually if swallows are streaking through an area it means the air is lifting enough to support our human-carrying wings.

SUMMARY

Lift in the atmosphere is a constantly changing characteristic because atmosphere itself is so mobile. Sometimes we can find up air by our skill and knowledge of conditions. Other times we simply blunder into it. In either case the more we know about the ways of the sky the better we can practice our tactic of delaying the inevitable: returning to earth.

The many forms of lift we have received here can occur in any combination. When we add thermals to the mixture we can have a confusing state of affairs unless we understand how each lift source behaves. Studying the principles involved, experiencing them in the air and thinking about what you experienced is the key to this understanding. The pay-off is making a prediction where lift should be then gliding out and actually finding it. When you do this on a regular basis you have truly become a creature of the air. Thermals are a particularly important part of this process because they are so widespread and variable. We turn to this important lift source next.

CHAPTER IX

Instability and Thermals

One of the most common and sought-after sources of lift in our universe of air is thermals. These bubbles and plumes of lift can extend to great heights and form stepping stones across the fluid sky. They can be elusive in weak heating conditions or they can be as abundant as dandelion puffs on a lawn when the sun and earth conspire to percolate the atmosphere.

Because thermals are so important to most of sport aviation, we expend some time in this chapter understanding how they are created, what conditions they favor and where to find them. In the next chapter we investigate thermal behavior in the sky.

We have previously introduced the concept of instability and lapse rate (see Chapter II). It may be a good idea to review that material at this time. We will build on the basic ideas and gain more insight into the workings of our flying environment as we learn the lift history of thermals.

THE BIRTH OF A THERMAL

A thermal is a collection of air rising through the general air mass because it is lighter than its surroundings. Thermals come in a wide variety of shapes, sizes and strength. If we want to be most effective at hitching a ride on these free elevators we must begin our study with their birth.

The sun heats the earth's surface on an almost daily basis. This heat is passed directly to the air above the surface as we found out in Chapter I. If heating is slow the warm air may rise in a light, continuous plume. In a faster heating process a bubble may form that remains on the ground for a period of time before it releases in a sudden rush. As an illustration of how this works, watch a pan of water coming to a boil. At first the bottom of the water is heated you will see convection currents—plumes—begin to rise.Then, as heating gets more intense bubbles begin to form on the bottom, breaking away to rush to the top surface. As the heating continues larger bubbles form in the hotter areas. A couple other things to notice in our model is the presence of downward moving currents in the early heating process, the general spherical shape of the bubbles and the overall mixing of the water. These features also occur in the air.

Figure 166 shows the effects of surface heating. In the first case light heating causes a slow circulation plume. This may occur in the morning, when a layer of cloud partially obscures the sun or in the evening when warm ground cover slowly releases stored heat.

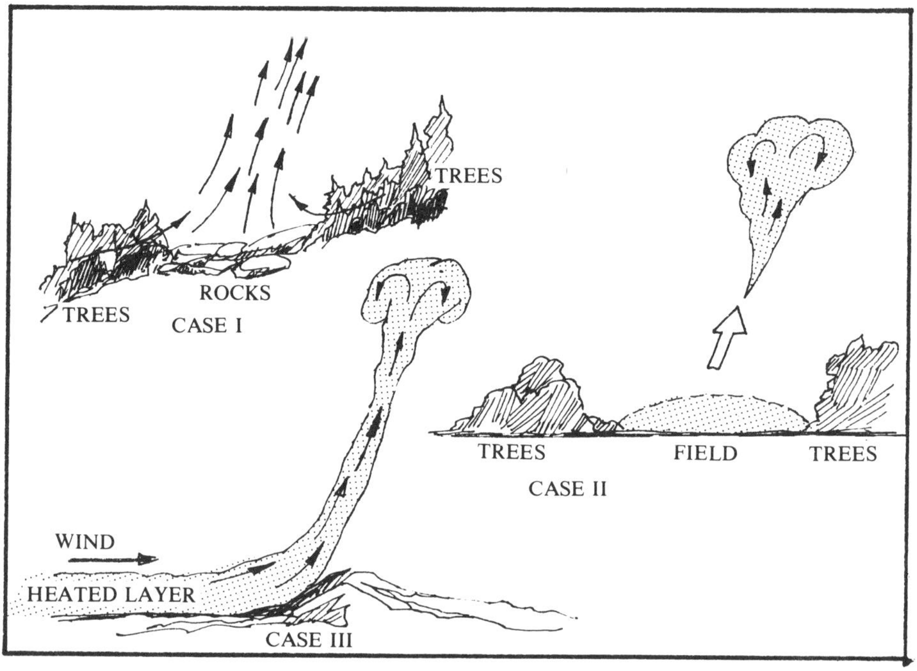

Figure 166 - Variations in Surface Conditions

Second case shows a warm dome growing on the surface. A dome such as that illustrated will be of limited extent in terrain where the size of the heated area is determined by the size of a field or other surface. Trees and other cooler ground cover often determine how large a heated dome may be.

When this dome grows at a fast rate it expands to move the air above it as shown in the figure. This fairly rapid expansion and the inertia of the heated air serve to keep it on the ground until some disturbance dislodges it. When the heated dome does release it consolidates into a bubble as we shall see below. This bubble process is the most efficient method nature has devised to carry heat aloft to relieve the imbalance of excess heat at the surface.

The limited area of heated air shown in case II occurs most often in greener areas such as Europe and the eastern portion of North America. In desert areas large tracts of undifferentiated surface heat up on an equal basis forming a broad layer of hot air. This potential thermal will rise at distinct trigger points and continuously feed into a tall column of rising air. This is case III on our illustration.

THERMAL TRIGGERING

A potential thermal may sit on the ground for many minutes as it is building. Although this is an unstable situation, the warm air must push up through the cooler air and a flow under the thermal must begin. If the warm air is expanding it may remain on the ground until a gust breaks it away or it becomes so large that the expansion slows and the cooler air pushes in from the sides. The sudden release of a thermal is called *triggering*.

Certain wind irregularities can serve as thermal triggers. For instance, a downdraft from a previously released thermal or a gust from a passing car can cause a thermal release. Many a pilot has achieved a low save when their ground crew raced by in a car to release a thermal. One site in Pennsylvania has a train that wends up the valley and releases thermals on schedule. Sailplane pilots have been known to dive at the ground to cause thermal release then zoom up to ride the thermal. This is almost like pulling yourself up by your own bootstraps.

A general surface wind serves to release thermals as it swirls around ground obstructions. Such a wind often limits the size of thermals because it triggers them more frequently. In open terrain a wind will blow the heated air along until it meets a hill or other rise which starts vertical motion and serves as a trigger as shown in figure 167.

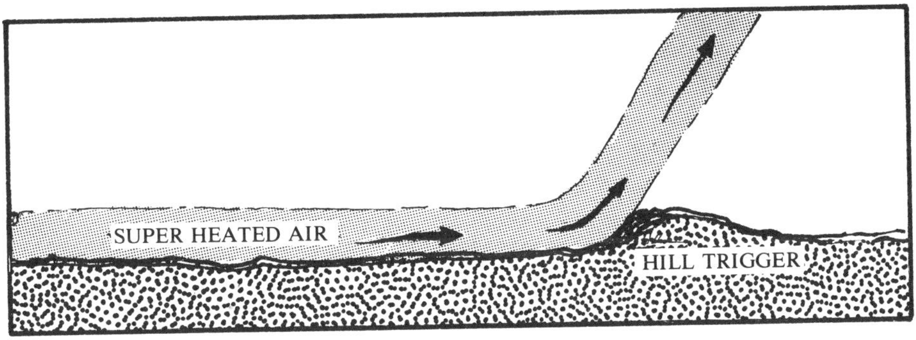

Figure 167 - A Hill Thermal Trigger

The passing of a cloud over sloping terrain can serve as a triggering mechanism. In extreme cases the ground can cool as much as 50 °F (27 °C) in several minutes when a cloud shadow stops the solar heating. A quick slug of cool air can form which slides downhill to trigger any potential thermals in its path.

When no wind is blowing any terrain irregularity can serve as a thermal trigger. A feature with an upward slope may have an upslope breeze established which will initiate thermal release. Figure 168 shows how hills, a tree, a pole or a tower with their rising convection currents will trigger thermals. Other irregularities such as buildings, plateaus and tree lines will trigger thermals in undifferentiated terrain.

LEE SIDE THERMALS

An important terrain effect is the blocking of wind by ground obstructions—hills, buildings and stands of trees. The downwind or lee side of such solid objects will experience very little wind disturbance if the wind is not too strong. Consequently this wind sheltered area will often allow a

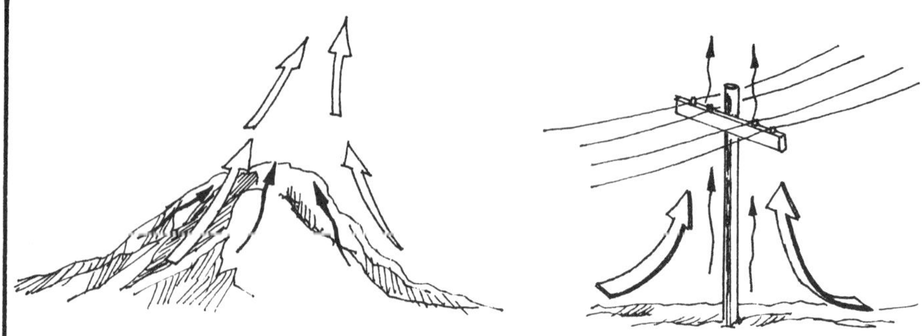

Figure 168 - Upslope Breezes Trigger Thermals

thermal to grow to considerable proportions before it releases.

Lee side thermals have a reputation in the soaring world for being bearers of healthy lift. However, in any significant wind the lee side of a mountain isn't our first choice of places to be. We need to be well above a mountain to exploit a lee side thermal safely. At that point we will most likely be encountering thermals from both sides of the mountain. Lee side thermals are not a dependable source of lift near a mountain because we can't safely spend time on the lee of a mounain searching for lift. However, look for wind sheltered areas on the surface as sources for good thermals.

THERMAL SOURCES

Intimately connected with thermal triggers are thermal sources. A thermal source is a point on the terrain most likely to produce thermals and consists of an area of good heating and triggering.

Good surface sources are those areas heated most rapidly by the sun. Bare ground, plowed areas, pavement, dry crops and weed fields are tops on our list. Any place you would feel burning bare feet on a hot summer day is a good thermal generator. Sand is readily heated but the sand particles trap a lot of air so the heat capacity of sand is low and they cool quickly in passing clouds. Stands of corn in autumn are excellent thermal sources since they trap a thick layer of air to be warmed. In a similar manner a town or city heats a deep layer of air due to the reflection from the sides of buildings and the great amount of pavement.

Rocky ground is a fine source of thermals if the rocks are small. Larger rock outcroppings have their own peculiarities. They readily conduct heat to their interior so their surface takes quite a long time to heat up. They don't come into their own as thermal sources until early afternoon. In the evening rock surfaces can be primary sources of thermals as they slowly release their vast amount of stored heat. Quarries are also good later thermal sources. However, if they are deep holes the thermals they grow aren't readily triggered.

HOUSE THERMALS

The terms *house thermal* and *resident thermal* refers to a thermal source near a particular flying site that is fairly reliable. This thermal may be a continuous column or more likely a regular succession of bubbles. Many a pilot has dove to a house thermal only to find disappointing sink. Nothing is guaranteed in the air except gravity and the rising cost of equipment. However, a house thermal is the most reliable form of thermal we have.

Essentially a house thermal exists over a good source. That it exists so close to a regular flying site is the reason for its reputation. Sometimes a rock outcropping, a quarry or a lone hill will be the source. At a mountain site a particular ravine or bowl may serve to herd thermals to the same place so it appears that they are from one source. No matter how a house thermal arises, it should be visited on a periodic basis if you are in need of a thermal in the area.

The latter point brings up an important matter that soaring pilots would do well to remember:

> **Thermal Reliability**
> *Good thermal sources tend to be consistent producers of thermals on a periodic basis throughout the day and every day.*

TERRAIN SOURCES

We have already mentioned that hills and other rising surfaces are good triggers. Here we will expand on that idea. High ground such as mountains or ridge tops are excellent thermal generators for a number of reasons. First, they are heated more readily by the sun because they are in thinner atmosphere which attenuates the sun's radiation less. Secondly, they often have slopes that directly face the sun as shown in figure 169. The figure

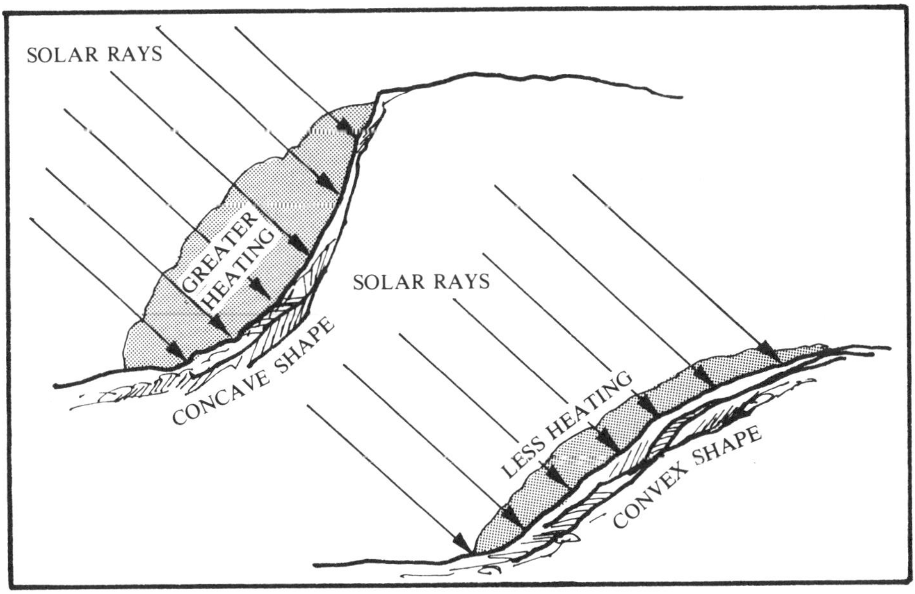

Figure 169 - Heating on Concave and Convex Slopes

also illustrates how a concave shaped hill heats the air more readily than a convex shape. A concave bowl also shares this property.

The third reason a mountain generates thermals exceptionally well is because the air overlying it is much cooler than that in the valley, while its surface temperature gets just as warm or warmer. Thus a thermal will originate from the mountain earlier than in a valley and be more bouyant or frequent during the warm part of the day. Finally, a mountain top will be above the nightime inversion layer that sets in as cool air slides down the mountain (shown previously in figures 101 and 135). Consequently the mountain will send off thermals earlier establishing it as a primary source for the day.

Over deserts and terrain where the surface is largely undifferentiated, the high points will be the thermal sources. Figure 170 illustrates this concept. The highest points are the better sources. One way to envision this is to turn the figure upside down and imagine a layer of water on the surface. Where it would drip off is a thermal source. You can perform this imagination trick while flying to help you locate the best possible thermal sources.

By the above analogy, flat pans and depressions would tend to be areas of sink. However, over a large flat area a depression may serve as a trigger source if it disrupts a normally steady wind flow that often exists in desert areas.

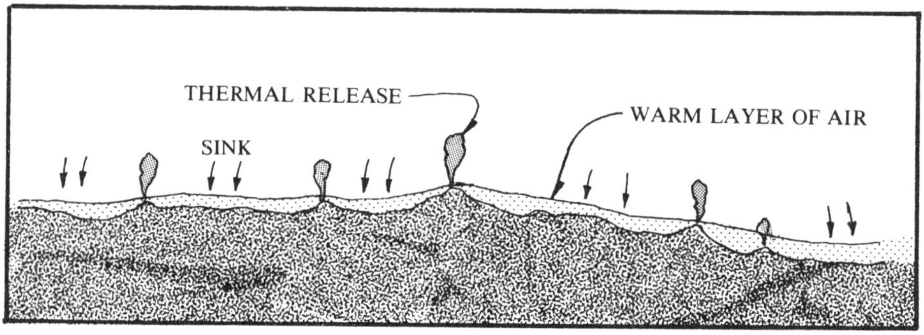

Figure 170 - Thermals Located at High Ground Points

WET GROUND SOURCES

We have learned that water is not a warm surface because of evaporation, its large heat capacity and the spread of any heating down from the surface. Thus it tends to be an area of sink and is normally avoided by pilots with a desire for altitude. However, there are some exceptions to this generality.

Water that lies in a thin layer in a marsh perhaps or after flooding will become nearly as warm as other morc solid surfaces. In this case thermals may readily be found, but they will tend to be light, large and difficult to pinpoint to a definite source if the wet area is large.

Ground that is moist after a rain is generally a poor producer of thermals because of the cooling effects of evaporation. However, the water vapor present does help this air to rise once it is lifted. On largely moist

Perfect thermal clouds over the Colorado plains.

areas look for thermals over higher or well-drained areas as these areas are likely to be dryer. In the more humid regions it is the areas that tend to be dryer that generate thermals. This is the case in northern Europe and the eastern half of North America. When droughts occur in these regions the improvement in thermal production and strength is dramatic.

Large bodies of water are natural heat sinks, but even these expanses can produce thermals in some situations. When cold northern air blows across a body of water it gets heated from below and enjoys the production of large and light thermals with a smooth disposition. This state of affairs is often experienced in autumn and winter near lakes and oceans and may be referred to as *lake thermals* or *water thermals*.

Snow reflects sunlight in the day and radiates off heat at night. As a result it remains very cold even after a succession of sunny days. However it too can produce thermals when very cold air moves over a snow-covered landscape. Snow thermals like water thermals are large and smooth and not at all uncommon as winter pilots will aver.

THERMAL CYCLES

We have learned that solar heating of the earth undergoes a daily and seasonal cycle. In Chapter I, figures 7 and 8 depict these cycles. The main point we learned is that maximum daily heating and thus thermal production does not occur when the sun is at its zenith, but a bit later due to a lag in the surface temperature compared to the solar radiation. The chart in figure 7 indicates that maximum surface temperature and thus maximum thermal production should be expected between 2:00 and 3:00 pm (14:00 to 15:00 hours).

These cycles times can be greatly altered, however, by terrain and cloud effects. For example, a west facing slope may not receive peak heating until four hours after that of a horizontal surface while an eastward facing slope may receive peak heating in the morning. A north facing slope in the northern hemisphere may only receive ample solar heating in the height of summer. This is expecially true at latitudes closer to the poles.

A layer of fog or clouds can greatly reduce surface heating and of course prevent thermals from forming. Morning clouds that dissipate later will naturally delay the normal thermal production cycle. But when the clouds do disolve the build-up of heat is rapid and thermals form quickly (unless the cloud is a thin layer that disappears slowly). Generally high cirrus clouds reduce thermal strength while spreading cumulus clouds stop thermals altogether if they remain over a wide area. Sometimes such cumulus build-up undergoes cycles as thermals rise, clouds develop, thermals get cut off, clouds dissipate, thermals reappear and the cycle repeats itself.

In general, on an annual basis thermal production goes with the sun. Peak solar heating produces peak thermal generation. Winter brings fewer and weaker thermals. This cycle is modified somewhat in temperate zones where cold fronts in spring and fall introduce unstable air from the poles and thus make these seasons the peaks for thermal soaring.

The daily cycle goes like this: morning heating starts the first stirring of the air. Light circulations develop that give way to the first thermals around 10 or 11 o'clock. Thermals then continue to build until 2 to 3 in the afternoon then they taper off and give way to late evening heat releases around 6 to 8 pm (18:00 to 20:00 hours).

Morning thermals start producing cumulus clouds.

There are often two *thermal pauses* that occur during the day. The first happens about ½ hour or so after the first thermals appear. The air seems to take a deep breath, thermal activity stops then returns with vigor. This early morning pause seems to be caused by the ground reaching trigger temperature and releasing a major fusillade of thermals that bring down a large volume of cool replacement air that takes time to heat. Once this air starts producing thermals they continue on a more regular basis.

The second thermal pause appears in the evening when regular thermal production wanes. Sometimes there is a period of about ½ hour occurring between 4:00 and 6:00 pm (16:00 to 18:00 hours) when not much seems to be happening. After this pause thermals produced by direct solar heating are rare and they are replaced by residual heating which we look at next.

EVENING THERMALS

Once the sun's smiling face begins to lose its bright disposition regular thermal production is reduced. Areas that turn to shadow and those that cool rapidly such as sand start producing sinking air. Other areas that were poor thermal generators or sink areas during the day come into the picture.

Forests and rock areas are particularly good places to look for an evening thermal. Fields of deep crops are also releasers of late afternoon heat. Finally, water comes into its own and displays its natural heat capacity by warming the air for hours in the evening. Deep water is assisted in its ability to produce thermals if a wind is blowing to help stir the water and bring heat up from the depths. Shallow water and all other evening sources are best in light or zero wind.

Thermals of the evening variety are never as strong, abundant, high rising or reliable as day time thermals, but weak lift is better than no lift. Besides, occasionally it is a joy to circle in a glassy bubble in a quiet sky. And sometimes we can be surprised by a rowdy little bullet released late in the day by some patch of ground storing it just for us.

In passing we should mention man-made areas such as parking lots and towns as being good places to look for evening lift. Also don't forget the fires and smoke stacks we mentioned in Chapter VIII.

At this point we summarize what we have learned about thermal sources:

Thermal Sources

DAYTIME	*EVENING*
•House thermals	•House thermals
•High ground	•High ground
•Heated areas such as: Bare ground, dry fields plowed fields, rocks (later in day), chalk areas, sand, quarries and dry areas	•Areas with residual heat such as: Rocks, towns, tall dry crops, forests (especially pines) and water.
•Avoid: wet areas, low lands, green areas, areas in long lasting shadows and blue holes.	•Avoid: close to high slopes and sandy areas.

NOTE: The above lists are ordered according to reliability.

THERMAL RISING

Once a thermal leaves the ground it undergoes a few changes. First it comes together to form its characteristic bubble or column shape as shown in figure 171. This process may require several hundred feet (100m) in a large thermal. As it is consolidating its form the thermal also accelerates to the speed appropriate for its bouyancy. This bouyancy is determined by its deficit of density compared to the surrounding air and its size. We discuss thermal bouyancy in Appendix IV.

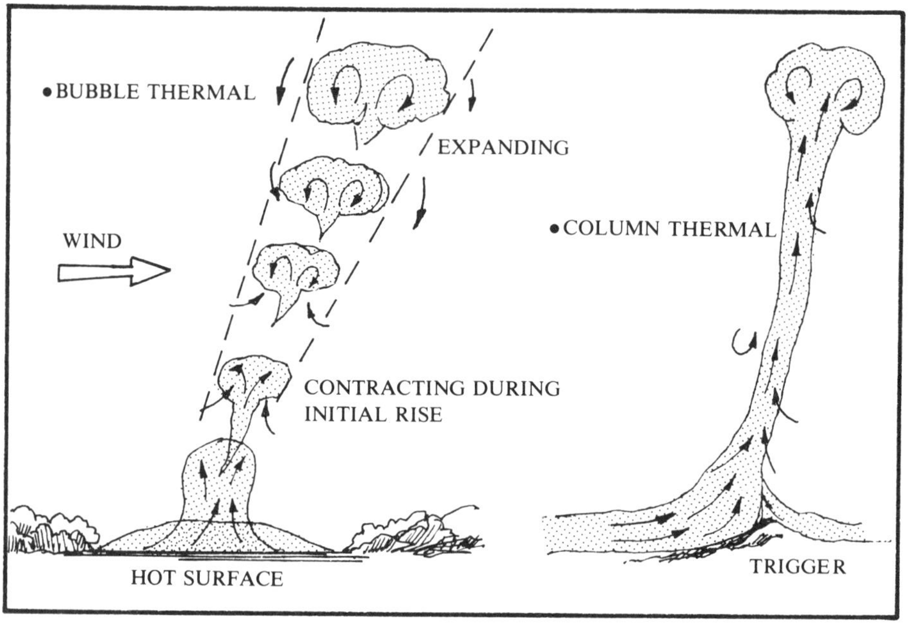

Figure 171 - Changes in a Rising Thermal

When a thermal rises abruptly air from around the area rushes in to replace the thermal air. If this air is also heated it will be entrained by the thermal and rise into it. A vast source of heated air as in case III of figure 166 will feed the thermal for several minutes creating a thermal column that stretches for thousands of feet (1000m)

If the supply of warm air is limited, then cool air will replace the thermal which will then be of limited size. The cool air will take time to heat then will release as another thermal. The time for a thermal to form in this repetitive process may be from several minutes to an hour or more depending on the strength of the heating.

Figure 172 shows how the air rushing in below the thermal can come from all directions in a light general wind. This in-rushig air can be quite vigorous in strong thermal conditions and can make landing in mid-day thermals a tricky affair. Switching winds called "light and variable" on weather reports are a sign of thermals. In a stronger general wind the direction won't change as much but the gustiness will increase.

When a thermal rises in the first 1,000 feet (300m) or so it may have an inflow of air from all sides. This general "convergence" tends to pull a

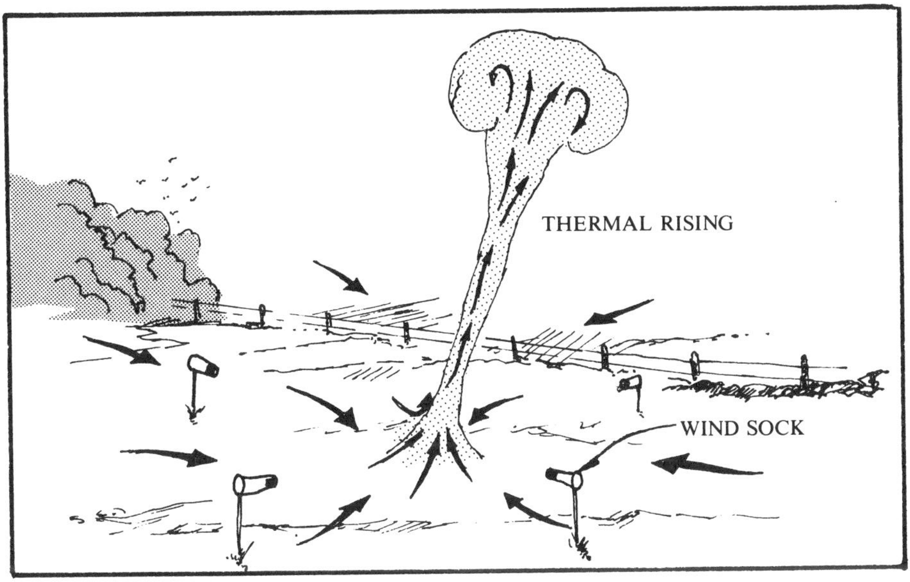

Figure 172 - Variable Ground Winds Under a Thermal

soaring aircraft towards the center so that less bank angle is needed to produce a given diameter of circle. Up higher the bank angle may have to be increased to maintain the circling diameter.

In general, thermals tend to be more turbulent close to the ground until they become more uniform up higher. However, thermals often rise into inversion layers that break up the thermal or contain shear turbulence. In windy conditions thermals may be so broken up that there exists a layer of mixed and turbulent heated air near the surface as shown in figure 173. This air may send off turbulent thermals at trigger points which will continue up as rowdy lift.

Figure 173 - Thermals in a Turbulent Layer

REAL LAPSE RATES

We have learned in Chapter II that unstable air induces a lifted or warmed parcel of air to keep on climbing. Our favorite thermals are precisely these warmed and lifted parcels. Let us see how typical lapse rates affect a thermal.

Figure 174 illustrates the lapse rates near the ground on an average summer 24-hour day. In the morning we see a considerable inversion layer near the ground. Remember, a lapse rate graph is simply a chart of the air's temperature at various altitudes. An inversion is a layer where this graph shows the air getting warmer or not cooling enough to be unstable as we go higher.

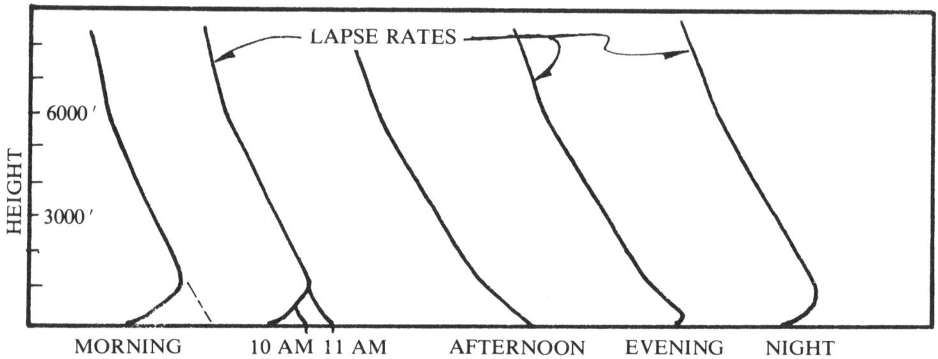

Figure 174 - Daily Lapse Rate Changes

The morning ground inversion is caused by cooling of the ground at night which cools the overlying air. Near mountains the nighttime downslope breezes can produce a deep layer of this cool air which represents a thick inversion (a 1,000 ft-300 m ground inversion layer is not uncommon). Evening thermals, clouds and wind can reduce this ground inversion by mixing the lower layers and reducing the radiation loss from the ground.

The point of the matter is this: thermals must continue to rise and die in this inversion layer until it disappears or they are strong enough to punch through it. Figure 175 shows how a thermal warmer than the surrounding air rises until it reaches the altitude where the air temperature equals to its temperature. At this point it stops its rise and mixes with the surrounding air. As this process continues the surrounding air from the level of thermal rise and below is displaced downward so that it eventually is warmed near the ground. Thus the lower layer begins to warm as shown.

As solar heating continues thermal temperatures rise and thermals climb higher in the inversion layer, continuing to warm the bottom of the layer. Eventually thermals are hot enough to rise past the inversion layer. When they do this they jump up rapidly in height as can be seen in the figure. The thermal temperature required to pass the ground inversion layer is known as the *trigger temperature*.

As can be seen in the succeeding lapse rates of figure 174, the ground inversion is eventually wiped out by thermal heating of the air. Later in the

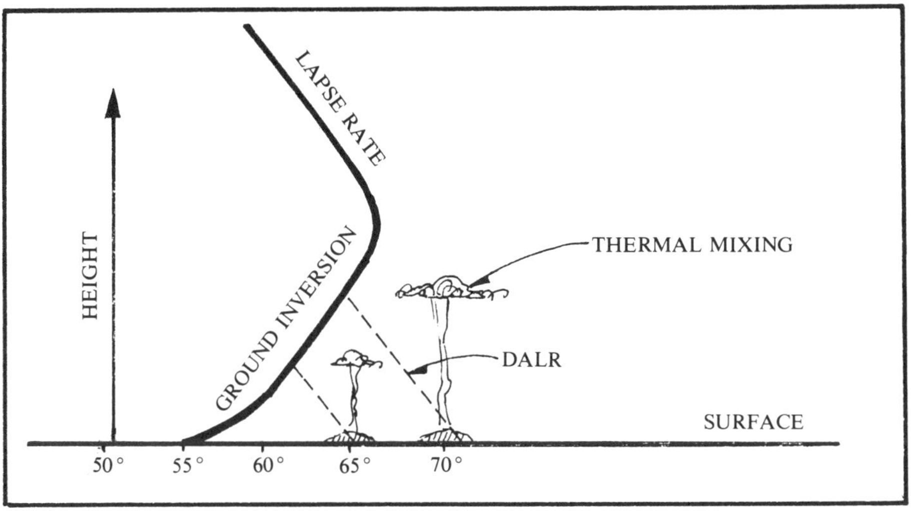

Figure 175 - Thermals in a Ground Inversion

day as surface cooling occurs it returns, of course. A very strong inversion (high temperature rise within a given altitude) caused by a clear, cold night will tend to hold off trigger time and good thermal production until later in the day. In such a condition heating may rise rapidly on the ground because the heat energy is trapped in such a low layer. It may feel like thermals should be popping, but nothing happens until quite a bit later. In Appendix V we investigate how to determine trigger times.

From the foregoing we can make the following conclusions:

Thermal Production

- Clear nights create a thick, stable ground inversion delaying thermal production on the following day.
- Clear days promote good heating and thermal production.
- Trigger temperature is the important factor in determining the timing of the initial usable thermals.

THE LAPSE RATE ALOFT

Besides a ground inversion, we have seen how inversions aloft also occur. Subsiding air in a high pressure system typically produces an inversion at around the 6,000 foot (2,000 m) level as shown in figure 176. Often different layers of air will lower at different rates and in stages. This action can produce two or more inversion layers. Also, the incursion of warm air aloft can produce an additional inversion as shown in the figure.

The multiple inversion layers have a profound effect on thermals. To see this look at figure 177. Here we have shown a typical lapse rate on a thermal day. Once the thermals rise past the ground inversion they rapidly increase their maximum height until they reach the layer of less instability. They then max out more slowly and meet a ceiling when they hit the inversion. Note that if the thermal does pass one inversion layer it often has another further up to contend with.

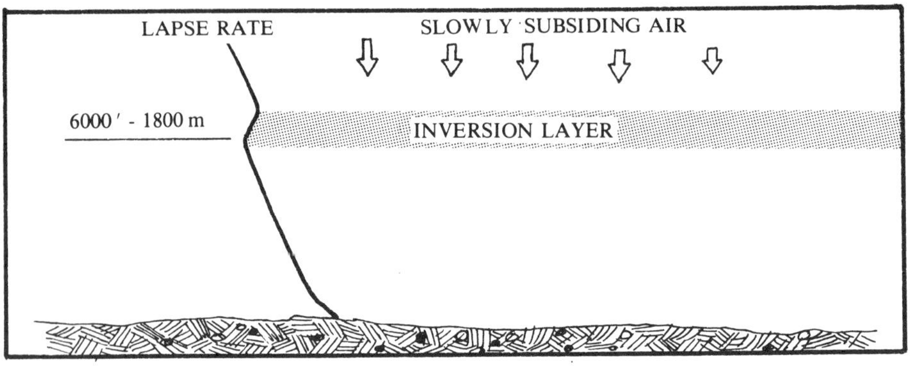

Figure 176 - Typical Inversion Under a High Pressure Cell

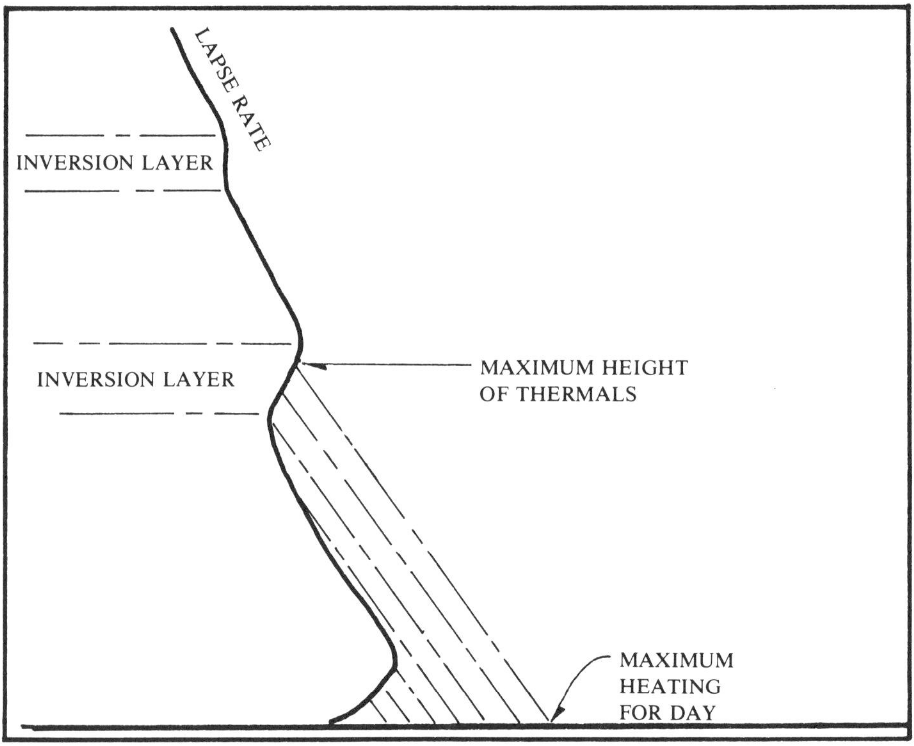

Figure 177 - Multiple Inversion Layers

Most thermals will stop at the inversion layer. Thus all the mixing goes on below this layer. As a result, dust, smog and general pollution stops at the inversion layer and can often be seen on the horizon as a brown line above which the air appears crystal blue. Occasionally a brown dome appears in this haze layer as a strong thermal pushes higher than normal. Identifying the inversion layer helps you know when to expect thermals to slow down and whether or not you have climbed above it.

In the next chapter we'll see about thermals that penetrate inversions. Here we should mention that on a strong thermal day the warming process in the layer below the inversion can eliminate the inversion. This will be noted by a sudden increase in the altitude achieved in thermals.

The inversion layer may not exist everywhere the same, for areas of good thermal production such as mountain chains may wipe it out while it is still thick in other nearby areas. Figure 178 indicates how an inversion will generally be higher over a mountain due to drifting of the air in ridge lift and upslope breezes. Also shown is how mountain-born thermals more readily bust up the inversion.

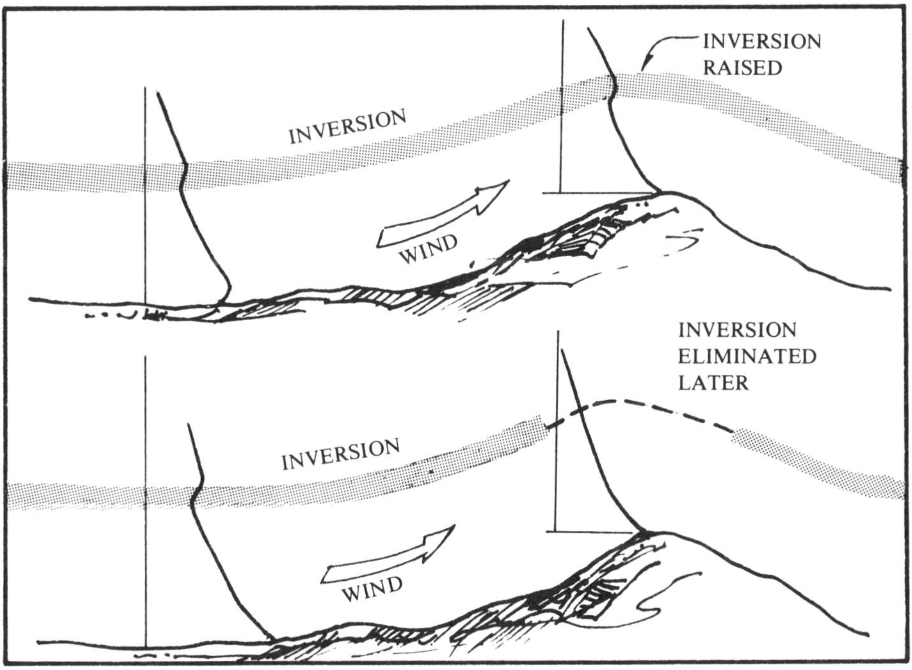

Figure 178 - The Raising and Elimination of an Inversion Over a Mountain

THERMALS AND LAPSE RATE VARIATIONS

The strength of the thermals on a given day depends on the lapse rate profile, the amount of solar heating and the moisture present. In Chapter XII we investigate the *thermal index* which takes into account these factors to provide a soaring forecast.

The trend in the lower thermal layer is for the lapse rate to approach the Dry Adiabatic Lapse Rate (DALR) or 5.5° F per 1,000 ft (1° C per 100 m). The reason for this is that thermals spread the heating up and down and bring the air to their temperature at each level. Of course, the lapse rate can be much different from that given above. Let's see what happens to thermals then.

A very stable ground inversion has been shown to stop thermals until later in the day. At times a much thicker layer of stable air can move into an area. This stable air will not have as strong stability as the ground inversion but it will dampen thermals. Figure 179 shows such stable air and how it is possible to have thermals of a weak nature even though the air is

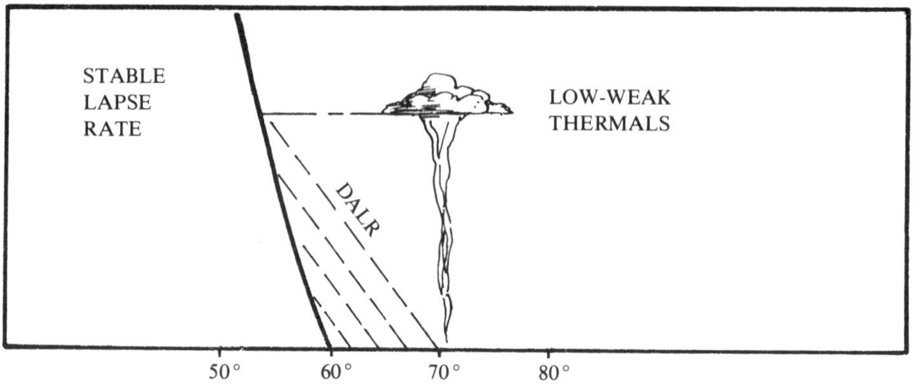

Figure 179 - Weak Thermals on a Stable Day

stable. Such thermals tend to slow down as they rise and can be quite turbulent as they erode away. These days tend to be hazy because thermals do not carry the moisture aloft.

We generally think of thermals as being warmer than their surroundings, but in truth the criteria for a thermal is simply that it be lighter than its surroundings. It has been found in more humid areas that thermals are often rising not because they are warmer but because they contain more water vapor than the surrounding air. This situation often occurs on those sultry, humid days that create thunderstorms in moist areas.

On clear, hot days it is common to get a layer very close to the ground that is superheated. This layer can be a few feet thick in green areas to several thousand feet thick in the desert. This layer is called the superadiabatic layer as mentioned in Chapter I (see figure 11). The superadiabatic layer has a lapse rate greater than the cooling rate of ther-

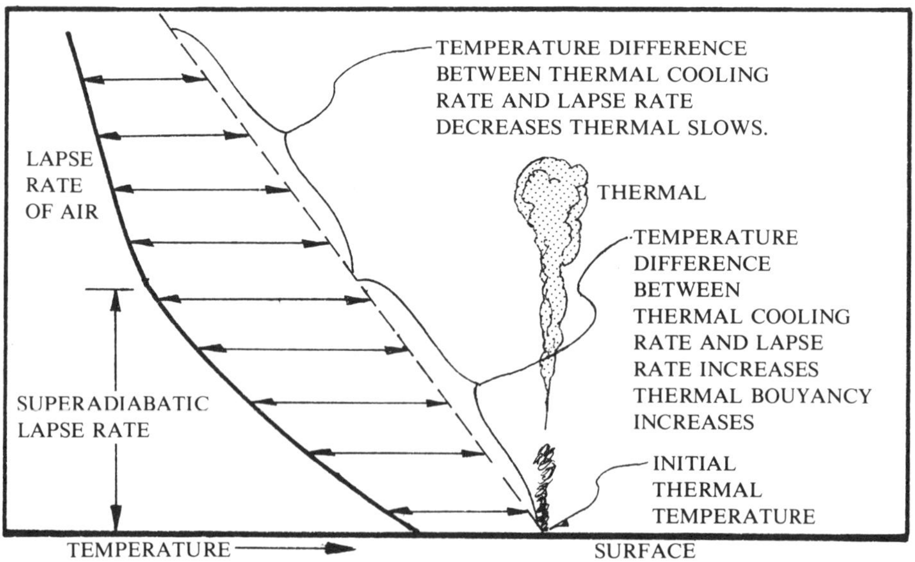

Figure 180 - Acceleration of a Thermal

mals, the DALR. As a result, the difference between the thermal temperature and that of the surrounding air is continually getting greater as the thermal rises through this layer. Thus the thermal accelerates upward. Figure 180 illustrates this principle.

Thermals rising in a superadiabatic layer tend to be of smaller diameter, punchy and strong. They are most commonly experienced in the dry, sunny areas of the world. They also give rise to dust devils.

DUST DEVILS

Tight cores of swirling wind will pick up dust, leaves and other debris to become a visible ground disturbance or towering column of brown dust in areas of bare ground. Such whirlwinds are known as *willy-willies* in Australia and *dust devils* elsewhere.

Dust devils occur when a thermal lifts off in superadiabatic conditions (see figure 181). The air rushing in to fill the area below the thermal usually has some turning motion due to Coriolis effect if it has been flowing for some time. When this air comes together its spin is exaggerated just as a skater spins faster when his or her arms are brought in. This spinning air would soon lose its impetus except for the accelerating thermal "stretching" the air vertically and bringing the rotating column tighter as it gets higher, much like a column of thick syrup gets thinner as you pull the spoon out of it.

From the foregoing we can make a rule:

> **Dust Devils**
> *Dust devils are formed when thermals rise in a superadiabatic lapse rate. Dust devils lie under the rising thermal, mark its track, size and often height as well as duration.*

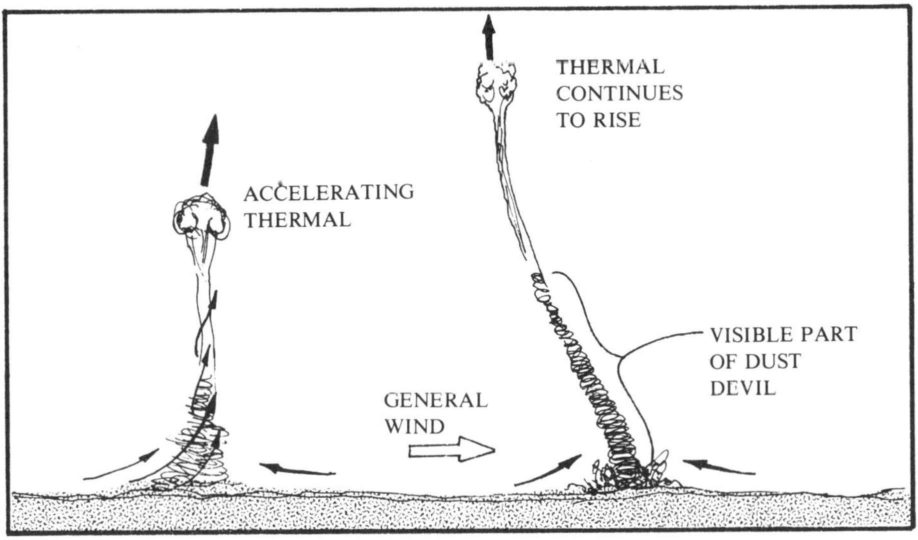

Figure 181 - Thermals Creating Dust Devils

Dust devils sometimes reach up into a thermal cloud, but usually stop well below this level, being typically only several feet to several hundred feet high (up to 100m). In some desert areas however, they can tower over several thousand feet (1,000 m) when fine dust and strong continous thermals abound. In these areas the height of the dust devil will indicate the minimum height of the thermal as well as its duration. However, at times the dust devil lasts past the production of usable lift as many unhappy pilots diving for a devil have found out. Watching the climb altitudes and rates as well as the duration of dust devils helps you judge the duration of the thermals creating them.

The vast majority of dust devils turn counterclockwise in the northern hemisphere and clockwise in the southern hemisphere. They are low pressure phenomena. The few devils that turn in the opposite direction are probably artifacts of rotation that began through turbulence or moving past a bluff. There is some conjecture that dust devil action spins the thermal air, and indeed, rotating thermal clouds have been seen on a rare occasion. It is likely that the air continues to spin above the dust although it probably stops its spin due to drag when the thermal leaves the superadiabatic layer. On this basis, it is reasonable to expect a better climb rate when turning against the flow of the dust devil (clockwise or to the right in the northern hemisphere) when in the strong lift of the superadiabatic layer. The reason for this better climb rate against the flow in spinning air is your rate of circling is slower so *less bank angle* is required to offset centrifugal force. Less bank angle gives you a better sink rate.

It is also important, to enter a dust devil thermal going against the flow for safety reasons. If you join the spinning air in the same direction as the flow you will experience a sudden strong tailwind which may stall you. If you enter against the flow you will experience an increasing headwind, as shown in figure 182, which will provide improved maneuverability.

A dust devil is a stable entity in that air from the outside cannot join the dust devil along the column and dilute it. Outside air can only enter it from below where the spin is slowed close to the ground as shown in figure 182. The air on the outside of the column is spinning and rising as shown while inside the column downward flow can occur due to lowered pressure. An example of this action can be seen in a stirred cup of coffee with up flow on the outside and a depression in the middle. The center of the dust devil is generally clearer than the sides.

The death of a dust devil occurs when the supply of warm air feeding the thermal is exhausted or some terrain effect blocks its progress. Dust devils will of course move up a steep mountain and are in fact quite common on heated slopes. A dust devil may continue a bit past the life of the thermal, but the devil soon looses energy and collapses. Witch doctors in Africa had a good business destroying dust devils by running through them, leaving the populace in awe of their demon-defeating powers.

The top view in figure 182 shows the track of a dust devil in relation to the wind. If a thermal moves with the wind or rises straight up above the ground wind layer it will be generally to the left of the dust devil track in the northern hemisphere and to the right in the southern hemisphere. This

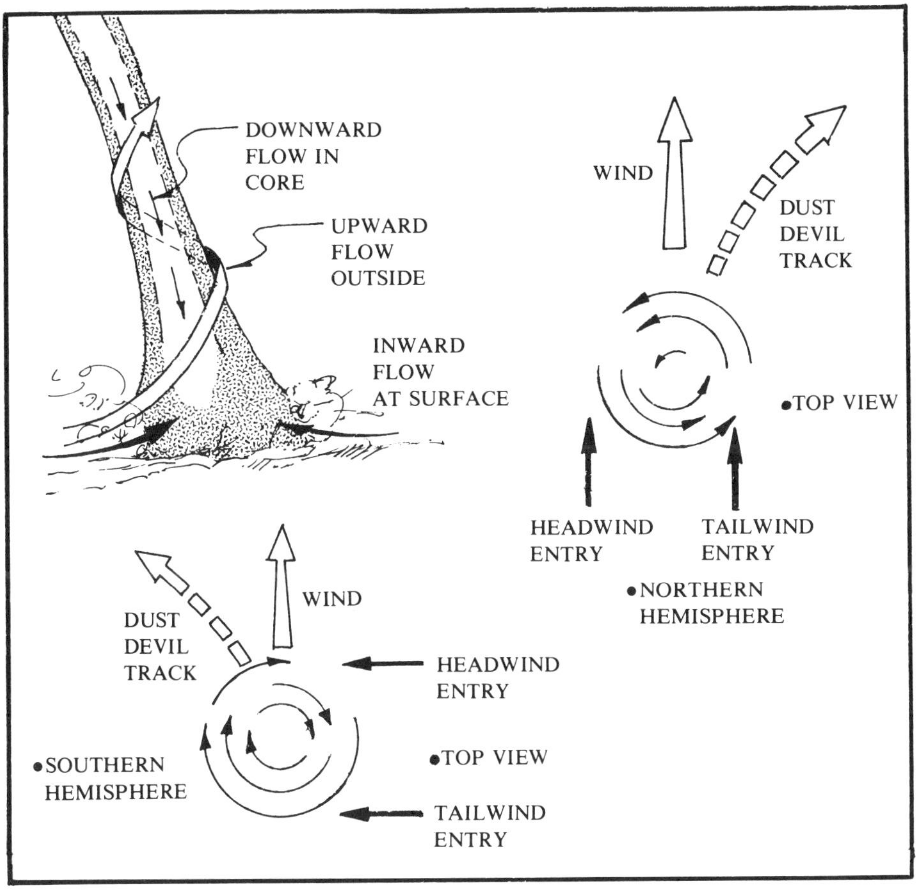

Figure 182 - The Nature of a Dust Devil

knowledge can help you locate thermals based on dust devils. Figure 181 shows how a dust devil snakes up to a thermal. Very tall dust devils can be seen to follow various undulating paths in different winds. The reason the dust devil travels at an angle to the wind direction is that the friction at the dust devil leading edge where it takes in the most air pushes it to the side.

Dust devil strengths can be quite variable according to their size and rate of spin. Indeed some dust devils have blown apart house trailers just like tornados. Although dust devils look like mini tornados, they are caused by ground conditions and rise from the surface while tornados develop from instability aloft and come from the clouds down. A circulating wind of around 15 mph (24 km/h) in a dust devil 100 feet (30 m) across is typical and perhaps reasonable for sport aviation purposes.

Using dust devils as thermal markers and sources of lift themselves is not without its hazards. Within the confines of the dust devil severe turbulence can be found (as well as a serious sanding of your leading edge). This turbulence has broken some aircraft and sent others out of control. These dire possibilities lead us to formulate the following dust devil safe flying rules:

Dust Devil Flying

- *Do not enter dust devils below 1000 ft above the ground.*
- *Do not enter dust devils below the top of the visible dust.*
- *Do not use excessively large and violent devils at lower altitudes.*
- *Use a turn direction opposite to the dust devil spin.*
- *Locate a thermal based on a dust devil to the left (northern hemisphere) or right (southern hemisphere) of the dust devil path.*
- *Newly formed dust devils are more reliable thermal markers than older ones.*

Dust devils are most prevalent and powerful in desert areas. Some of these monsters can be ½ mile (1 km) or more in diameter. In greener areas dust devils are more rare, shorter lived and lower in extent. Part of this reason is the lack of dust to carry aloft. This author once flew in a thermal in Pennsylvania at 5,000 feet up with scores of corn leaves circulating in the thermal like a flock of hawks. We call this a leaf devil. On another occasion we witnessed a dust devil created on a rock outcropping in New Hampshire, that had no dust to pick up but made a sound on the rocks like fizzing fireworks.

One other matter we should mention is water devils which occur when dust devil type swirls move over the water. These are usually short-lived and do not rise very high but they indicate good thermal conditions.

IDEAL THERMAL CONDITIONS

Air masses moving into an area play a great role in the stability and thus the thermal prospects. Warm fronts and warm air masses in general are not condusive to thermals because their load of humidity cuts down surface heating by scattering the sunlight. The humidity itself accepts heat directly from the sun and warms the air before thermals can develop.

Cold air masses are generally good thermal producers. This is because they usually bring clear, dry air and become unstable when their under surface is heated. This isn't always the case as we have seen in the discussion of the sea breeze air mass which is stable. But cold fronts from the poles are almost always bearers of thermals.

In the eastern US and northern Europe such fronts are welcome for the fine soaring they bring. Unfortunately they are also driven by high pressure systems and thus the trailing air mass is gently subsiding. The vigorous thermals push up through this sinking. air, but they are slowed slightly. The real problem is that high pressure dominated air masses create inversions due to the subsidence of the air and thus a lid on thermals. For this reason it is normal in the eastern US for thermals to stop in the inversion around 6,000 feet (2,000 m) above sea level and 12,000 foot cloud bases are a rare, glorious sight.

On the other hand, desert areas are in prime soaring form when a low pressure system sits over the area. The slightly rising air in the low reduces the stability aloft and aids thermal progress. It is not unusual for thermals to rise above 20,000 feet (7,000 m) in these areas because an inversion is usually not present. Most lows in the desert are heat lows (see Chapter IV).

Lows are not often thermal producers in moister areas because their rising air creates clouds and rain. Pilots in green areas must settle for highs and lower altitudes. In moister areas dryer conditions are sought after. On the other hand in the desert a little moisture is desirable because the added humidity in the thermals helps make them lighter so they rise better higher up. Moister thermals also produce clouds which are great thermal indicators at altitude.

We summarize here:

Good Thermal Conditions

- *Clear skies and bright sun*
- *Light to moderate winds*
- *Cold front, high pressure systems and dry days in moist, green regions.*
- *Low pressure systems and some moisture in desert regions.*

LIFT IN A THERMAL

Once an ideal thermal leaps into the sky and organizes itself it ideally takes on the shape of a mushroom turning itself inside out like a smoke ring as shown in figure 183. The air rising in the core or center of the thermal is moving upward about twice the rate of the top of the thermal. Thus it is possible to be near the top and climbing slowly while other pilots are climbing up to you from below. It is not always their better thermaling skills at work in this situation, but their position in the faster rising air.

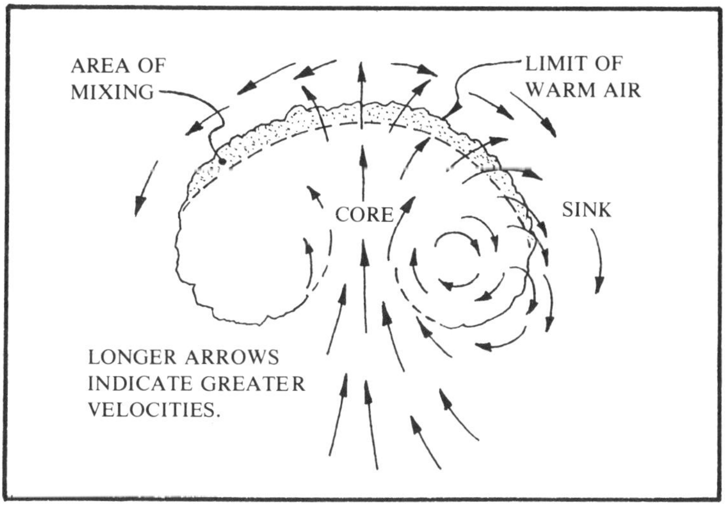

Figure 183 - Cross Section of an Ideal Thermal

As the thermal rises it pushes the air above it up and out of the way creating sink and turbulence along the sides of the thermal. An area of turbulent mixing occurs at the leading edge of the thermal as shown. This sink and turbulent area are often what announces the thermal to a searching pilot.

As our ideal thermal rises it continues to expand as it takes in more air and encounters lower pressure. It is fed from below as long as the supply of warmed air lasts and also pulls in air from the sides which may aid the thermal strength if it is a warm residue from a previous thermal or dilute the thermal if the air is cold. Some vortices and calves of the thermal are left behind in its wake as shown in figure 184.

It is probably a sure bet that the ideal thermal exists in nature judging from the thousands of pilot reports depicting textbook lift patterns in the thermal. However, there are also many occasions when cores are elusive, multiple and varying in strength. We'll look at the variety of thermals in nature in the next chapter.

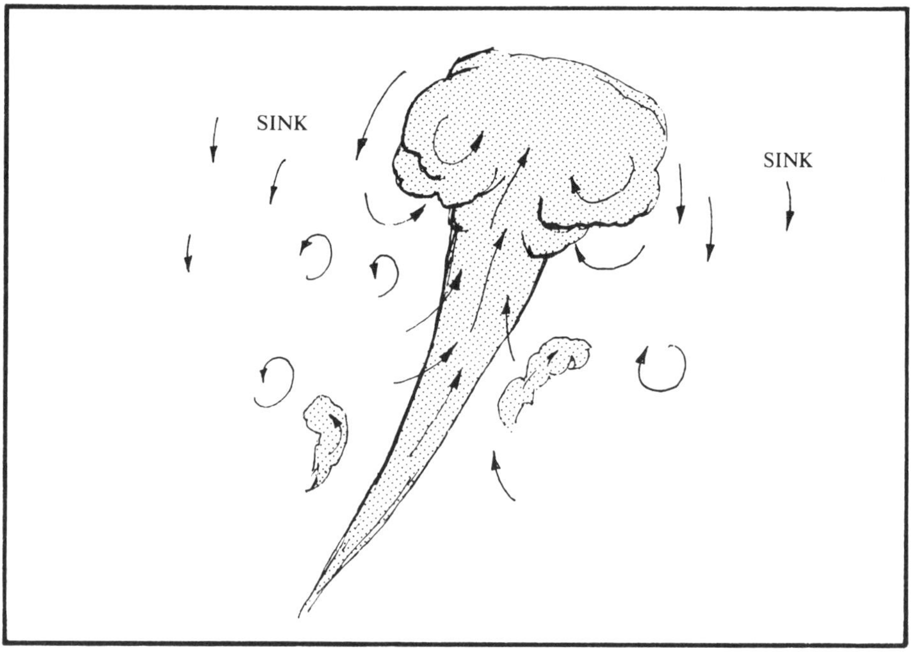

Figure 184 - Mixing, Sink and Lift Around a Thermal

THERMAL SINK

In unstable conditions we know that lifted air wants to continue rising. We should also know that air given a downward push wants to keep moving down since it continually remains cooler than the surrounding air in an unstable lapse rate. This sinking air acts like a negative thermal.

In good thermal conditions sinking air will be abundant. Usually the stronger the thermals the stronger the sink. However, because thermals inhabit typically 1/10 of the sky or less, the sinking air is usually more spread out and not as organized into strong vertical slugs.

Interthermal sink is usually strongest higher up where thermals are larger and more able to start a wider area sinking. If thermals are organized by a mountain, other terrain effect or streeting action the sink can be also more organized and widespread. Sometimes the best policy when immersed in sinking air for a long time is to turn 90° to your course in hopes that you were flying along the long axis of a elliptical sink area and can thereby escape the sink.

SUMMARY

We seek to prolong our adventures aloft by hopping a free ride whenever we can. One of the best vehicles a soaring pilot can find is a thermal. These conveyances are like hot air balloons rising to the heavens. The only trouble is they are invisible for the most part. Thus we have to study their behavior so we can make the best guesses possible as to how, when and where to find them.

Thermals are abundant and found practically everywhere at various times. They are variable in all their properties: strength, turbulence, size, duration reliability and height. Only experience, study and a little luck will afford you the ability to find the best thermal in the conditions at hand. We now have a good background in the basics of thermal behavior. Next we learn the deeper secrets of thermal lore.

Given the principles we have learned, can you guess the cause of these two arching cloud towers over Niagara Falls? *Answer on page 268.*

CHAPTER X

Thermal Lore

You will find no person on earth who exhibits signs of pure rapture more than one who has landed afer an extended flight in pure thermals. These gifts of nature reward the pilot who finds them and successfully exploits their lift with a high, cool vista and a warm sense of acccomplishment. Thermal flying is like sailing or fly fishing in that a minimum amount of equipment stands between the participant and the environment. The combination of chance and skill is what determines success or failure and therein lies the joy.

We have gained a solid understanding of the mechanics that generate and perpetuate thermals in the previous chapter. Now we turn our attention to the behavior of thermals in the sky. Our goal is to gain experience and knowledge of thermal lore so we can minimize chance and maximize our skill. Thus we are rewarded with more rapturous times on high.

THERMAL SIZES AND STRENGTHS

Thermals are will-'o-the-wisps. We can't see them and they linger only long enough to catch our fancy and show us their magic. All we can tell about them is their general diameter and upward velocity. Even these factors are suspect in a given thermal when someone else joins us and thermals up past us in a nearby core that we didn't know was there!

We *will* make generalities though to further our overall picture of thermals. From the experience of flying through countless thermals we can say that a thermal of 150 feet (50 m) in diameter is a fairly large one. Most of the thermals we encounter are 100 feet (30 m) or less in diameter. Some thermals—especially those in weaker conditions—can be quite large of course. However, when we find lifting areas larger than say 300 feet (100 m) in one direction we should suspect a source of lift other than pure thermals.

Lets imagine a thermal with a hundred foot diameter. If we assume a spherical shape and work out the volume we have over 500,000 cubic feet. At sea level air weighs .076 pounds per cubic foot so our thermal weighs over 19 tons! No wonder it can carry our feather-weight craft aloft. Incidently, the volume in this spherical thermal can be made up by a layer of air a bit over 5 feet (1.6 m) deep on a field 300 feet (100 m) square—not a large thermal generating area at all. If our thermal had twice the diameter the volume goes up eight times and it weighs more than 150 tons. This great mass gives thermals a will of their own in the free air as we shall see.

Thermal strengths can be essentially defined by their upward velocities. These can be highly variable from near zero to several thousand feet per minute or fpm (17 m/s) in thunderstorms. Typically we encounter thermals from 200 to 700 fpm (1.1 to 3.9 m/s) in humid climates with occasional climb rates a bit over 1,000 fpm (5.5 m/s). In desert condition climbs from 500 to 1,500 fpm (2.8 to 8.4 m/s) are common in midday with greater strengths occasionally. The strongest velocities occur at levels where the lapse rate is most unstable as shown previously in figure 180.

One meteorologist has found that the strength of thermals is directly related to the height they reach. It also can be generally said that dry thermals are weaker than those that contain enough moisture to produce clouds. Thus we can form the following chart for average rates of climb over several minutes:

Thermal Strengths

DRY THERMALS	
Maximum Thermal Height	Average Thermal Strength
3,000 ft (1,000m)	330 fpm (1.7 m/s)
6,000 ft (2,000m)	500 fpm (2.5 m/s)
10,000 ft (3,000 m)	700 fpm (3.6 m/s)
MOIST THERMALS	
Cloud Base Height	Average Thermal Strength
3,000 ft (1,000m)	375 fpm (1.9 m/s)
6,000 ft (2,000m)	600 fpm (3.0 m/s)
10,000 ft (3,000m)	780 fpm (4.0 m/s)

The average thermal velocities listed above do not take into consideration your aircraft's sinking rate. This must be subtracted from each value. In desert conditions we should expect stronger thermals, and may wish to add to these values.

We can make a further generality and state that the stronger thermals are, the more turbulent, tighter and longer lasting they are. Weaker thermals tend to be more benign, often wider but less reliable. This matter has to do with the conditions in which the different strength thermals rise.

THERMAL HEIGHTS

The maximum heights that thermals reach on a given day at a particular place depends on one of several things: the height of an inversion layer, the height of cloud formation or the height of the dry adiabatic lapse rate layer. Here we have outlined the causes of thermal demise as shown in figure 185.

In the first case we see an inversion layer stopping the thermal climb.

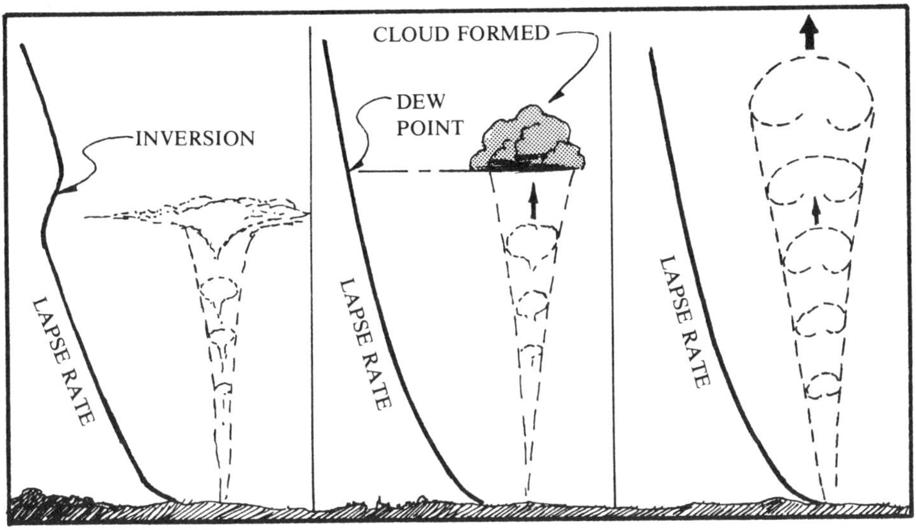

Figure 185 - The Death of a Thermal

When the thermal reaches this inversion layer it often becomes turbulent as it disorganizes. If many moist thermals are stopped at the inversion layer the humidity in the layer can increase until cloud is formed—usually of the stratocumulus variety. Inversion layers can be anywhere from very close to the ground to well above normal thermal reach.

Some thermals may punch through an inversion layer if it isn't too thick. These thermals will be the strongest in the sky. In order to ride the thermal through the gauntlet of the inversion layer it is necessary to locate the best core and be patient as it slows and disorganizes in the inversion. Often it will come together again to reform into a more coherent thermal rising out the top of the inversion.

In the second case shown in the figure the thermals reach the dew point level and form cumulus clouds. When this occurs much mixing with the surrounding air takes place as latent vaporization energy is released during condensation. The thermal loses its bouyancy and identity due to this mixing.

The height of the dew point depends on the temperature profile of the air and the humidity of the thermals. Since mixing near the ground spreads the humidity fairly evenly, thermals generally contain the same relative moisture and thus create clouds with bases nearly the same level. Different or changing cloud bases mean a different air mass is entering the area.

The final case whereby thermals stop their climb occurs when they enter neutrally stable air and gradually get weaker and weaker as they continue to mix with their surroundings. This is the situation with dry thermals when an inversion layer is not present as shown in the figure.

THERMAL DURATION AND DAILY VARIATION

Thermals can be passing fancies or semi-permanent plumes. Again we can only generalize, but say that thermals providing lift for ten minutes or less is the norm. It is rare in fact for one thermal to carry us from very near the ground to cloudbase, even in desert conditions where thermals last longer for cloudbase is usually much higher there. In truly weak conditions we have to pass from thermal to thermal like stepping stones to gain altitude.

In the morning thermals are gradually becoming more abundant, larger and longer-lived. By midday a steady state is normally reached whereby half the thermal clouds are building and half are dying at any point in time. A thermal cloud typically lasts for about 20 minutes and is fed by two or three thermals. One thermal usually will not form a lasting cloud. Thus we have a thermal duration of 6 to 10 minutes in normal conditions.

Figure 186 illustrates the typical daily progression of cloud sizes, base heights and shapes. Note how the bases rise as the air dries out. Also the bases do not come down in the evening as thermals weaken. In very moist conditions the bases do not rise because the surface air does not dry.

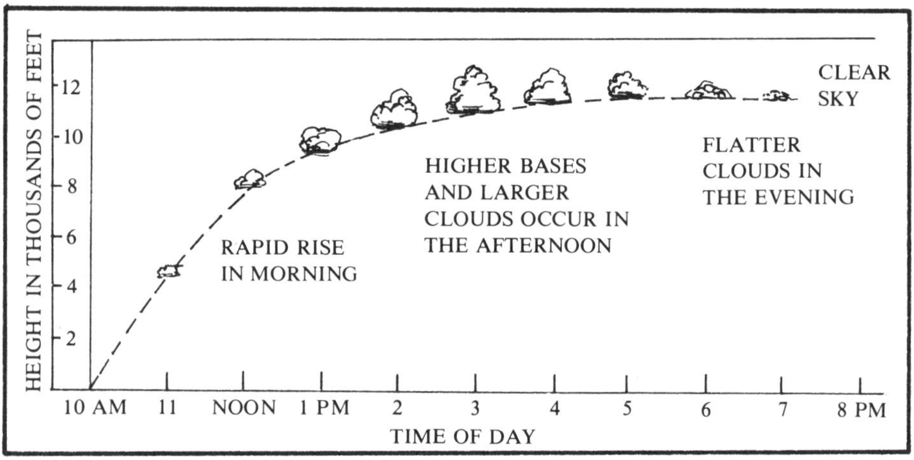

Figure 186 - Daily Thermal Height and Size Changes

THERMALS IN WIND

We have mentioned that wind tends to trigger thermals more frequently. Wind also moves the thermal and allows it to pick up more warm air than it normally can if it is rising above one spot. However, stronger winds will distribute heat away from the ground and prevent cohesive thermal build-up. In this case thermals may originate from above the ground as a couple of warm volumes coalesce then entrain other warm patches.

When a thermal rises in wind it will tend to float with the wind, but its great mass as mentioned earlier causes it to move slower than the wind due to inertia. Figure 187 illustrate how the thermal's horizontal motion will always be lagging behind that of the wind as it rises into stronger winds.

In a strong wind shear or in erratic winds at different levels a thermal may get broken apart as shown in figure 188. When this occurs the thermal may organize again above the shearing level. In weaker shear the thermals may tilt and be highly angled downwind as shown in figure 189. This state of affairs complicates thermaling because a pilot must constantly pay attention to the core location and avoid drifting too far from a safe haven.

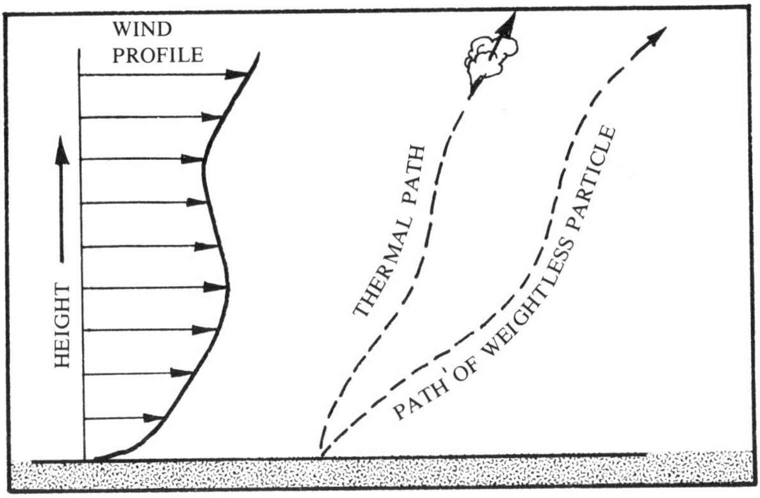

Figure 187 - Thermal Drifting in Wind

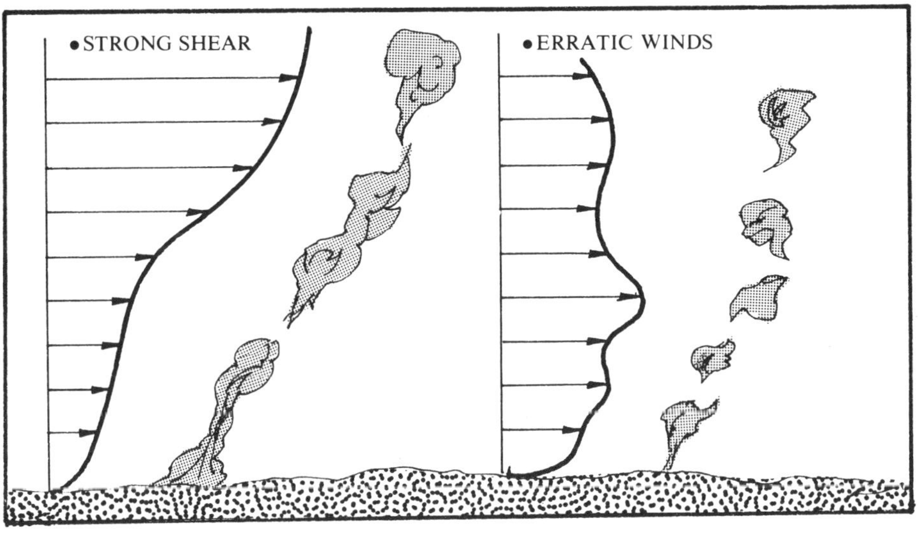

Figure 188 - Wind Breaking up a Thermal

In lighter variable winds thermals may get pushed this way or that by the wind or even move around as they pick up influences from other nearby thermals and join together. Such snaky thermals are commonly found in Brazil and other tropical areas this author has experienced.

Occasionally we will even find a thermal to move *upwind* if it rises from a ground wind drifting it in the opposite direction of the upper winds. Such elusive thermals demand all a pilot's thermaling skill and attention in order to stay in them, let alone maximize their potential. Birds, other gliders and airborne fluff or debris can greatly aid in the location of these wishy-washy thermals.

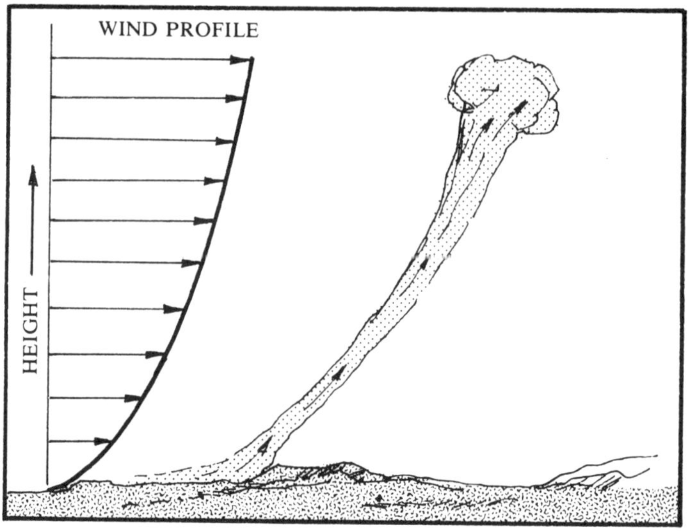

Figure 189 - Thermal Tilt in Wind

THERMAL TRACKS AND CYCLES

Wind tends to collect thermals on the mountains. The reason for this is the wind drifts the thermal toward the mountain at which point it rises above the mountain as shown in figure 190. This is another reason why the high ground is a good source of thermals.

When thermals climb up a slope we can often see their tracks in the trees or watch their progress through the ground cover. At times, however, virgorous activity in the trees at mountain peak level is caused by fast moving sink and should be avoided. Usually this occurs in stronger horizontal winds. In lighter winds tree tops rustling identifies a thermal presence.

When thermals rise near a slope they tend to hug the terrain because the inflowing air is blocked on the slope side. This effect causes thermals to

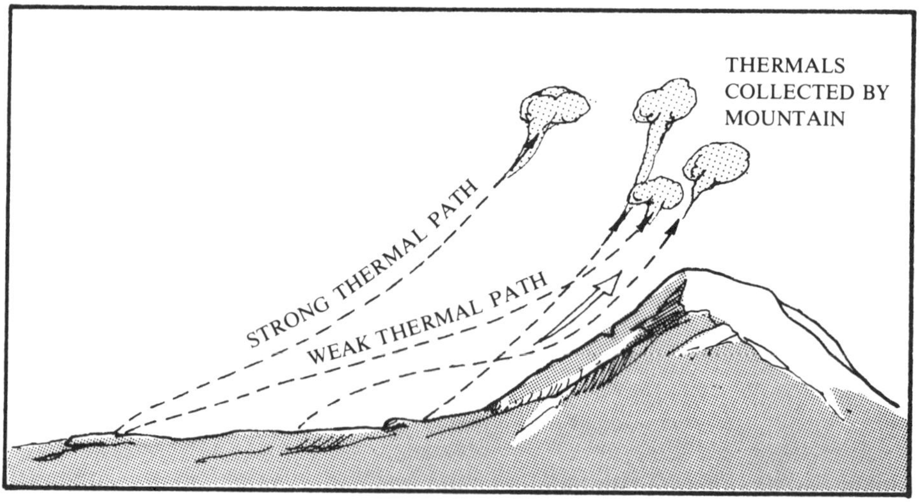

Figure 190 - Thermals Collected by a Mountain

ride up the walls of canyons and ravines. In addition, the combination of upslope breezes and the tendency for the thermal to ride close to the terrain can result in the lower portion of the thermal riding up the slope in front of the higher portion as shown in figure 191. The important point here is that thermals that show themselves as disturbances on the crest may be located *upwind* of the crest.

As thermals approach the top of a hill or mountain they often reduce, stop or reverse the incoming wind as they suck up warm air like a big vacuum cleaner (see figure 191). Consequently one sign of an approaching thermal is this change in wind. It may take several minutes for this wind cycle to occur as thermals pass through. Sometimes they can be heard approaching as they rustle through the brush. On high mountains the approach of a thermal is frequently accompanied by a dust devil as superheated air lifts off. These cycles are often very regular and it is useful to time them to predict thermal approach.

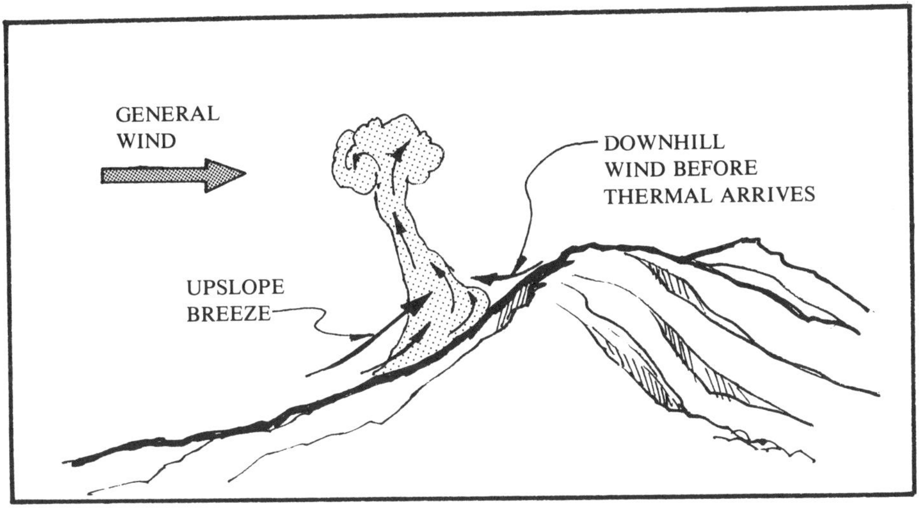

Figure 191 - Thermal Tilt Near a Slope

THERMAL TYPES

In the last chapter we discussed thermals as light circulation, bubbles or columns and described the ideal thermal. Here we look at variations on these themes.

The first and most common difference we find between nature and the ideal thermal is the presence of multiple thermals. It seems that thermals rise most readily in paths already taken by previous thermals. This is due to the general area of sink caused by the bubbling thermals. When a secondary thermal rises it usually catches up to the preceding thermal because of favorable motions in the wake. This action has been demonstrated in visible fluids.

Once two thermals are joined together they may combine their efforts or somewhat maintain their own identity. This can be seen in the many thermals that exhibit multiple cores. In some cases four or five cores will be rising at once with areas of sink or lesser lift in between as indicated by a group of gliders thermaling together. In such a multiple core situation it is desirable to be in the strongest core but you may not know there is a stronger core if other gliders or birds are not around. Figure 192 depicts a hypothetical shape of a multiple core thermal. Such thermals seem to develop most often when ground wind, large expanses of heated air or an inversion alter thermal production from the tidy little point source of the ideal model.

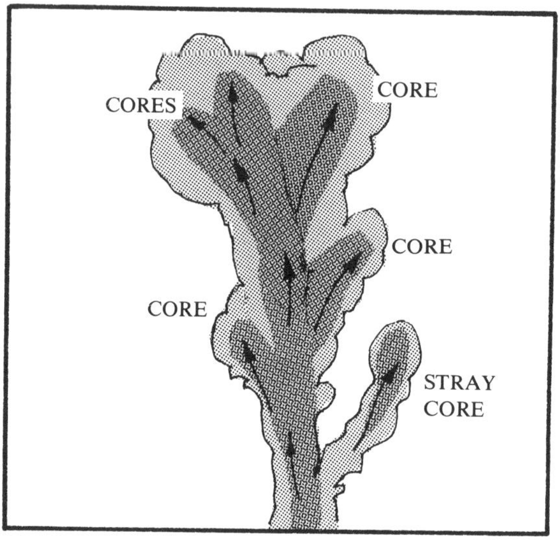

Figure 192 - Multiple Cores in a Thermal

Often thermals appear to be oblong or elliptical shaped with a long axis oriented towards the wind. Multiple cores can be found in such thermals as well. The reason for the oblong shape may be the picking up of extraneous bits of warm air as the thermal moves horizontally in wind.

When thermals are found to be elliptical most of them will be elliptical throughout the day. In this case flying directly upwind or downwind will present the most lifting air and help you find the core.

We have previously mentioned evening thermals and how they are generally weaker, shorter lived but often wider. Figure 193 shows the differences in daytime and evening thermals in a side view.

THERMAL SPACING

All the thermals that the ground produces do not go to cloud base. Indeed, many of them stop part way up as they lose bouyancy through entrainment of the surrounding air. Others join together with nearby thermals to rise together. As figure 194 shows, the higher we go in the atmosphere the fewer but more powerful thermals we encounter. Only the stronger and longer lasting thermals rise to the top.

The spacing of thermals is closely related to their maximum height. Thermals tend to be from 1½ to 3 times their maximum height apart. Down low we have more thermals spaced closer. Higher we encounter fewer thermals further apart but they are stronger. The spacing numbers given above do not vary greatly on a given day but change within this

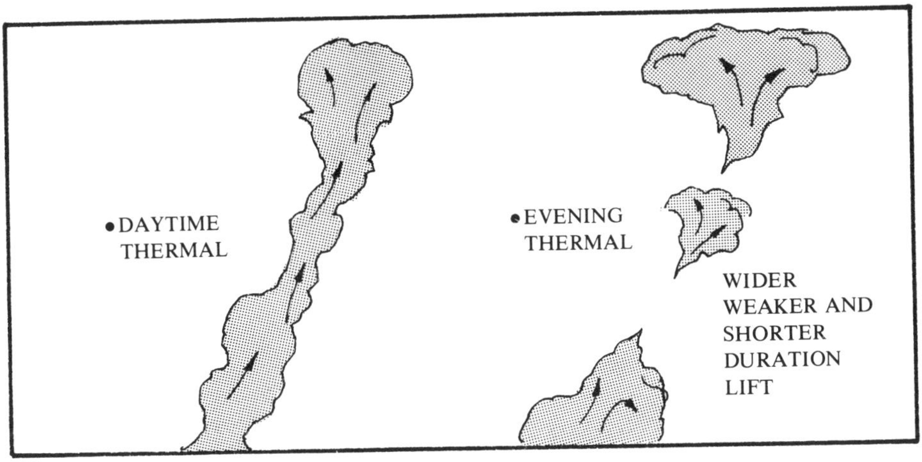

Figure 193 - Daytime and Evening Thermals

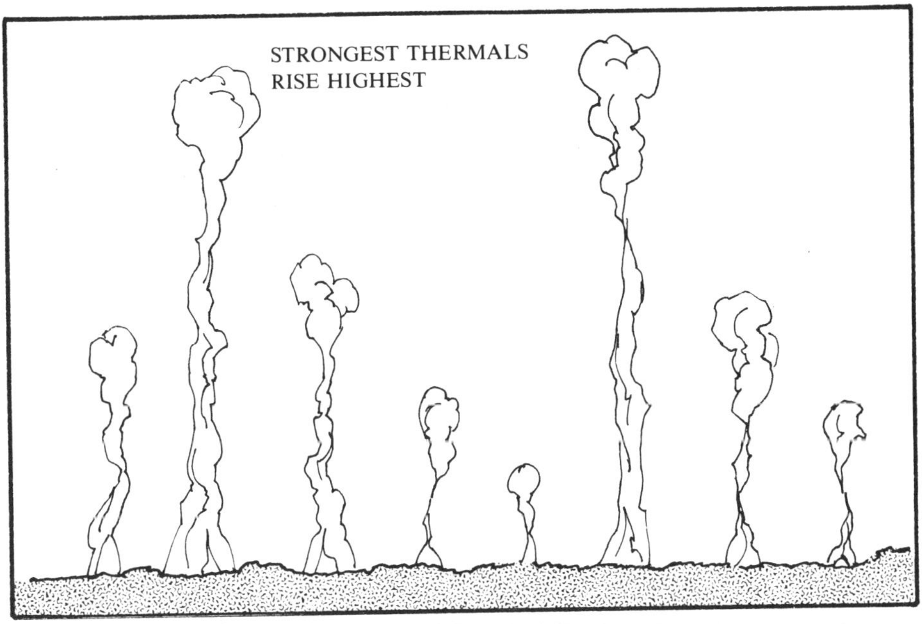

Figure 194 - Thermal Differences With Altitude

range according to the day's conditions.

Several other related numbers are: Thermals tend to take up about 1/10 of the sky. At higher altitudes they are fewer but wider. Clouds on a typical thermal day occupy about ¼ of the sky. It may appear that clouds cover much more than this, but this is because the sky is obscured by the vertical development of the clouds as shown in figure 195. Thermals are typically 1/3 the diameter of the cloud they are feeding which accounts for the greater coverage of the sky by the clouds.

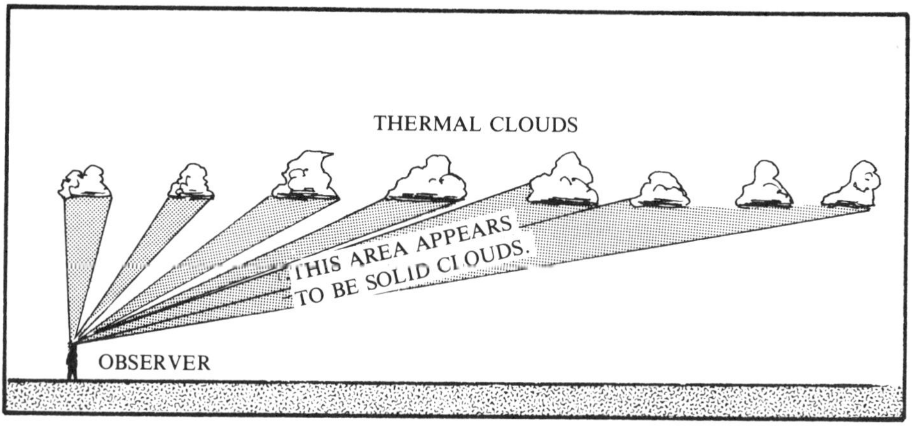

Figure 195 - Thermal Clouds Appear to Cover the Sky

THERMAL PATTERNS

Nature has shown us how to create adjacent cells that use the minimum amount of connecting lines with no wasted space. These are hexagonal honeycombs. We can see such honeycomb shapes approached in cracking mud, frost heaves and "mackeral sky" clouds. Thermal patterns over undifferentiated desert or water will have this honeycomb shape also as shown in figure 196. Here we see lift in the center of the open space and sink all around. In fact, mackeral sky or altocumulus clouds are precisely caused by thermal circulation initiated when a lifted layer of air reaches the level of instability and autoconvects. In the absense of any trigger points high in the atmosphere, the hexagonal pattern appears.

Over the years naturalists have observed a curious pattern in the flight paths of seagulls and other water birds. Under certain conditions the birds soar far out to sea using circular flight paths as if they were riding ther-

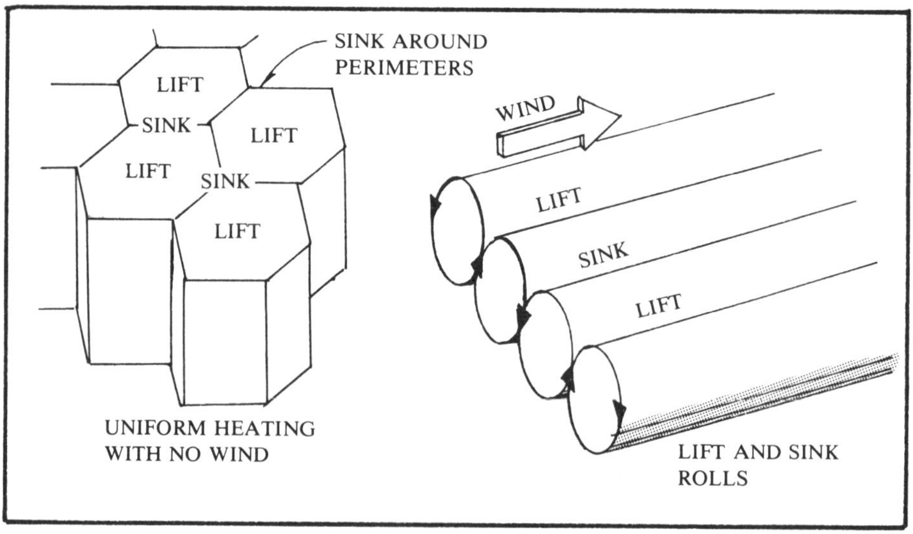

Figure 196 - Thermal Patterns

mals; at other times the birds would follow straight-line paths as if they were riding corridors of lift. After some research and experimentation it was found that in the first instance the birds were indeed thermaling in regularly spaced convection cells of the type described above. The size and height of these convection cells (sometimes called Benard cells) depends on the amount of heating and the height of the heated layer.

When a wind begins to blow these cells begin to tilt. At about 15 mph (24 km/h) the cells are lying on their side and have become horizontal rolls as shown in the figure. These rolls are very important to sport aviation and are associated with thermal streets. The experienced seagulls use the lift between the rolls to achieve their long, straight soaring flights.

THERMAL STREETS

We use the term *thermal streets* to denote any row of thermals. These rows can be created by several mechanisms. The first is a continuously pumping point source such as a hill or quarry. Several of these point sources in an area can give a thermal cloud pattern as in figure 197. Here we see cloud rows of various length and spacing reflecting the effectiveness of the point source in producing thermals, its positioning and rate of heating.

Rows of clouds or thermals formed in this manner are called *cloud streams* or *thermal streams* depending on whether a cloud is formed or not. Such rows can extend 3 to 15 miles (5 to 25 km) downwind from the point source depending on the wind velocity and how quickly the clouds dry out. It is likely that such a stream is useful for only about half its length as the rest of it is eroding away.

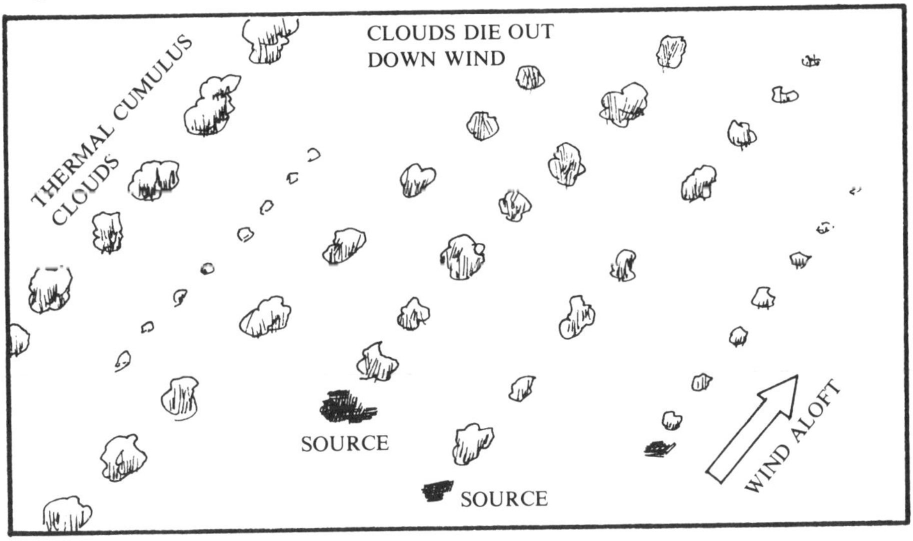

Figure 197 - Cloud Streams

Another form of thermal or cloud row is seen along mountain chains or ridges as in figure 198. These rows can be properly called streets, but they do not form in the classical method seen below. Streets along mountains tend to be stationary since they are formed by thermals and convergence over the mountain peaks. This author experienced a flight above the Pennsylvania ridges with rows of clouds along each of the several ridges in view in the morning. No wind was apparent and ridge lift was non-existent. However, we were able to climb up in thermals to about 4,000 feet over the top and fly along the ridge. A half hour into the flight the clouds disappeared as the air dried. However, the thermal streets still remained above the ridges as we demonstrated by flying north for over two hours in pure thermals. Pilots who ventured out into the valley found no useable lift.

Figure 198 - Streets Above Mountain Chains

CLOUD STREETS

True *cloud streets* are created in the free air and take part in the rolling action described above. In figure 199 we see such rolls and the organized cloud streets above the lift areas with rows of sink or sink streets in between. The parallel rolls are not real entities but are representations of the air circulation. A particle of air ideally follows a spiral path along a roll as shown.

The conditions favorable for cloud street production is a wind blowing steadily in the same direction and increasing to 15 mph (24 km/h) or more within 2/3 of the height of cloud base as shown. It is also desirable to have a stable layer above the street so that no one cloud can grow exceptionally large and offset the uniform street patterns.

When such thermal streets are dry they are known as *blue streets*. These clear-sky streets occur more often than most pilots expect in areas prone to streeting. Searching directly upwind or downwind for another thermal when leaving a thermal is a wise idea for this reason.

Cloud streets occur most often in high pressure dominated air masses. This is because of the stable layers that high pressure subsidence produces aloft. Since these inversion layers are not as frequent in desert areas such as the American Southwest, true cloud streets are not common there.

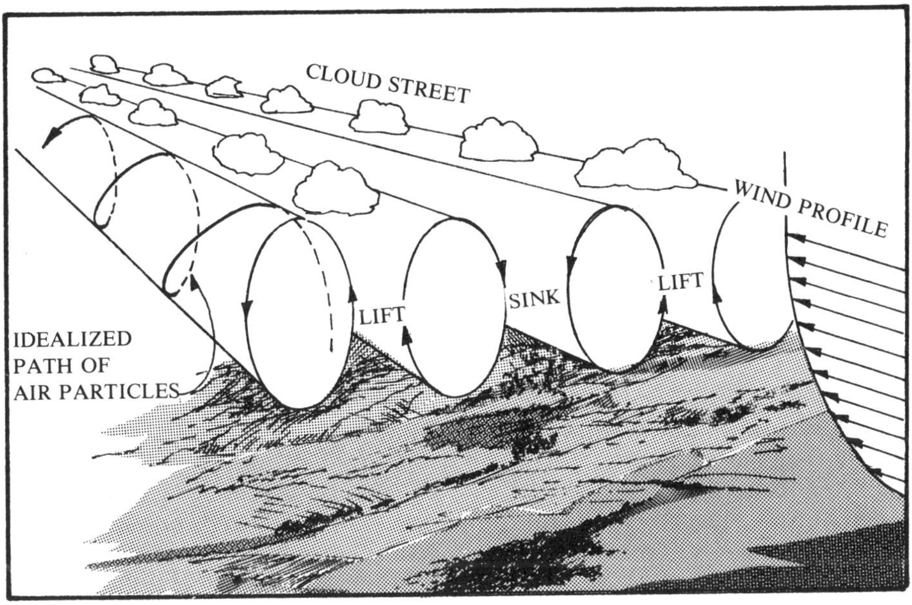

Figure 199 - Cloud Street Mechanics

Elsewhere cloud streets are common and should be expected whenever cumulus thermal clouds are around. It may be difficult to detect streeting action from the narrow viewpoint we have of the clouds in the air. The trick is to use the clouds shadows to discern any lining up of the clouds.

We expect that streets would occur after every cold front passage if it weren't for the random effects of the terrain on air mass heating. As it is, cloud streets are difficult to relate to any particular ground thermal source, but it seems that some strong sources alter the street's pattern and may indeed be the determining factor in street placement.

Weak thermals organized in streets.

CLOUD STREET BEHAVIOR

Cloud streets can be solid lines of clouds in moist conditions or sparse dots of clouds in drier air. Generally the thicker the clouds are along the street the more continuous the lift is. The spacing of the individual streets is usually two to three times the cloud height. Thus a cloudbase of 6,000 feet (2,000 m) gives us a street spacing of from 2.25 to 3.5 miles (4.5 to 7.0 km). This spacing depends on the lapse rate and is important when trying to cross streets.

It is common for a line of streets to stop for a space or entirely while others continue on. It is not unusual for a single street to be 50 miles (80 km) or more in length and the whole streeting region to be hundreds of miles long and wide. Such long streets usually curve with the isobars aloft.

Streets are not steady-state creations for the clouds within them regularly die and are replaced. Also, the streets themselves frequently die and are reformed to the side. Timing and good fortune are needed to fly rapidly changing streets. Sometimes the sun lines and shadows are "in phase" with street production as shown in figure 200. In this case the solar heating is directly under the lifting air and the streets are invigorated. Other times the shadows may be under the streets so they eventually die or shift to the side. The moving sun in relation to the direction of the streets can readily change the lift patterns.

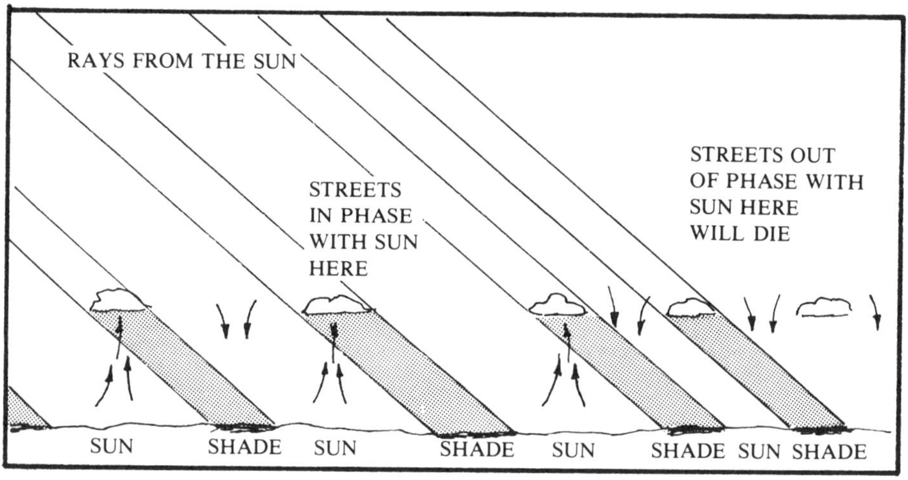

Figure 200 - Streets in Relation to the Sun

When flying in street conditions the ideal is to select a healthy street and fly underneath it, remaining as high as possible. Sometimes you can fly straight in abundant lift. Other times you have to step from thermal to thermal. In any case the sinking air between thermals in a street is much less than the sink between streets. The sinking air between streets can contain very strong downdrafts which have been known to reach all the way to the ground.

When crossing streets it is prudent to start as high as possible and take the shortest path to the best cloud in an adjacent street. You should expect to lose 1,000 feet (300 m) at a minimum in this venture. The strong sink between cloud streets can totally eliminate healthy ridge lift as described in

Chapter VIII. There is generally less sink between blue streets than cloud streets because the latter feature stronger circulations.

Sometimes cloud streets drift sideways when a crossing wind exists at the inversion layer. In this case it is possible to ride up the front of the barrier produced by the cloud and even above the visible cloud into the clear air wave aloft. This matter was also discussed in Chapter VIII.

Here is a summary of cloud street concepts:

Cloud Street Flying:

- *Use the cloud shadows to detect streets and locate the best street.*
- *Fly along the street as much as possible.*
- *If your route requires crossing streets, do so perpendicular to the street and aim for a good cloud. Expect to lose abundant altitude.*
- *Look for blue streets along mountains and in the free air. They are common in post-frontal conditions.*
- *Be aware that waves can exist above the streets, especially when the streets drift sideways.*

THERMALS AND CLOUD CHARACTER

We have seen that clouds based on thermals begin first in the morning and remain longest over the best thermal sources such as mountains. However, except for cap clouds over islands and high mountains, thermal clouds undergo a continuous decay and rebirth process. Figure 201 shows the typical life process of an isolated cumulus cloud. Normally it takes more than one thermal to build such a cloud that lasts twenty minutes or so from the first wisps to final dissoluting. As long as thermals feed it will grow. When thermals stop the cloud dries and dies.

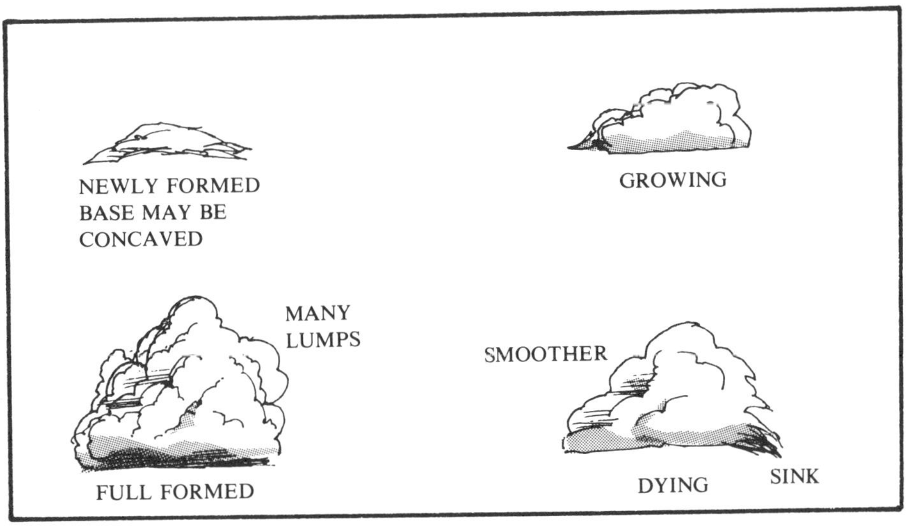

Figure 201 - Cumulus Cloud Life Cycle

Building thermal cloud in center of photo. A dying cloud is visible in the upper left.

In figure 202 we show a schematic of typical cloud shapes as they age. Note how the point of the triangle shape is upward in the growing cloud while it is downward in the dying cloud as drying takes place near the bottom first when the thermal lift stops. Drawings 1 to 3 represent growing clouds. Four and 5 show cloud decay. Be careful not to mistake the 5th drawing for the 1st.

Other signs of feeding or productive thermals clouds are dark, flat bases, sharp outlines and swelling or boiling cauliflower type heads on the clouds such as seen on the 3rd drawing. Signs of dying or disintegrating clouds are wispy outlines of the cloud, especially near the base, a poorly defined bottom and a reduction in size. Color is also an indicator of cloud health. A growing, robust cloud will be gray or white, bright or dark according to how it catches the sunlight. A dying cloud appears slightly off-color and may take on a yellowish or brown hue because the small moisture particles evaporate first when it begins dying and this changes the reflectivity.

The general strength of thermals and length of production can be cor-

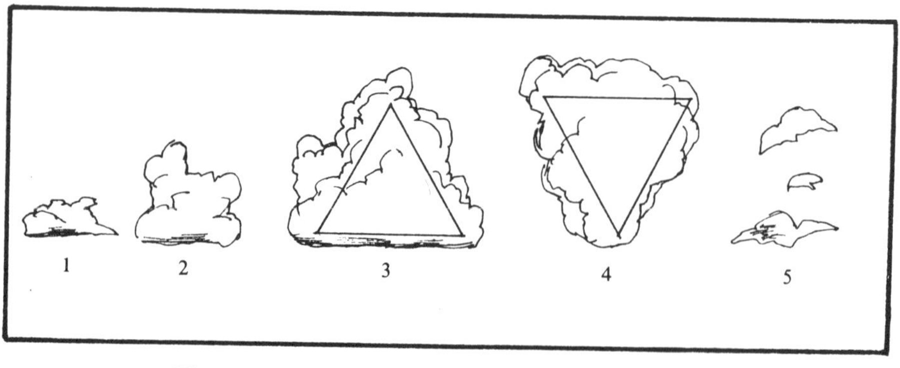

Figure 202 - Cloud Shapes as a Thermal Ages

related to how high and large a cumulus cloud builds. Stronger thermals surge higher in the cloud and longer lasting thermals make it larger. However, the character of the air has something to do with this, for many flat thermals are caused by the blockage effect of an inversion layer. Figure 203 shows the development of a thermal created cloud with and without an inversion layer present.

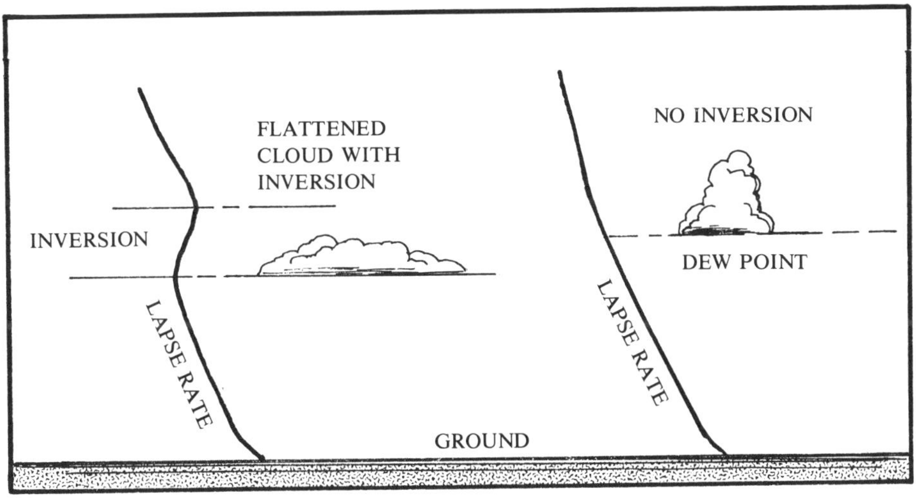

Figure 203 - Thermal Clouds With and Without an Inversion

When too much moisture builds up at the cloud level, either because the thermals are moist or an inversion layer holds the moisture in a narrow band, the clouds will spread and turn into stratocumulus forms as mentioned earlier. Thermals may still be found in these conditions if some sun is allowed to peep through or the cold air is moving over warmer ground. In that case the darker areas of the base (and areas of towering cloud if they can be seen) are the places to find thermals as shown in figure 204. The over-spreading of cumulus clouds is known as overdevelopment or O.D.

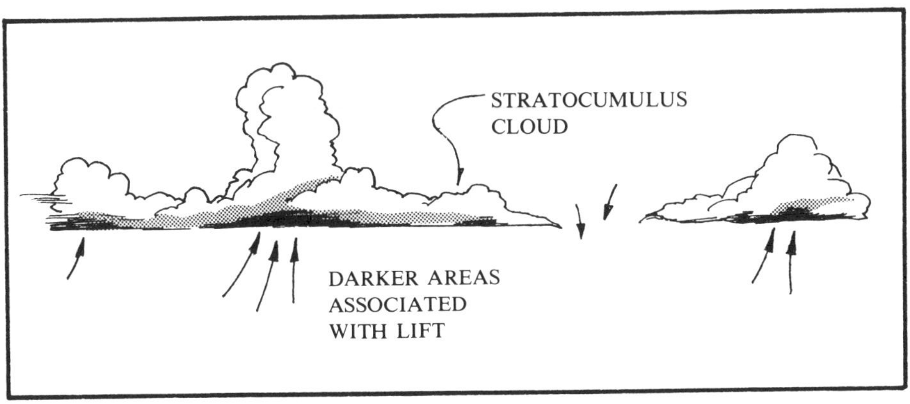

Figure 204 - Lift Under Spreading Cumulus

Spreading cumulus in wind. Darker bases indicate better lift.

NEAR CLOUD FEATURES

Close to thermal clouds the lift can get suddenly stronger in a phenomenon known appropriately as *cloud suck*. Many a thermaling pilot has found themselves fighting to stay out of the clouds. Cloud suck seems to occur most commonly in low pressure weather and especially in humid conditions.

At times a haze or veil can be seen below a growing thermal cloud. This haze is a sign of vigorous climb beneath the cloud as the air in the thermal is rapidly cooled by its rise. Some of the pollution particles that promote condensation take on water long before the dew point is reached and become visible when they gather enough moisture, thereby causing the haze. Such subcloud haze occurs beneath thermals above towns and other pollution sources and are generally signs of good lift.

Often the opposite of cloud suck occurs. That is, lift lightens exceptionally below the thermal cloud so that it is difficult to climb the last few hundred feet (100 m) to cloud base. The cause of this frustrating state of affairs is two mechanisms. First, clouds are not always fed by ground thermals. They frequently suck some of the surrounding air into the base and thus are self-sustaining as this air condenses and releases heat of condensation (see figure 205). It will be difficult or impossible to climb under such clouds. The bad part is that not all healthy-looking cumulus clouds can be depended on to have a useable thermal feed.

The second mechanism is: the general sink produced by rising thermal warms as it compresses and often produces a layer of stable air just below the cloud level. This *subcloud layer* as it is called can be 500 feet (150 m) thick and is shown also in figure 205. Once the thermal reaches this layer it can be disorganized and turbulent as in any inversion. Only by staying tight in the best core can we expect to continue climbing.

WIND EFFECTS ON CLOUDS

We have seen previously in figure 36 how the wind alters the cloud

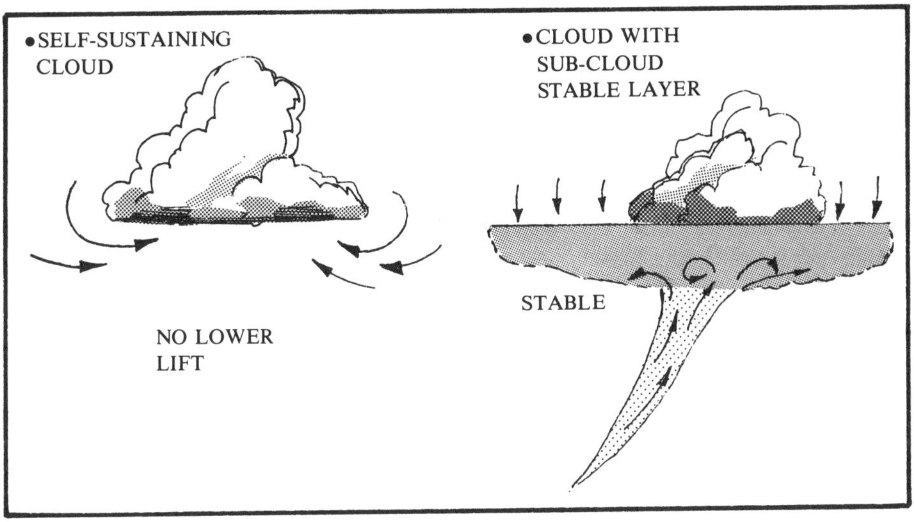

Figure 205 - Self Feeding Cloud and Sub-Cloud Layer

shape. Such a wind can create continuous thermal production as it causes the downdrafts to be located on the downwind side of the cloud and the updrafts on the upwind side. In figure 206 we see how downdrafts can generate a new thermal and distort the shape of the cloud. Figure 207 shows the frequently viewed appearance of a cloud over a continuous thermal source in wind. Here the thermals enter the cloud on the upwind side and are blown downwind as their cloud tower erodes. They are then replaced by a new thermal.

An important feature of clouds in wind is the barrier effect. We mentioned this before in Chapter VIII when discussing waves produced by streets. Here we'll expand on the idea to note that whenever a thermal is rising in a wind that is increasing aloft it moves slower than this wind and thus acts like a hill (single thermal) or ridge (a row of thermals) to create

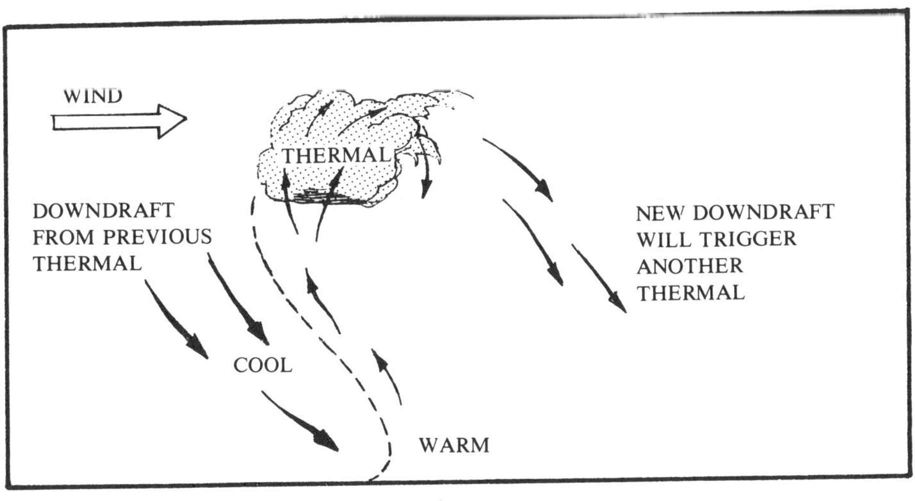

Figure 206 - Thermal in Wind Triggering a New Thermal

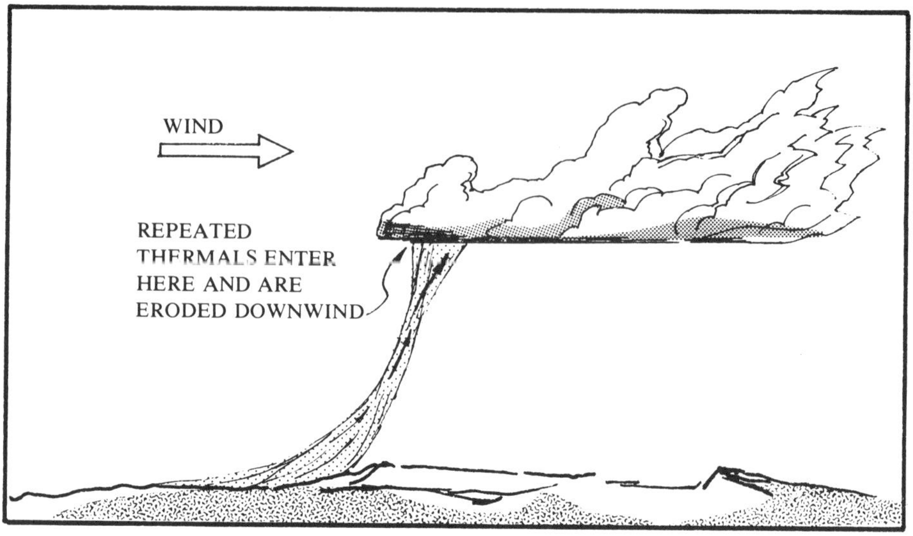

Figure 207 - Repeated Source Thermal in Wind

lift in front of the thermal as shown in figure 208. After the thermal forms a cloud is the easiest time to find this lift.

To exploit cloud barrier lift we must search in the upwind side of the cloud. As cloud base is approached we move toward the front edge of the cloud and hope for lift. It helps if a fresh thermal meets you near cloud

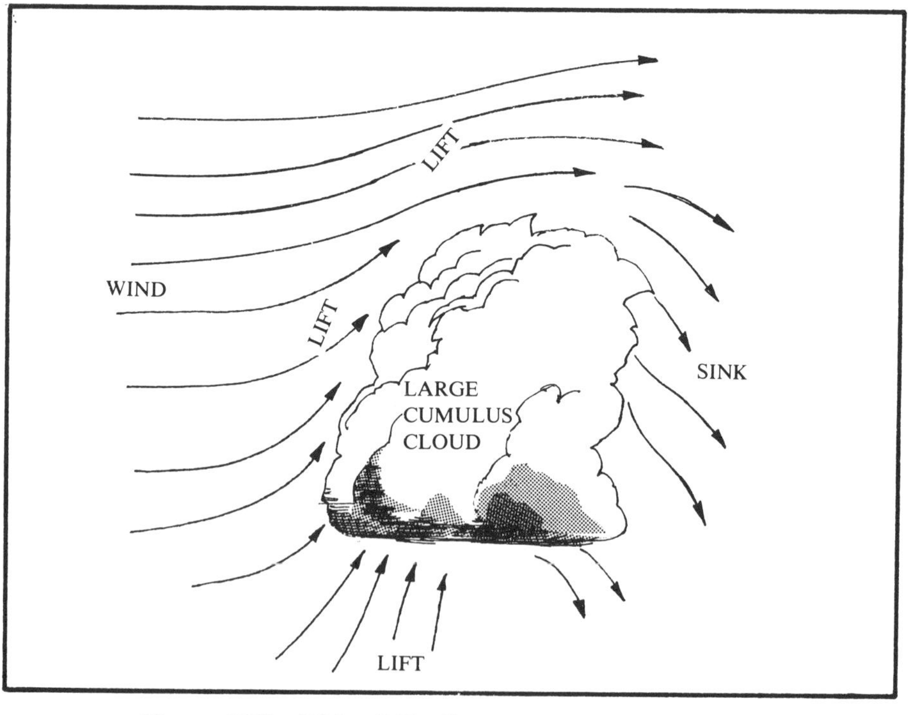

Figure 208 - Ridge Lift Created by a Cloud Barrier

base to boost you up the cloud. Many pilots have known the special thrill of climbing up the wall of a cumulus cloud in smooth lift. Such lift is light but rewarding and not all that common so we should savor the experience when it occurs.

High winds shredding the downwind side of clouds.

LOCATING THERMALS

Finding thermals is like a treasure hunt: sometimes the clues are wide open and apparent; other times we only get hints as to their presence. Frequently we have no guidelines at all and must trust to dumb luck and blunder around until we find a core. However often thermal clues do appear but are either subtle or complicated so we need to understand them if we aren't going to overwork our guardian angel.

We are going to divide the sky up into thirds from the earth's surface up to cloud base or the maximum height of thermals. We will discuss thermal clues in each third.

LOWER ONE-THIRD

Close to the ground we mostly rely on terrain features to locate thermals. We have previously seen that high ground is our best bet. Valleys between long ridges or mountains tend to be sink holes during the heat of the day. Ridges and mountains collect thermals drifting on the wind and produce thermals readily because of upslope breezes and sun-facing slopes.

Also we have learned to look for types of ground cover that heats readily, recognizing that different thermal sources produce at different times of the day. For example, rocks, towns and to a lesser degree trees and water release their heat at night. Even mountain lift is variable as the sun moves and heats different slopes.

At low altitudes smells, floating debris and other flying creatures locate thermals. Farm smells, smoke and factory smells are occasionally found in the air indicating the presence of a thermal or the wake of an old one. For instance, a particular chemical factory makes a particular odor in Pennsylvania that always signifies a thermal.

In 1988 this author was flying near Bright, Australia and encountered airborne fertilizer that caused excessive tearing and near blindness. The experience was unpleasant but the fertilizer sure pointed out a good thermal! Other debris more welcome are seed fluffs, leaves and even paper that gets picked up by strong thermals.

Birds, butterflies and other pilots are good visible clues to thermal location. Hawks, vultures, eagles and ospreys are frequent thermal partners, but this author once shared a thermal with two migrating great blue herons. Monarch butterflies in migration are frequent denizens of thermals in some seasons.

Close to the ground thermal tracks on trees up the slopes of mountains so adorned can locate thermals. Often the thermal process is cyclic, so knowing the cycle length and where the thermals last appeared can help locate another. Figure 209 illustrates how it is often best to return to your starting point to find another thermal once one has departed.

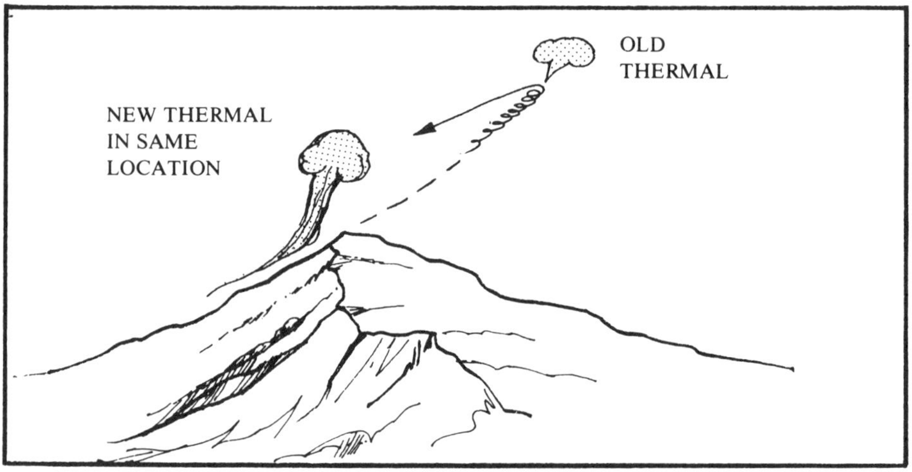

Figure 209 - Drifting Thermals and Repeating Sources

MIDDLE ONE-THIRD

When flying in the middle area of the airspace we have defined, it is important to use both terrain and cloud clues as well as in-air clues such as smells and thermal markers. When flying cross-country we set our sights on a distant goal and alter our path to take advantage of favorable cloud formations. Looking for patterns in the sky helps correlate the apparent lift with the terrain effects. It is no use flying to a mountain if it is not producing a cloud while all the others are!

Cloud shadows can be discerned from this middle altitude and they can help you locate the best clouds. This author once flew 50 miles (80 km) preceding a cloud that was creating a downdraft downwind that triggered off thermals. All the time climb rates were slow and we remained in the general level between the ground and the cloud.

UPPER ONE-THIRD

At this level we are mainly keying in on the clouds if they exist. We previously discussed signs of good cloud feed. To that we'll add the following points: A raggedy appearance indicates a dying cloud while a hazy appearance often means that ice is being produced (the cloud is above the freezing level). Ice formation doesn't particularly relate to cloud phase (except with thunderstorms as we'll see in the next chapter) or thermal strength.

Look for the best lift in clouds on their upwind portion, under their highest build-up and their darkest base areas as shown in figure 210. It is common to encounter multiple thermal cores under a cloud—some may be better than others. A concave or bulged upward base is a sign of strong lift. When flying towards a cloud watch for swellings on the top and sides as indications that it is still growing. Once the growing stops the thermal feeding the cloud has died.

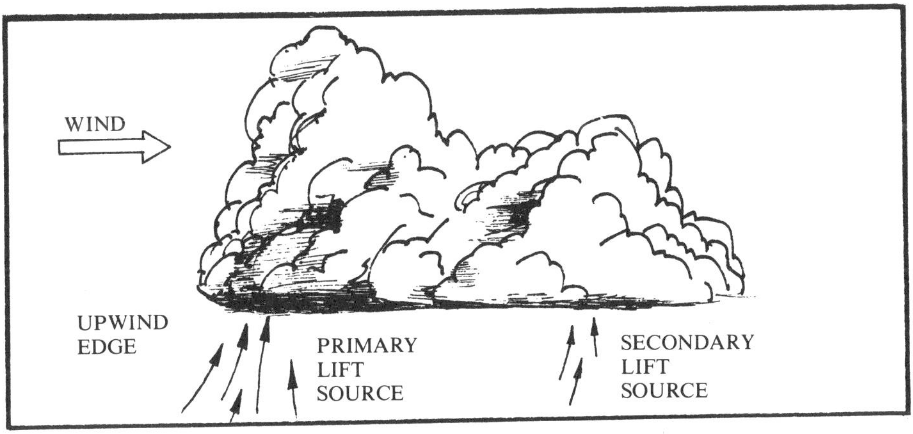

Figure 210 - Locating Lift from Cloud Appearance

Here is a summary of clues to good thermal producing clouds:

Cloud Thermal Clues

Look For:
1. Large puffy clouds of a white fluffy hue.
2. Growing clouds with swells or bulges.
3. Sharp outlines and flat bases.
4. Newly forming clouds.

Aim For:
1. The darkest and thickest looking clouds (except in thunderstorm conditions, of course).
2. The darkest area of a cloud base.
3. A concave (upward dished) base area.
4. The area with the highest cloud development.
5. The upwind area of the cloud.

Avoid:
1. Wispy clouds or areas of clouds.
2. Diminishing clouds.
3. Off-color or "dirty" clouds.

When no thermal clouds are around we must rely on terrain guidelines no matter what height we achieve. At higher altitudes, however, we should look farther ahead to pick likely thermal sources. Also, we should be aware of the possibility of cloud streets and carefully choose a track parallel to the wind if possible once we leave a thermal.

BLUE HOLES

All too frequently on thermal days we encounter areas of no clouds. These are aptly called *blue holes*. Lift in blue holes is likely to be weak, sporadic or non-existent. Invariably, it is better to skirt around a blue hole than go through it if possible. If your track takes you through a blue hole you'll probably end up using terrain features to locate lift since you'll most likely sink for a long way before you find a useable thermal. For this reason, it is wise to aim for a likely thermal source from the beginning of your blue hole crossing. Expect to find only a few weak thermals in the blue expanse even though the whole area is in sun. Working everything you can is usually the way you'll get cross the dreaded area.

Blue holes are often what we can term the nonproductive areas over valleys. Blue holes also exist over flat terrain and can be related to an upper air wave effect, poor thermal production of the surface or the fact that some areas begin thermal production earlier than others and thus suppress thermals in the late-starting area due to subsidence. Blue holes are common in areas with light winds. In these situations the blue hole rarely fills in so waiting for a change usually doesn't pay off.

SUMMARY

In this chapter we have discovered more about thermals. If anything, we can say they are complex beasts. They live and breathe for only a short time, but in that period they make their presence known by leaping into the sky and sprouting a crown of cumulus. We find that each thermal, like a star or snowflake, is unique. We ride them in wonder, awe and respect but also with a little pride for it is our understanding and skill that guarantees the most success.

Thermals are lift opportunities. They do not occur for simply no reason. There are natural principles involved governing the presence or lack of thermals. Soaring pilots especially must understand these principles, but non-soaring pilots should also learn about thermals if only to avoid their turbulence and downdrafts.

We have completed our study of lift and benign influences of the air. We next turn our attention to overgrown thermals and what these growing problems portend for pilots.

CHAPTER XI

Thunderstorms

We have all marveled at the spectacular show a thunderstorm presents as it marches across the sky, booms through a valley and strikes with violent bolts of sound and fury. Our marvel turns to abject fear, however, when we are confronted with the same spectacle while flying. Thunderstorms are a serious threat to all aviation and when we aviate with small, bouyant craft we are especially vulnerable to the whims of these storms.

Like most of the deadly surprises found in nature we can greatly reduce the direct threat of thunderstorms with a little knowledge, vigilance, caution and a healthy amount of avoidance. In this chapter we will cover the basics needed to understand how thunderstorms work, then explore the methods of judging their potential danger and escaping their long reach.

WHERE AND WHEN

Thunderstorms, most simply, are overgrown thermals. In certain conditions a thermal keeps getting bigger and bigger, devouring every speck of warm air in its vicinity then moving on to find more. Eventually the storm dominates the weather in its area then creates its own weather.

Thunderstorms appear in almost all areas on the globe at some time, although certain localities are more prone to thunderstorm production. For example, the lower Midwest of the United States with its overabundant supply of Gulf of Mexico moisture experiences frequent thunderstorms. Also the northeastern United States sees plenty of thunderstorms when cold fronts push under the warm humid air of summer. These areas with the most frequent thunderstorms also tend to suffer the most violent ones. Europe too is assailed by thunderstorms but they are less vigorous than those in North America because of the general toning down of temperature extremes in the overall maritime climate.

While some areas rarely see a thunderstorm and others like some tropical locales get one every day, most of us have seen enough of them that we are aware of their threat if not their causes. It has been estimated that at any given time on the earth over 2,000 thunderstorms are occurring which add up to 45,000 a day.

In dry regions thunderstorms tend to occur during the day when the powerful heating sends up untold tons of air. Often such storms are related to a disturbance aloft or an incursion of moist air. In more humid regions thunderstorms can occur day or night when a cold front moves through the area, lifting the moist air. Since fronts tend to slow down during the night, they can cause thunderstorms to linger for hours in an area after nightfall.

THE CAUSE OF STORMS

Thunderstorms develop from normal thermal conditions when the air is *sufficiently unstable, laden with moisture and some triggering mechanism is present.* The first two criteria seem to be the most important, for while cold fronts or mountains help trigger thunderstorms, they are not necessary and the normal thermal process can build into an isolated thunderstorm given enough instability and moisture. The instability may be caused by radiative cooling aloft, surface low pressure influence, a trough aloft or extreme surface heating.

Humidity is necessary for thunderstorm production because it is the release of latent heat of vaporization into the cloud when water vapor condenses that provides the energy for the storm to build. This latent heat also fuels tornados, hurricanes and other strong winds. Water vapor is also important because humid air in the lower levels absorbs heat which adds to the instability near the ground.

A THUNDERSTORM LIFE CYCLE

We typically divide the life of a thunderstorm into three parts: building stage, mature stage and dissipating stage. Let us look briefly at the important features of each stage.

Spreading cumulus building to thunderstorm proportions.

BUILDING STAGE

A thunderstorm begins with thermal build-up that overdevelops. This type of overdevelopment is not the spreading kind, but tends to climb vertically. Indeed, if an inversion layer or even a layer of dry air is present aloft thunderstorms are not likely to develop since the vertical extent of the clouds will be limited. The difference in a normal strong thermal cloud and a thunderstorm cloud is the latter keeps building because of the plentiful supply of humid air near the ground being forced aloft.

As the thunderstorm cloud grows in size it reaches a height where it becomes a "heat pump" just like a fireplace chimney. Cooling takes place at the top where cloud is evaporated while warming continues below. The result is strong updrafts that suck up warm air from below and to the sides and pump it aloft. Above a certain size such a cloud is self-stoking and continues to grow into the mature stage unless the supply of moist air at the ground is cut off.

Figure 211 shows the thunderstorm's building stage. At this point updrafts will be light to strong depending on the size of the storm, the

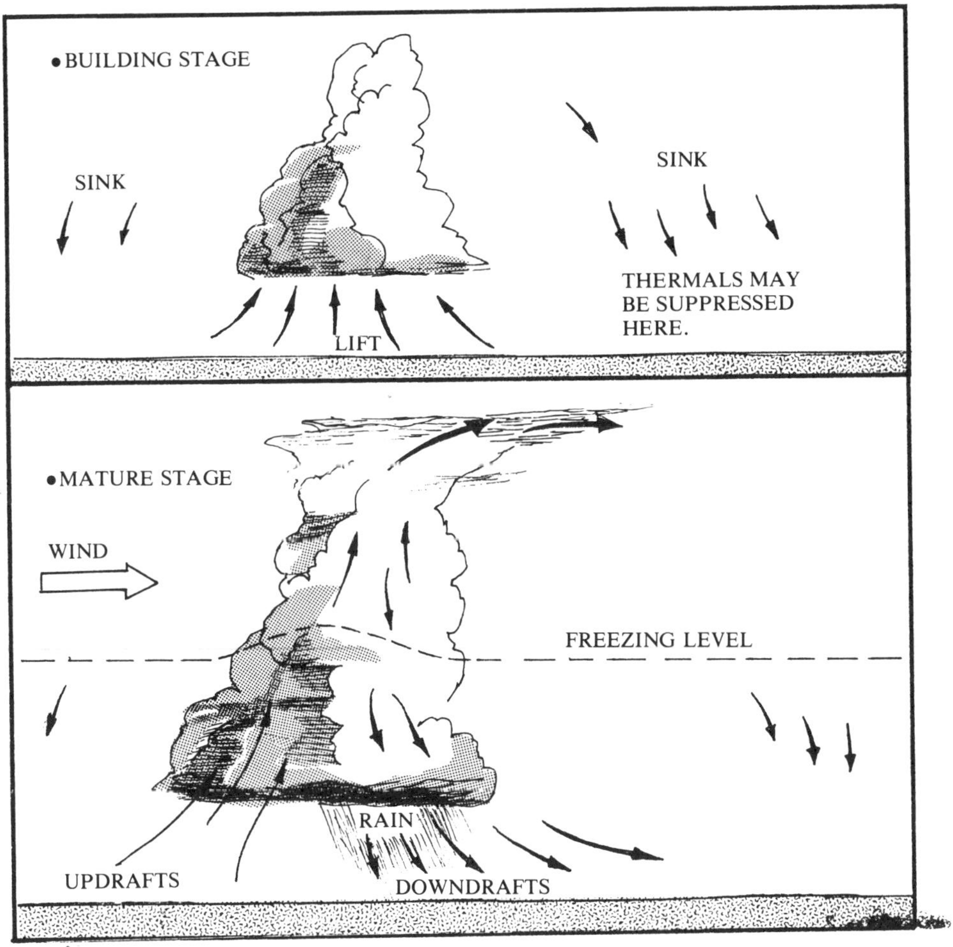

Figure 211 - The Building and Mature Stages of a Thunderstorm

strongest being that of the best thermals in the region where the storm is developing. At this stage the thunderstorm doesn't influence the local winds too much but it may begin shutting down lift for quite some distance around it due to the wide area of sink it creates.

MATURE STAGE

The second stage of a thunderstorm occurs when the upper reaches of the cloud climb well past the freezing level as shown in the figure. Ten or 15 minutes after this occurs, ice crystals grow to hail and they begin to fall when they are large enough to overcome the updrafts. Continuous upward gusts reaching 2,000 fpm (10 m/s) can blow this hail back up causing it to grow until it is of window shattering proportions. It has been estimated that hail of baseball size at the ground must have been formed in vertical wind currents of at least 70 mph (112 km/h)!

At the mature stage a thunderstorm typically reaches 25 to 35 thousand feet (over 10 km). Some monsters extend to the tropopause and top out at 50 to 60 thousand feet (15 to 18 km). If a thunderstorm top reaches the jet stream its top will be blown away to form the characteristic anvil head as shown in the figure. This action results in a sudden increase in cooling due to evaporation which speeds up the maturing process. One sign of thunderstorm maturity and the attendant danger is the appearance of an anvil head. Thunderstorm dangers which we cover below are mostly present in the mature stage.

As the thunderstorm continues its development the cloud darkens with moisture to become a true cumulonimbus. The falling hail creates downdrafts that eventually reach below the freezing level. Hail and rain then fall as droplets coalesce into larger drops. The downdrafts get more powerful and eventually overcome the updrafts. Rain and hail fall to the ground. The up and down motion in the cloud transports electrical charges and lightning begins at this time. Downdrafts and gusts are frequent. The updrafts share the cloud with the downdrafts but are now confined to the upwind side of the cloud as shown in the figure.

Dissipating thunderstorm. Note falling rain. Gust front has just passed.

DISSIPATING STAGE

As downdrafts continue to fall they bring cool air from aloft that spreads out around the storm. This and the cooling effect of precipitation tends to stop the heating of the ground, the updrafts are eliminaed and the thunderstorm collapses. Eventually the supply of moisture in the cloud is exhausted by blowing away at the top and falling rain as shown in figure 212. Lightning and downdrafts can continue in this stage.

The whole cycle as described takes a half hour to an hour or so with about 20 minutes spent in the mature stage. Some storms seem to last much longer, but it is common for multiple cells to occur and these longer-lasting storms are probably serial storms.

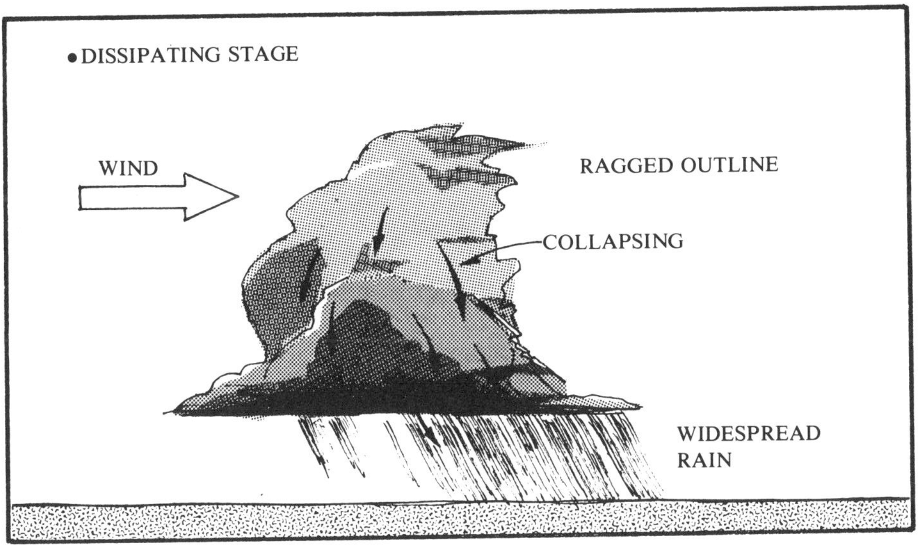

Figure 212 - The Dissipating Stage of a Thunderstorm

THUNDERSTORM VARIATIONS

Not all thunderstorms are created equal. Some are of course, more powerful than others and some are much slower to develop. Here we look at several possibilities.

Isolated Thunderstorms are those that develop in the middle of an air mass through convective heating of the ground, convergence under a low or the inflow of warm, moist air. These storms can occur day or night and can be some of the most severe in humid areas. Storms that develop over mountain peaks aided by mechanical lifting are also classified as isolated thunderstorms, but they can join up in chains over continuous mountains. These storms are most frequent in the afternoon and early evening because of the lift mechanism involved.

Imbedded Thunderstorms are storms concealed within a larger area of clouds, usually stratus type as shown in figure 213. At times the storm can be obscured by a general haze layer instead of clouds. These storms are often created in warm fronts due to the lifting of the warm air as the front progresses. Imbedded storms tend to be less severe because warm front lifting is slow and the excess of clouds reduces surface heating. However,

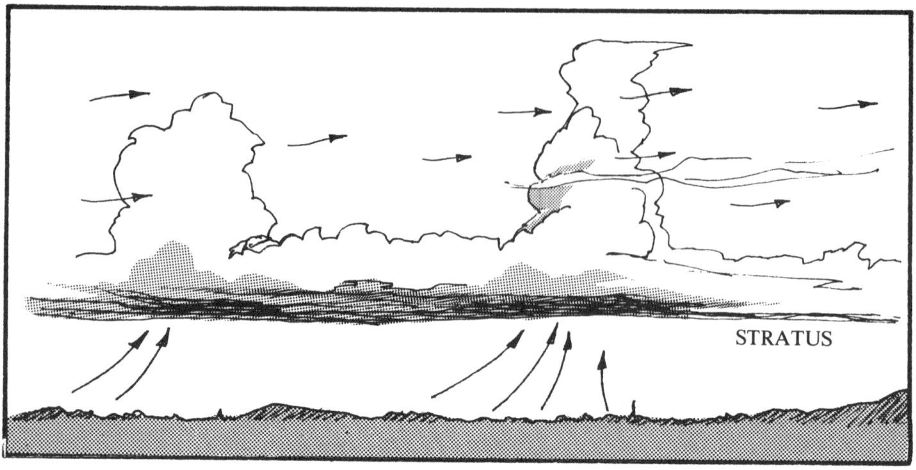

Figure 213 - Imbedded Thunderstorms

they can represent a grave danger to pilots for they can't be readily detected as can the larger more isolated storms. Often the sound of thunder is the first thing that tells us an imbedded storm is in the area. This may be impossible to hear in the air. We should expect imbedded storms any time haze or clouds obscure the sky and weather reports declare a chance of thunderstorms.

Squall Lines are a solid wall of thunderstorms working together like a football offensive line to march along pushing past anything in their way. These lines of storms are separate entities but are so close as to appear joined. Squall lines are cold front type conditions and the vigorous lifting in such a front guarantees that this type of storm is severe.

Squall lines can precede a cold front by scores of miles and appear to be caused by pulses that emanate from the front. Other squall lines are developed when a collapsing thunderstorm sends cold air spreading in front of it to act like a miniature cold front and lift the warm air existing at the surface. Squall lines can extend from 10 miles (16 km) to over 200 miles (320 km) in a fast moving front.

Thunderstorms that are produced by cold air wedging in under warm air have the particular property of arising very quickly. However, because they are constantly fed by warm air on the downwind side, as shown in figure 214, they tend to last longer and be very severe when they mature. Note that the initial downdrafts are on the windward side of the storm as opposed to the downwind side with the isolated storm shown in figures 211 and 212.

High Level Thunderstorms often occur in drier areas where the dew point is above 15,000 feet or so. The storms in this case are formed by low pressure disturbances aloft and are most active during the afternoon although they can be found day and night. Their distinctive feature is that the rain they drop rarely reaches the ground because it has so far to fall that it dries in the process. This evaporation cools the air through which the rain is falling and adds to the vehemence of the downdrafts. Rain falling in a veil partway to the ground from one of these storms is known as *virga*.

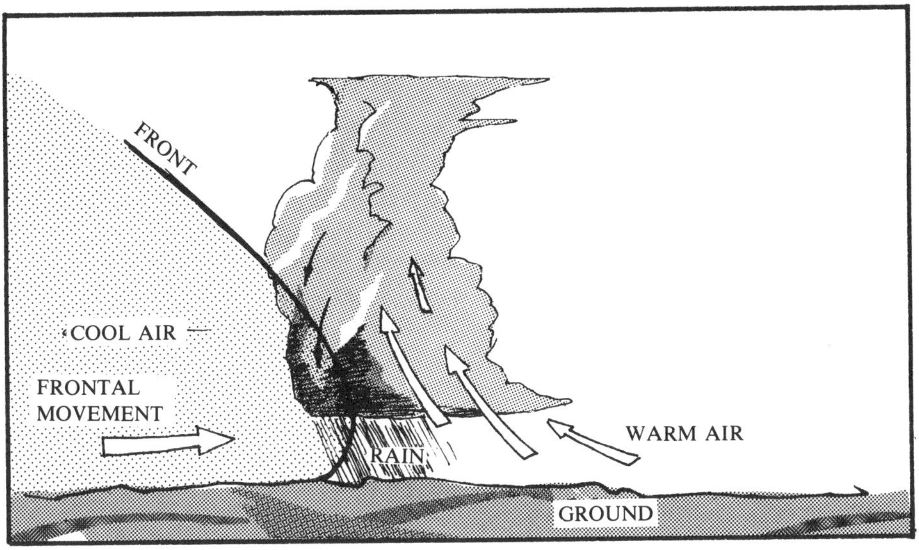

Figure 214 - Thunderstorms Due to Cold Front Movement

Two other thunderstorm characteristics should be noted here. The first is tilting of the thunderstorm by high winds aloft (see figure 215). As a result of this tilting the updrafts and downdrafts do not collide and the storm can build longer or last longer as building and collapsing occurs in cycles. Such a tilted thunderstorm is a sign of severity and high winds should be expected.

The second feature we'll note is the projection of a bench or foot from the thunderstorm. This bench is shown in figure 216 and is an indication that the thunderstorm is producing high winds. The bench usually precedes the thunderstorm and is an area of strong updrafts and tur-

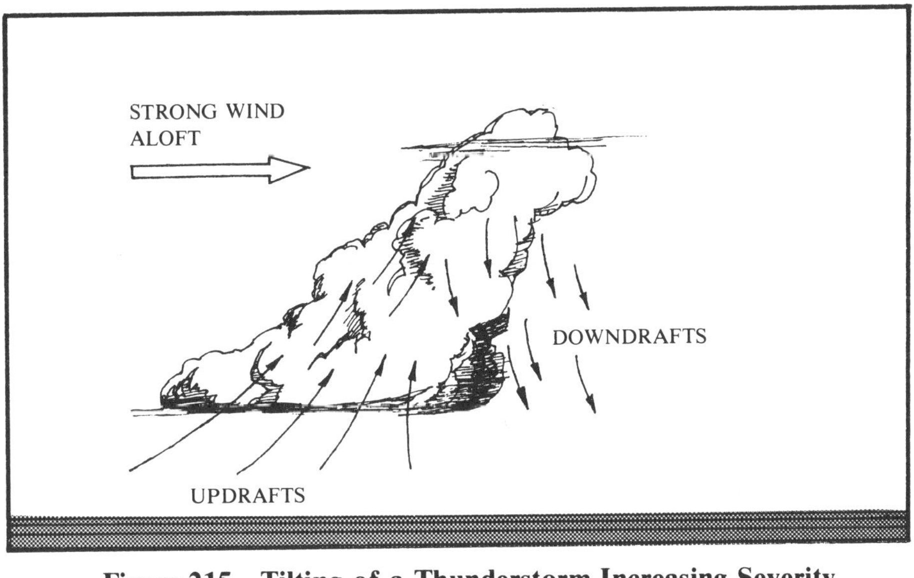

Figure 215 - Tilting of a Thunderstorm Increasing Severity

Mammata clouds beneath a protruding bench. These clouds are formed when the cold stable downdraft air fills in below the bench and mixes with the cloud layer.

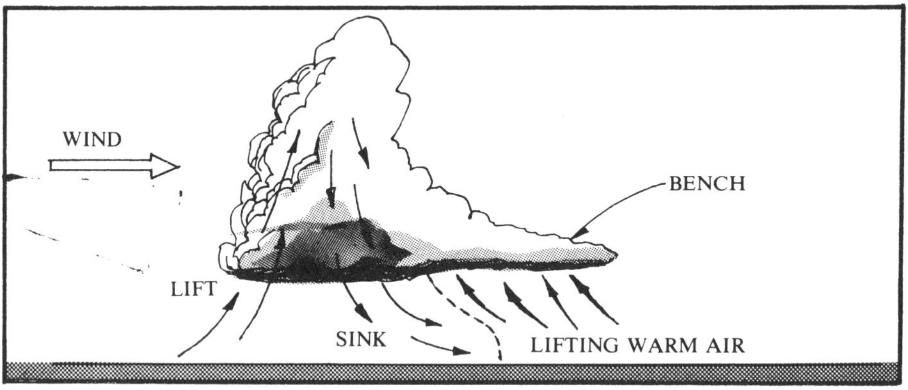

Figure 216 - Production of a Thunderstorm Bench

bulence as a gust front (see below) lifts the air. This author once had the dubious experience of skirting a thunderstorm below such a bench and flying 20 miles (30 km) in a straight line at a high rate of speed without losing altitude.

THUNDERSTORM DANGERS

Thunderstorms present real dangers to aviators. We'll review the dangers here then later see how to avoid them. They are:

Cloud Suck

The tremendous updrafts in a thunderstorm that can reach an incredible 100 mph (160 km/h)! can suck any aircraft up into the cloud. Then your troubles really begin. Once in the cloud you may suffer severe disorientation known as *vertigo*. Without full instrumentation (including a turn-and-bank indicator) a pilot in a turbulent thunderstorm will not be able to maintain control of the aircraft because there is no sure way of telling directions (magnetic compasses are useless once turbulence starts them swinging) or even which way is up. High forces from uncoordinated controls can break your aircraft.

In addition, as the storm carries you higher you may suffer *hypoxia*, which is a lowering of oxygen in the blood stream, until you become impaired, unconscious or a fatal statistic. If hypoxia doesn't do you in, freezing to death or *hypothermia* may. Remember, the temperature drops 5.5° per 1,000 feet as you climb. At 20,000 feet that is way below freezing..

The typical updrafts that a sport pilot can get caught in are not the maximums quoted above, for no one would be foolish enough to fly into such a monster storm, but it is easy to get caught in lift strong enough to make an escape difficult.

Turbulence

All the updrafts and downdrafts in a thunderstorm create considerable turbulence due to shear. All we have to do is think of the velocities involved and you can imagine the severity of the turbulence. Thunderstorm turbulence can (and has) tear apart airplanes.

Hail, Rain and Snow

Precipitation often comes all the sudden in a thunderstorm because the process of creating ice crystals above the freezing level is spontaneous and self supporting. Also, the sudden collapse of an updraft can allow the rain or hail to fall in a gush. The rain and snow can create severe visibility and icing problems. Hail can damage the aircraft and injure the pilot. Imagine flying in a sky full of golf balls.

Lightning

The processes in a thunderstorm cause a separation of charges that induce large voltage potential differences from place to place. When this potential builds up sufficiently it discharges to relieve the imbalance. This discharge is lightning. Such a huge spark expands the air and moisture explosively and we have thunder. Since thunder only occurs when lightning flashes, thunder is a good sign of a mature and dangerous thunderstorm.

The light from lightning travels to us almost instantaneously while the sound of thunder travels at the speed of sound. Since they are created simutaneously, we can judge the distance to the lightning and the storm by timing the difference between the lightning and thunder. The rule is: *there is 1 mile for every 5 seconds or 1 kilometer for every 3 seconds difference between the arrival of a lightning flash and thunder*. If a lightning flash is oriented perpendicular to you the thunder will arrive all at once in a loud clap. If it is on a radial away from you the thunder will arrive over a period of time as different parts of the lightning bolts' thunder arrive at different times. Echos and reverberations will also stretch out the duration of thunder. Thunder can never be heard more than 25 miles (40 km) away and rarely more than 10 miles (16 km). Thus, if we hear thunder we know the storm is close enough to affect us.

Cases of lightning striking an aircraft are rare since the aircraft must be in the direct line of the lightning. It is probably not possible for an aircraft to hold enough charge to attract a strike. However, corona discharge (the bleeding off of ions visible as a glow) known as St. Elmos fire can occur. When lightning does discharge there are typically electric currents flowing in various parts of the storm. These currents can be felt as anything from light twinges to painful shocks but are not in themselves dangerous or indicative of an imminent personal lightning strike.

Thunderstorm lightning tends to have preferred routes of discharge as shown in figure 217. The paths are numbered in order of their frequency. It can be seen that lightning occurs most often inside the cloud (1 and 2) which is a sign of a building storm. Ground strikes are a sign of a collapsing storm.

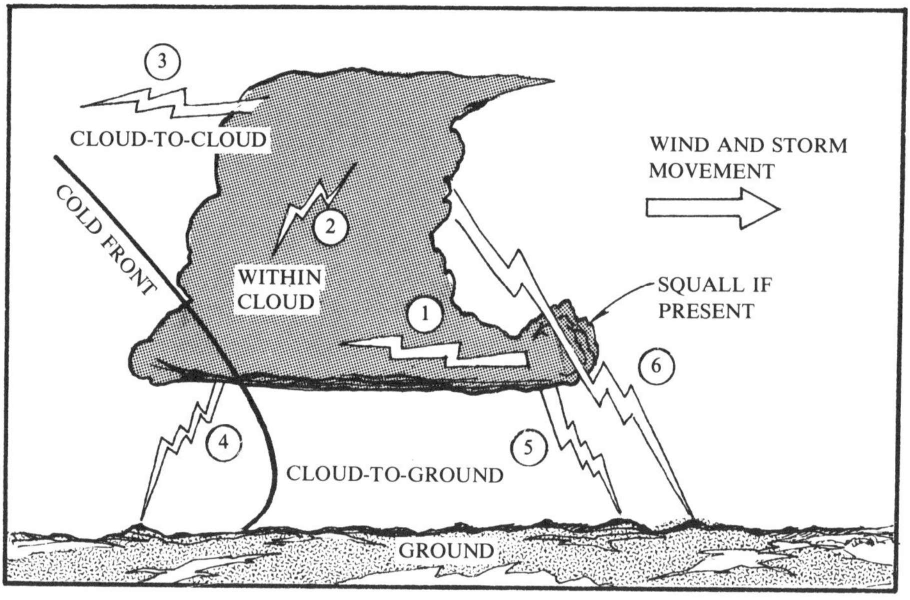

Figure 217 - Lighting in Thunderstorm

On the ground or in flight below the cloud the greatest danger is on the upwind side. Lightning also tends to occur most readily just above and below the freezing level. Lightning is not apt to begin until the cloud has climbed well above this level.

If you are on the ground when lightning threatens, avoid tall objects and metal objects (including aircraft). Seek shelter inside a car (not a convertible) or building. If you are in an open field, kneel down with your head on your knees and your hands on your head. Do not lie flat because you will increase your chance of encountering a strong potential and having a current pass through vital organs.

High Winds and Downbursts

The chance of encountering high winds is perhaps the most real danger that thunderstorms pose to aviators. We have already mentioned the incredible updrafts that can occur. Downdrafts can be up to 50% stronger than the updrafts. This important matter is discussed below.

Before we leave the list of dangers we should mention that a real problem with thunderstorms is that they can suddenly expand in any direction. You may be flying several miles from a rather mild storm and in a matter of minutes you may be part of it. This occurs most readily in humid conditions, and has happened to so many pilots that the only sure way to avoid such a fate is to give thunderstorms a wide berth.

DOWNDRAFTS AND GUST FRONTS

When precipitation begins in earnest from a storm it can drag literally tons of air down with it. The downmoving air allows the rain or hail to fall even faster than it would in still air so very high velocities can be reached. When this slug of cold air and water reaches the ground it spreads out like a tomato splattering on a wall as shown in figure 218. If a wind exists the spread is mostly in the downwind direction as shown. The leading edge of this cold, dense air is called a *gust front*.

The gust front acts like a miniature cold front, plowing under the warm air ahead of it, creating turbulent lift, clouds and shear turbulence. The gust front may advance in pulses as succeeding downdrafts give it impetus. Gust fronts typically extend 10 miles (16 km) or so downwind from a storm, but may remain in the confines of the storm if it is drifting very fast in which case they help feed the storm with the warm air they lift. Some gust fronts have been known to rush over 50 miles in front of a large thunderstorm. Downslopes aid their progress.

A gust front is no friend to aviators because it brings a shift of wind up

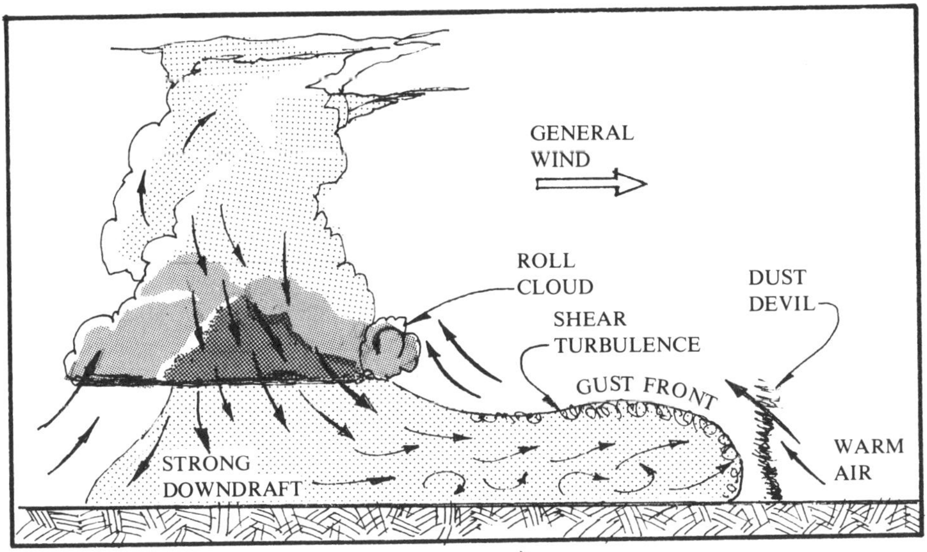

Figure 218 - Thunderstorm Gust Front

to 180° (typically 45°), an increase in velocity commonly around 30 mph (48 km/h) but occasionally several times this amount, and often severe turbulence. We can sometimes see the progressing edge of a gust front as it stirs up dust or creates dust devils. Over greener areas these signs are absent. Sometimes a telltale roll cloud exists above the gust front as shown in the figure.

The downdrafts that cause a gust front are themselves worthy of caution. In their most severe forms they are called *fallout, downbursts* or *microbursts*. They can slam an aircraft to the ground in the worst situations. Even though the air spreads out when it hits the ground, there is no cushioning effect for an aircraft when it falls faster than the air since it cannot change its direction as fast as the air.

When rain is falling without reaching the ground we must suspect the worst sort of downdrafts and gust fronts for the evaporation process creates cooler and thus heavier air as mentioned. As a generality, downdrafts that reach the ground occur 5 to 10 minutes after the peak updraft and the peak intracloud lightning. When precipitation is visible below the cloud, expect downdrafts and gust fronts to soon follow.

JUDGING THE STORM

It may be smart to set a policy of remaining on the ground when a remote possibility of thunderstorms exists. However, many potential good flying days will be missed with this plan. Also, some days produce thunderstorms by surprise. Therefore it is wise to learn how to distinguish the severity and immediate danger of a thunderstorm. By offering this information we are not condoning flying in or near a thunderstorm by any means. We merely believe in being prepared.

The factors to consider when judging a storm are:

Judging Thunderstorms

1. The rate of cloud build-up.
2. The extent of cloud vertical development.
3. The height of the base above the ground.
4. The cloud size (horizontal dimensions).
5. The cloud darkness, shape, position and movement.
6. The humidity of the surrounding air.
7. The presence of lightning and precipitation.

Cloud Build-Up

The faster a storm cloud grows, the sooner it will reach the mature stage. Powerful vertical currents are an indication of the energy contained in the local thunderstorm heat engine. With rapid build-up expect sudden collapse and powerful downdrafts.

Cloud Height and Base Height

A towering thunderstorm cloud is likely to encounter winds aloft that evaporate and cool it so that the eventual downdrafts are more vigorous. With greater cloud base height the rain beneath the cloud has more

Fractocumulus clouds near a spreading thunderstorm.

distance to evaporate and cool the surrounding air to increase its downburst potential.

Cloud Size

The more widespread the cloud the more area it influences and the more likely it is to produce dangerous winds. Also, a wide thunderstorm is more difficult to escape.

Cloud Darkness, Shape, Position and Movement

The darker the cloud, the more laden with moisture it is and the more imminent the lightning, thunder and fallout. When an anvil head forms the storm intensifies. When a bench forms it is a sign of high winds. Tilting of the storm cloud is an indication of a severe storm. A moving thunderstorm drifts over new ground and can pick up more moist air than a stationary one. Thus its build-up can be more rapid. A stationary thunderstorm is often self-limiting as it devours all the available heated air around it.

Humidity of the Air

In humid air conditions thunderstorms tend to be less severe although they are more often imbedded and therefore may impose a greater risk. More humidity means a lower cloudbase and a chance of precipitation before they get too large.

In dry areas storms build much slower and become much larger before they drop their load of rain. This author has witnessed a towering cumulus in the western US climb to over 30,000 feet (9 km) then gradually dry out due to lack of moisture aloft. This is not generally the case, however, and while drier areas exhibit fewer thunderstorms the storms that do develop are longer lasting and more powerful than those in more humid areas.

The occurence of lightning in a storm is a sign that it has neared the mature stage. It will soon be followed by rain and gust fronts as noted earlier.

The first six items in our chart above are the means for judging the storm severity. The first and last items help determine when the most severe danger is in store.

ESCAPING A THUNDERSTORM

We are all human and we all err, or worse yet we are tempted to push our luck. More innocently we can be blithely thermaling up beneath a benign thermal cloud that rapidly builds into a thunderstorm. For whatever reason we can imagine for encountering a thunderstorm, our knowledge of how it operates will help us escape its clutches.

The sooner we recognize the presence of a storm the better we are able to get away. If we encounter widespread lift of 200 fpm (1 m/s) in humid areas or 500 fpm (2.5 m/s) in dry areas we should suspect thunderstorm buildup. This is the time to look above for excessive cumulus build-up and start moving to the front edge of the cloud (upwind side). This places us in the best possible escape position.

When it becomes time to run from a thunderstorm by far the best direction to go is to the side (across the wind) as shown in figure 219. This escape direction is best to avoid the gust front downwind from the storm and the extensive rain and lightning on the upwind side as well as sudden growth of the storm cloud that most frequently occurs along the downwind edges. Unfortunately a thunderstorm can grow sideways in humid air. In this case the best escape route is across and downwind as shown in the figure.

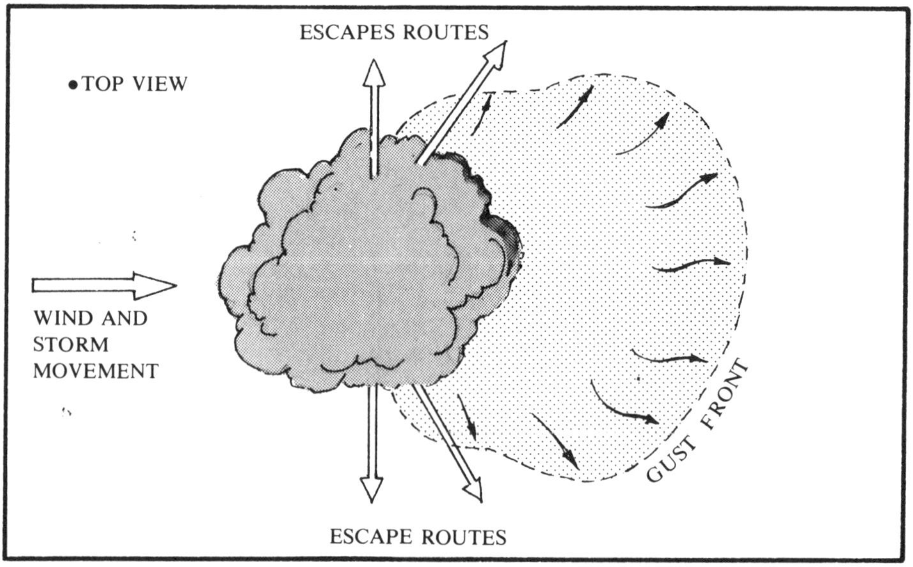

Figure 219 - Escaping a Thunderstorm

In the unhappy case that you get sucked up into a storm cloud the best thing you can do is fly at a moderately fast speed straight for the nearest edge of the cloud you spotted before the ground disappeared. Sometimes it takes several minutes to clear the clouds. Some pilots never do before dire consequences occur.

TORNADOS

Sport flying with any sanity should never occur when any chance of tornados exist. We only mention them here because they are intimately connected with thunderstorms. Tornados can be spawned by severe storms that are produced by convection, but they are most often associated with cold frontal passage.

Recent research has shown that tornados occur when large thunderstorms experience strong wind shear so that they are tilted and the downdrafts do not destroy the updrafts. The wind shear also causes a rolling action to the wind which becomes a vertical rotating mass when lifted in the middle by the updrafts. Finally tightening of the rotation through stretching as the column rises (the same mechanism as in a dust devil) increases the rotation velocity. Other factors occur to consolidate the rotation and bring the tornado down in its characteristic shape.

Tornados can contain winds of over 200 mph (320 km/h) and tend to skip along the ground in a northeasterly direction. Their great winds cause the damage for which they are notorious.

Tornados occur most often in the midwestern US where warm Gulf of Mexico air meets cold Canadian polar air to create spectacular storms. The rest of the world is not immune from tornados, however, for they are found in all states and countries.

Large thunderstorm in the building stages.

SUMMARY

We now have a good understanding why thunderstorms earn the respect of pilots worldwide. They possess incredible amounts of energy and make no bones about shedding it on an innocent countryside. Any flying creature that happens to be in the way will suffer. Ideally, pilots will avoid thunderstorms. However, every year a number of unfortunate souls have close encounters of the worse kind. We can only hope that our study here will stop such occurrences.

Thunderstorms come in many sizes and types. It is important for every pilot to know the general characteristics of the conditions in which he or she flies. Flat lands or mountains, dry or humid country, in the path of fronts or not—these are the parameters that determine the type of thunderstorms you will encounter. Once you recognize their style you can learn their behavior and increase your safety in the air. A safer pilot is a happier pilot.

CHAPTER XII

Watching the Weather

It wasn't until the 19th century that civilized man realized that certain cycles shorter than the seasons govern our weather. At that time the movement of fronts and the forces that drive them was postulated. Gradually study and technology increased understanding until we could actually make somewhat accurate predictions as to when our baseball games would be rained out. At first the lack of long-range communications and the necessary global view prevented the acquisition of the true nature of the weather. Now with satellites enveloping the earth and weather stations everywhere linked in a microwave network, we can paint the big picture. Our weather is no longer a mystery even if it isn't 100% predictable.

In the past couple of decades the weather information available to the individual has increased dramatically in most developed countries. Pilot weather services in particular have been expanded with the advent of home computers and more sophisticated telephone systems. In this chapter we review such sources of weather information as well as explore predictions based on our own observation of maps and the environment.

The more we learn about the weather, the more we can guesstimate its future behavior. It is this skill that sometimes delineates the adequate pilot from the accomplished pilot. The information in this chapter will help us put on the final polish to our pilotage skills.

WEATHER CYCLES

One of the most important things a pilot can do to become weather wise is to frequently view weather maps and correlate what is predicted with what actually occurs. After a few months of this it should be possible to do a fairly good job of predicting the general conditions day by day. After a few seasons it should be possible to pick and choose the best flying days.

Watching the weather over a long period of time brings out the appearance of cycles. The weather can't be expected to be regular, but at times six or seven day cycles get established that can succeed each other for a month or two. For example, a cold front will come through on Thursday followed by a warm front Sunday then another cold front the next Thursday and so on. Recognizing these cycles when they occur can be a great aid to the armchair weatherman.

In areas characterized by frontal passage, it is beneficial to understand the nature of post-frontal conditions. For example, after a warm front we expect southerly winds in the northern hemisphere and warm, humid air. After a cold front we expect northerly or westerly winds and clear air north of the equator. Figure 220 shows how the cold sector air conditions vary behind a cold front. Watching the progress of the high in relation to where we intend to fly gives us an idea of what to expect.

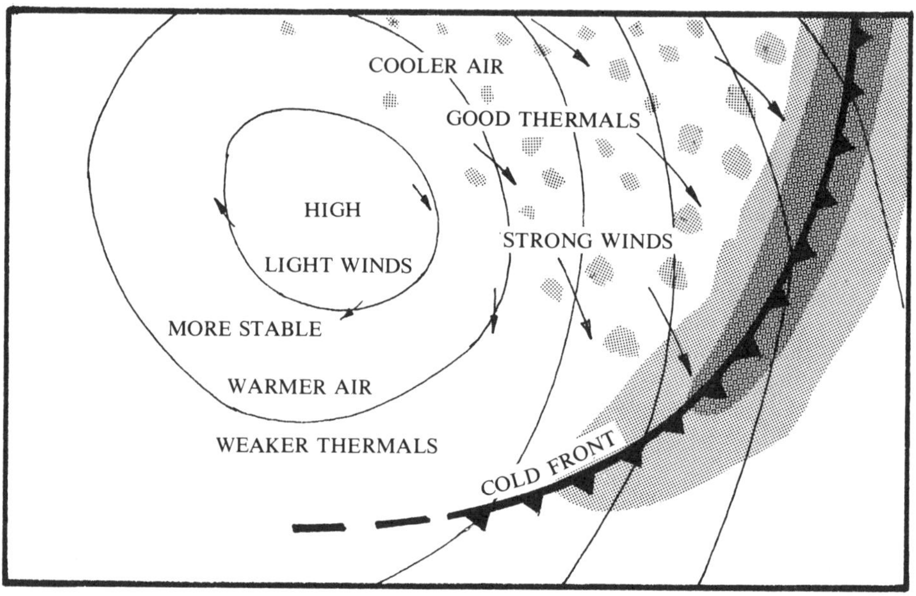

Figure 220 - Post-Frontal Conditions

We can apply this logic to other areas and other weather systems. The main points to remember are: *the weather is somewhat cyclic and the large-scale characteristics are repeatable.* When pressure systems and fronts are in the same general positions they were at a previous time we should expect about the same weather conditions. Remembering the results of past weather situations helps us understand the condition at the present. We should also recall that winds are generally light under a high and stronger under a low pressure system.

WEATHER AND THE BAROMETER

For the serious student of the weather a barometer is an invaluable prediction aid when used with a wind indicator. Pilots quite often use altimeters with barometer functions built in, so they are already equipped to predict the weather. The chart below gives you enough information to cover almost any situation. However, local effects such as sea breeze and valley winds must be considered.

This chart is designed for the northern hemisphere at sea level. For the southern hemisphere change all wind directions that read north to south and vice versa. Leave west and east directions unchanged. For altitudes above sea level, add 1/10 of an inch to your barometer reading for every 100 feet in elevation you are above sea level (add 10 mb for every 100 m above sea level).

Barometer/Wind Table

BAROMETER (REDUCED TO SEA LEVEL)*	*WIND DIRECTION*	*CHARACTER OF WEATHER INDICATED*
30.00 to 30.20, and steady	westerly	Fair, with slight changes in temperature, for one to two days.
30.00 to 30.20, and rising rapidly	westerly	Fair, followed within two days by warmer and rain.
30.00 to 30.20 and falling rapidly	south to east	Warmer, and rain within 24 hours.
30.20 or above, and falling rapidly	south to east	Warmer and rain within 36 hours
30.20 or above, and falling rapidly	west to north	Cold and clear, quickly followed by warmer and rain.
30.20 or above and steady	variable	No early change
30.00 or below, and falling slowly	south to east	Rain within 18 hours that will continue a day or two.
30.00 or below, and falling rapidly	southeast to northeast	Rain, with high wind, followed within two days by clearing colder
30.00 or below and rising	south to west	Clearing and colder within 12 hours
29.80 or below, and falling rapidly	southeast to northeast	Severe storm of wind and rain imminent. In winter, snow or cold wave within 24 hours
29.80, or below and falling rapidly	east to north	Severe northeast gales and heavy rain or snow, followed in winter by cold wave.
29.80 or below, and rising rapidly	going to west	Clearing and colder.

*30.20 inches of Hg = 1022.73 mb
30.00 inches of Hg = 1015,96 mb
29.80 inches of Hg = 1009.19 mb

The reason why the above chart works is that the barometer detects the approach or departure of pressure systems and fronts. The wind direction indicates where you are in relation to the highs and lows which further defines the expected weather. For example, the first entry on our chart with a high and steady barometer and westerly winds implies a high pressure cell is just to the southwest bringing its normally fair weather.

If your barometer is changing faster or slower than indicated in the chart but otherwise the barometer level and wind agrees, then the predicted weather will occur sooner or later respectively than the chart indicates. We must always take such predictions with a grain of salt of course, for the weather is a complex machine and one little nuance can throw the whole thing out of wack. A pressure cell can intensify, a front can stall or a volcano can blot out the sun. These matters are beyond our control and for the most part they occur rarely, so our chart is fairly reliable.

WEATHER WISDOM

Long before science provided some enlightment about the workings of the weather, the common man observed patterns and clues in nature that provided predictions with various degrees of reliability. Some of these we included in Chapter III. Others such as using insects to tell the temperature or severity of the weather are either of no use to us here or inaccurate. The few we include here are based on sound meteorological principles.

Weather Signs

Look for unsettled weather and rain when:

The barometer is falling.
The nighttime temperature is unusually high.
The clouds move in different directions at different levels.
A halo appears around the sun or moon.
Webby, disordered cirrus appear.
Thunderstorms develop in a westerly wind.
Summer afternoon clouds darken.
Cumulus clouds develop very rapidly.
Clouds (stratus or cumulus) become lower.

Look for steady precipitation when:

The weather is unsettled (see above) and the wind is south or southeast (north or northeast in the southern hemisphere) with the barometer falling.
The wind is southeast to northeast with the pressure falling.
Thunderstorms develop in a south to southeast (north to northeast below the equator) wind.

Look for clear weather when:

The barometer rises or remains steady.
The wind shifts into the west or northwest (southwest).
The temperature falls.
Cloudiness decreases after 3 or 4 pm.
Morning fog breaks within two hours of sunrise.
There is a light breeze from the west or northwest (southwest).
There is a red sunset.

Most of these general rules are self-explanatory given our understanding gained in previous chapters. We'll add here, however, that a halo around the sun or moon is caused by high, thin cirrostatus clouds refracting the light. These high layer clouds are a sign of an approaching warm front. As the front gets closer the clouds get lower which makes the halo appear smaller. The closer the halo is to the celestrial body, the sooner the rain will come.

WEATHER INFORMATION SOURCES

As we enter the 21st century technology has provided much greater access to information for the average person. Weather information is a dramatic example of this for in many areas the public can display current weather maps and other data on their computer screens through a subscription service. Pilots have additional sources of information through the weather services maintained specifically for aviation.

The following is a list of weather information sources in the order of their importance to pilots:

1. Flying weather service from airports and government maintained stations.
2. Personal computer services available upon subscription.
3. Continous broadcast weather radios.
4. Maps appearing in newspapers or at weather stations (at airports or universities).
5. Weather reports on the general radio.

Comments: Each of these items have a different accessibility in different countries. For example: no. 1 requires a pilot license in some countries while no. 3 is not available in many countries. You should inquire within the aviation community as to what is available in your area. Pilots are invariably up-to-date on the available weather services.

In the US, for example, the government runs a pilot weather system accessible by phone. It is called IVRS and is easy to use once you know the system. By pushing various code numbers on your telephone you can get a recording of any weather item you wish for any area of the country. To obtain a pamphlet explaining IVRS, call your local airport or Flight Services listed in the telephone book.

Computer systems can tap into the government weather service in the US. This system is called DUAT. Check in your country's aviation or computer magazines for access to this and similar systems.

Items numbered 3 to 5 above are designed for general public consumption. As such they are not detailed enough for pilots. However, with knowledge and practice we can ''read between the lines'' and get some useful information. It is important to know the general code words used in these reports. For example, in the US *light and variable* usually means thermal induced changeable winds with very little general wind. *Scattered clouds* or *partially cloudy* usually means thermal based cumulus clouds and good soaring. *Breezy* means winds over 20 knots and gusty. In some countries you can get a list of what the various standard terms mean from your weather service.

WEATHER MAPS

We have seen the way fronts, pressure systems and isobars are depicted on the weather maps in Chapters IV and V. These maps are typically produced every three hours and are sometimes available for the surface conditions, 850 millibars (mb) (about 5,000 feet or 1,500 m), 700 mb (about 10,000 feet or 3,000 m), 500 mb (about 18,000 feet or 5,500 m). The surface map tells us what weather we should expect in addition to winds near the ground while the upper air charts let us know what the wind conditions are aloft. Figure 221

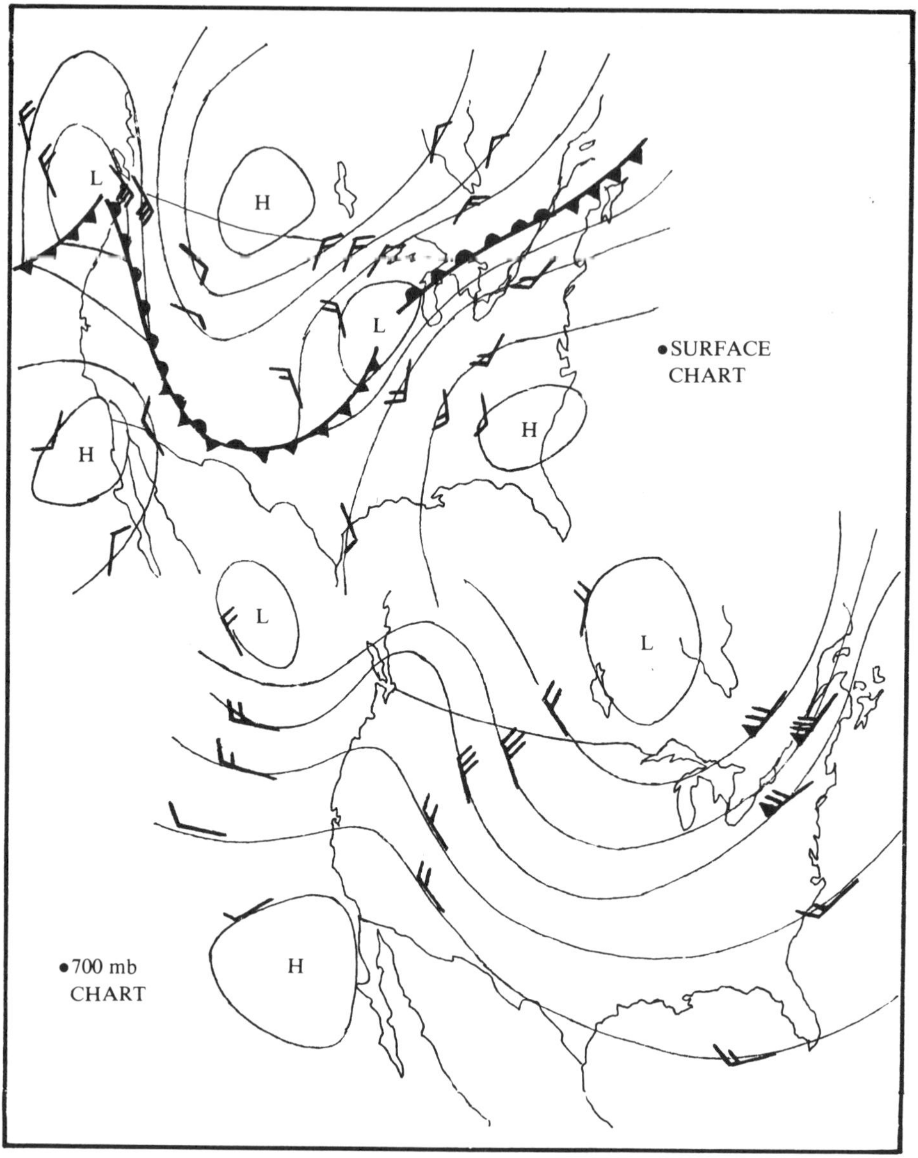

Figure 221 - Surface and 700 mb Level Charts

shows a typical surface chart for the US with that at the 700 mb level at the same time. Note how matters get simpler as we get higher.

From these charts we can gather the pertinent information: the presence of pressure systems and fronts as well as expected precipitation (rain occurs near fronts and in lows). We can also tell the wind direction and speed from the position and spacing of the isobars (see Chapter V).

The winds are very important to pilots and are usually shown directly on the weather charts. The way they are depicted is with an arrow and short "feathers" as shown in figure 222. The arrow points to the direction the wind is blowing. Our example thus indicates a northwest wind.

The marks on the tail of the arrow indicate the wind speed. A short mark is 5 knots, a long mark is 10 knots and a solid triangle is 50 knots (remember a knot equals 1.15 mph or 1.85 km/h). The arrow shown is indicating 25 knots. The 700 mb chart in figure 221 shows these arrows following the isobars.

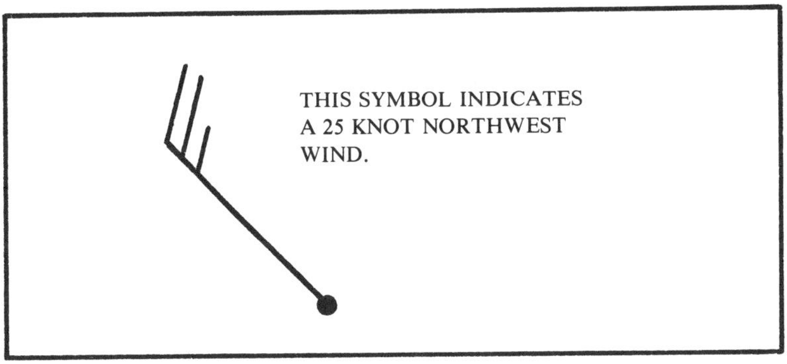

Figure 222 - Weather Map Wind Symbol

On surface charts each reporting station is placed on the map with a wind arrow and other symbols around it. These clusters of data can be very useful and are interpreted as in figure 223. Here we see the main features of the local weather depicted. We should learn to recognize the symbols for the wind (see above), the cloud cover and type and precipitation type. Also the temperature and dewpoint are useful to allow us to calculate cloud base height (see Chapter III). We should also note the barometer reading and trend in order to predict the near term weather changes.

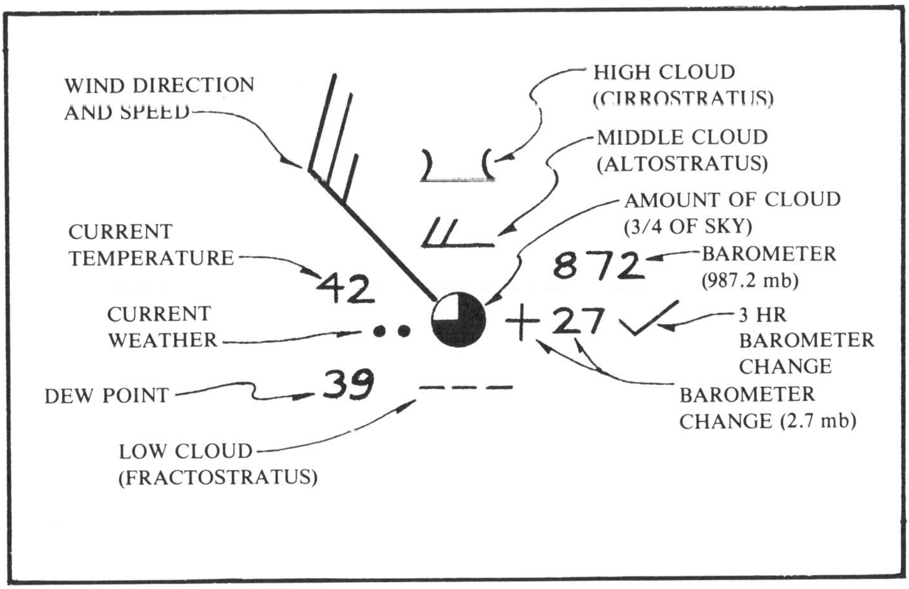

Figure 223 - Weather Map Symbols

The cloud symbol chart appears on page 38 in Chapter III. The symbols for various forms of precipitation appear below.

Weather Symbols

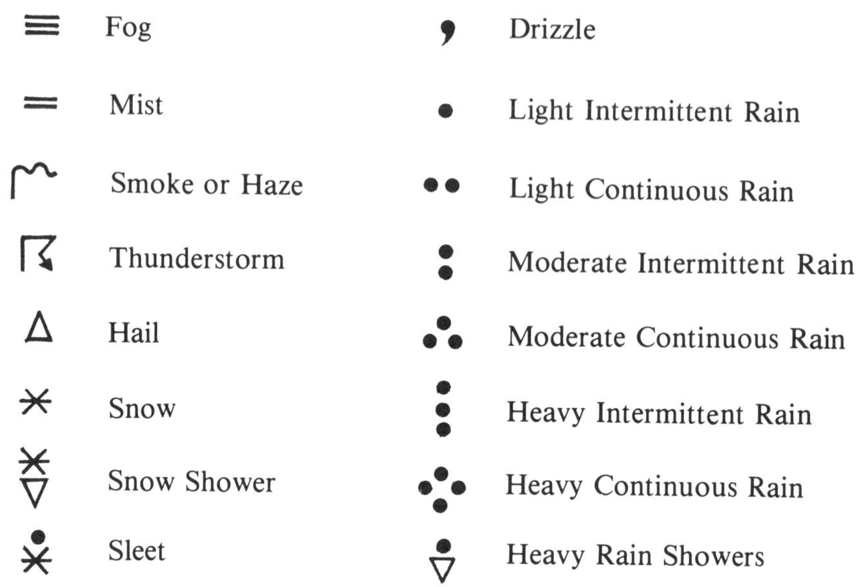

SOARING FORECASTS

Pilots who rely on up air to acquire airtime often want to know what their prospects are before they venture aloft. For this we use such matters as the thermal index, the lifted index, the K index and the trigger time. We will explain each of these.

The Thermal Index is a measure of instability using the expected maximum ground temperature and the current temperature aloft. Any temperature at altitude can be used but typically those at 850 mb (5,000 ft) and 700 mb (10,000 ft) are readily available. Now we take the expected ground temperature and subtract from it the temperature change we expect a parcel of air to experience as it rises to our given altitudes. As long as cloud isn't formed, this temperature is 5.5 °F per 1,000 feet (1 ° C per 100 m) multiplied by the altitude difference (in thousands of feet) between the ground and our chosen altitude. Then we subtract the calculated temperature at altitude from the measured temperature. This final value is our thermal index. An example will make matters clear.

Let's assume that the forecast high for the day is 85° F. The temperature at 850 mb or 5,000 feet is 52° F and that at 700 mb or 10,000 feet is 39°. Also let us assume that our altitude is 1,000 feet MSL. For the first calculation we are going from 1,000 feet up to 5,000 feet for an altitude difference of 4,000 feet. Thus we expect a parcel to cool 5.5° F times 4 (in thousands of feet) or 22° F. Subtracting this value from 85° gives 63° which is the calculated temperature at 5,000 feet. Now subtracting 63° (calculated) from 52° (measured) gives −11° F. This is the thermal index up to 5,000 feet.

Continuing onto the 700 mb level we see a difference of 9,000 feet from our ground altitude to 10,000 feet. This gives 5.5° F times 9 or 49.5° F as the temperature drop in 9,000 feet so we have 85° − 49.5° = 35.5° as the calculated temperature at 9,000 feet. Subtracting 35.5° (calculated) from 39° (measured) gives us +3.5°. The plus sign indicates stability since the actual air is warmer than a lifted parcel would be. Thermals will not reach the 10,000 foot level.

Negative values of the thermal index are needed in order for good thermals to develop. The more negative the index is the stronger the thermals. The amount of instability indicated by the thermal index depends on the altitude for which it is calculated. For this reason it is important to use the same altitude from day to day in your calculations so you can learn what thermal strength relates to what index value.

The lifted index is a special form of the thermal index. While the latter can be calculated for any altitude, the former is always calculated for 18,000 feet (5.5 km). Since the lifted index is at one altitude, we can readily assign it values. These are:

Lifted Index Rating

Lifted Index	*Atmospheric Stability*
10 and above	Very Stable
5	Stable
1 to 4	Marginally Stable
0	Neutral
−1 to −4	Marginally Unstable
−5 to −9	Unstable
−10 and below	Very Unstable

The lifted index is published every 12 hours in the US and is available from aviation weather services. It is most useful in the high desert areas where inversion layers do not occur. In moister areas with lower cloud bases, calculating a thermal index will give a more accurate indication of thermal strength in the lower levels.

The K index uses the above calculation method in addition to the moisture present to predict the probability of thunderstorms. The table below gives the values and their meaning.

K Index Rating

K INDEX	*THUNDERSTORM PROBABILITY*
15-20	Less than 20%
21-25	20-40%
26-30	40-60%
31-35	60-80%
36-40	80-90%
Over 40	100%

We can combine the lifted index and the K index to get an overall view of the weather and soaring prospects. This appears in the following table:

Forecast From Indices

LIFTED INDEX	*K INDEX*	*PROBABLE WEATHER*	*SOARING FORECAST*
Zero or negative (unstable)	High (humid)	Instability; showers or Thunderstorms.	Turbulence; may be dangerous; clouds may prevent soaring.
Zero or negative (unstable)	Low (dry)	Limited cumulus activity; little if any precipitation.	Bumpy but not dangerous; good for thermal soaring.
Positive (stable)	High (humid)	Stratified cloudiness; steady precipitation.	Smooth air; no thermals.
Positive (stable)	Low (dry)	Predominantly fair.	Smooth air; weak thermals, if any.

The final factor in our soaring forecast is the trigger temperature. As mentioned in Chapter IX, a ground inversion often exists to stop thermals until they are strong enough to punch through the inversion or the inversion itself is eliminated. The time at which this happens is very important for it determines when usable thermals will first arrive at a mountaintop. As shown previously in figure 174, the approach to trigger temperature is a gradual one. If we know this trigger temperature, the morning lapse rate and the rate at which heat is supplied to the surface, we can calculate the trigger time. The trigger time information is also often available from weather services. We show how to calculate it in Appendix V, given the lapse rate profile.

READING AN AREA

Like Sherlock Holmes, the sport pilot should be concerned with details when preparing to take a flight. He or she must take into consideration both the land forms and the weather. Pilots flying a familiar area will most likely have a good idea of what to look for and will be most concerned with what the winds are doing. Often a simple weather report of wind speed and direction will be sufficient to determine whether or not a well-known area is suitable for flying.

The wind velocity can be checked on your way to a flying site, although it is most important to reevaluate the judgement at the point of flying. Essentially this factor will be continously watched by an aware pilot. Note that when mountain flying, slopes that face the prevailing wind will be most reliable while other slopes will be less predictable.

Going hand-in-hand with the wind velocity is the stability of the air. This can be obtained from weather reports, calculations (see above) or observation. Stratus type clouds indicate stable conditions, while cumulus clouds are indicative of instability. Bright, sunny days are usually unstable to some degree. If the air feels cold and the sun is bright, instability should be expected. Smoke remaining close to the ground and general haze denote stable air while smoke rising to great heights and good visibility signifies the opposite. Soaring birds and rapid changes in wind direction are signs of instability and thermals.

We can largely reckon the amount of turbulence present by noting the instability and the strength of the wind. Variations in wind socks, wind meters, water surfaces and foliage can fortell gusts that are the effects of turbulence. Shear turbulence can be evaluated by noting any large differences in wind velocity at different places or altitudes.

Wind gradient should be taken into consideration by noting the difference in wind at the surface and thirty or forty feet up. A significant change should warn of extra required speed on landing. At the same time, changes in wind direction aloft should be noted to determine probable thermal drift.

The foregoing is concerned mainly with safety factors. The experienced pilot will also look for ways to capture lift. Checking slope directions with respect to the current wind direction will greatly enhance our ability to utilize ridge lift and find thermals drifting up a mountain. The location of cumulus clouds and potential thermal generators should be noted at this time as well. The possibility of convergence should always be considered by looking for terrain effects and cloud clues. Finally, we should do the same thing for waves to be ready for their abundant lift if they appear.

SITE EXAMPLES

In order to practice our ability to ''see'' what's happening in the invisible air, let us examine several hypothetical situations. The pictures in figures 224 to 226 indicate many terrain effects and conditions. The drawings consist of widely varying weather—the first being unstable in lighter winds, the second stable with light wind and the third stable with higher wind.

Looking at figure 224, we see the smoke in the valley at A gives an indication of wind speed and direction (light and along the valley) as well as the bouyancy of the air. The smoke breaks apart from the light turbulence and expands as it rises. Not much wind gradient is expected since not much wind exists. Thermals are present as can be told from the isolated cumulus clouds at B.

Rough air is expected everywhere due to the thermals, but strong turbulence is only encountered near the arid hot patch C and the towering cumulus D. The thermals above C are small and intense while the thunderstorm D creates gusts which reach the ground and spread. As the thunderstorm marches up the valley it can bring stronger winds and trigger of dust devils in front of it.

Lift is expected at points marked E as these correspond with good thermal producing surface areas. In time the thunderstorm advance may shut down these thermals long before the storm arrives. Other good thermal spots are likely to be over the high ground F. Note the probable areas of sink low over forested areas labeled G and over water H. Flying in the area beyond the town is not considered safe due to the threatening storm.

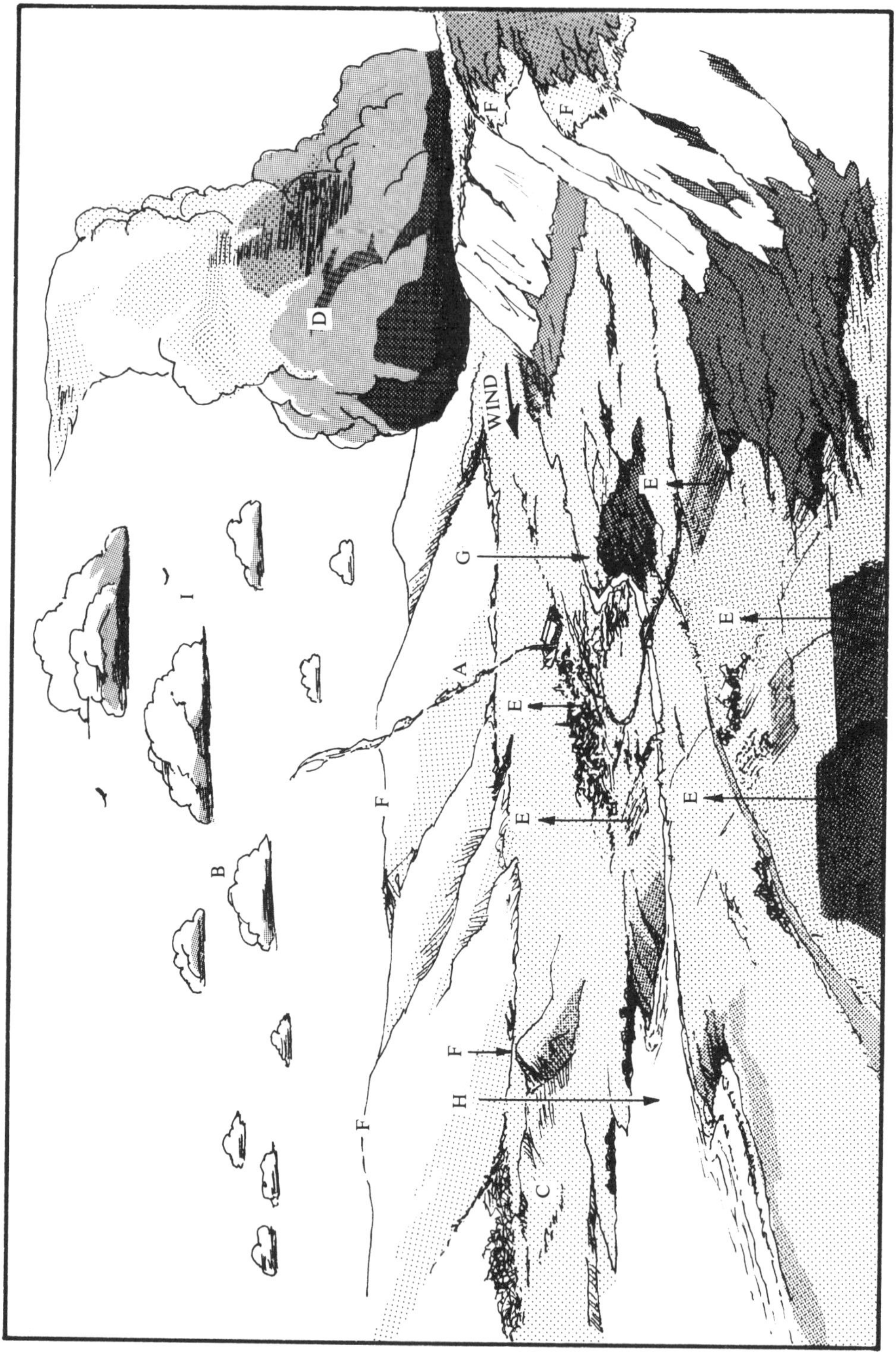

Figure 224 - Conditions on an Unstable Day

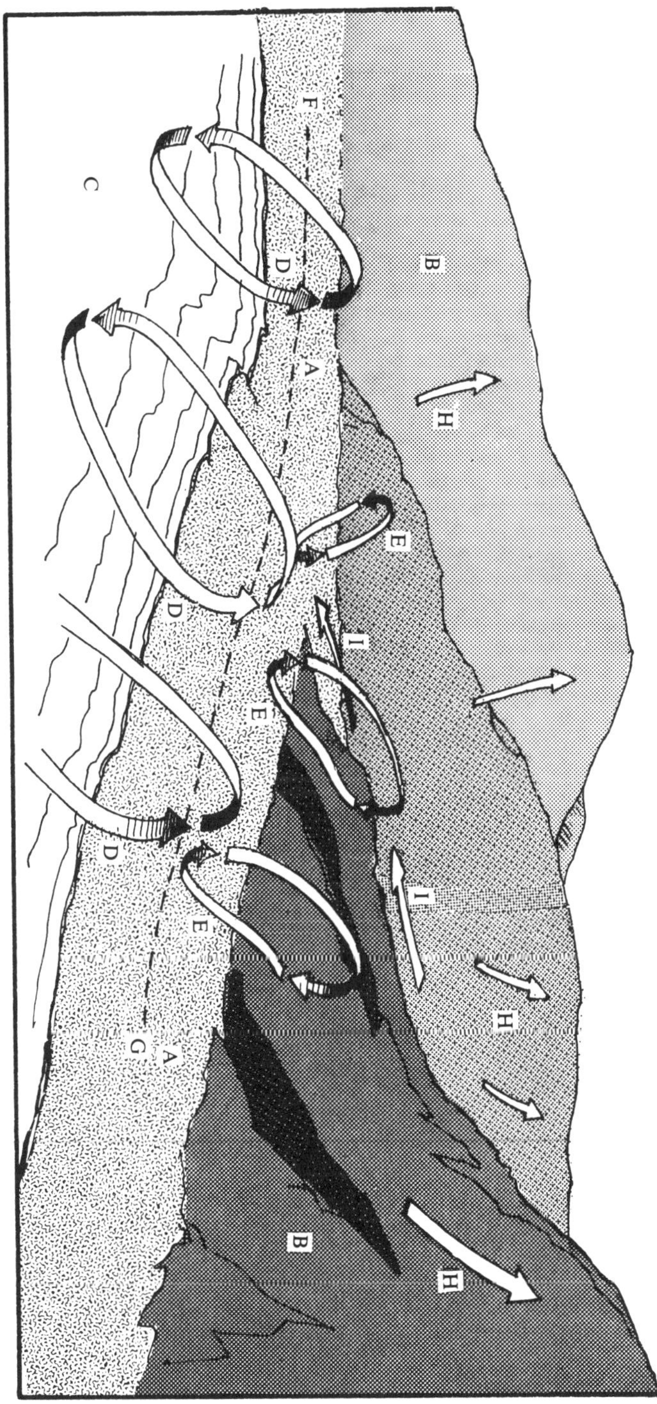

Figure 225 - Conditions on a Light Wind Stable Day

Figure 226 - Conditions on a Windy Stable Day

Finally, notice the birds soaring at I, identifying thermal location.

In figure 225 conditions are cloudless and stable. This produces a large difference in temperature between the beach area A and the forest B or water C. Thus local circulation in the form of sea breezes D and flow from forest to beach E is set up. Lift above the line F-G where these two flows meet can be expected as this is a convergence zone. However, due to the heating of the valley, upslope wind (anabatic flow) is present on all the mountain slopes H so the flow from the forest to the beach will be eliminated as the day progresses. We must note that these circulations can exist in unstable conditions as well, however, the presence of thermals can make the circulations more erratic. Downslope winds spewing from the canyon at I should be expected in the evening.

Viewing figure 226 we can see a very different situation. The wind is blowing strongly which can be told by the smoke in the valley at A. The wind gradient and direction change is shown at B. Normally a change in wind direction is not conducive to wave production, but here we show the wind channeled by the valley.

The air away from the ground is expected to be smooth, but here we see turbulence behind ground objects C. Special attention should be given to possible strong turbulence behind hills D and trees in landing area E. The stratus clouds at F can block the sun and increase stability, as well as accompany a high level shear. Such an abrupt cloud deck can produce a heat front in light winds.

Ridge lift can be expected all along the hills on the right, but attention should be paid to the gaps G. The presence of waves is indicated by the lenticular clouds at H and they can possibly be encountered when flying at the elevated point I. A rotor exists as indicated by the cloud at J and should be carefully avoided. The lee side of the ridge K should not be approached too closely in significant wind.

SUMMARY

In this chapter we learned a little more about the vastly complex weather system around us and how we can readily access pertinent weather data. We left out one important point, however: there is no substitute for the information gained by talking to other pilots. This is especially true if you are new to an area since each locale has its own weather peculiarities, at least on the small scale.

Ours is an experiential pastime. We learn from doing, thinking, reconsidering and exploring. All sport aviators must be students of nature to excel and progress in safety. Observation is the key to this progress. Much can be learned in a single day of watching clouds, wind effects and birds. You are not being lazy if you are loafing in the interest of science.

One of the beauties of aviation is that it provides excitement and relaxation away from the hustle of everyday life. The essence of flying at its best is peaceful communication with the environment. The pilot that understands his or her aerial surroundings truly finds a home in the air.

Description of photo on page 212:

The two columns in the picture are rising towers of mist from the American Falls (left) and Horseshoe Falls (right). The pounding, churning water creates mist and water vapor which is lighter than the surrounding cold, dry, dense air. Thus the mist rises just like a thermal and this remarkable photo is a visual simulation of a large thermal, especially with the more cohesive column on the right. The mist columns form an arch because of the circulation created by the cold sinking air over Goat Island in the center.

APPENDIX I

DENSITY CHANGES IN THE ATMOSPHERE

Aircraft performance is greatly affected by the air's density. The less dense the air, the faster all speeds become. This is particularly important during takeoff and landing.

The density of the air changes according to variations in temperature, humidity and pressure as well as gain or loss of altitude. Altitude changes are the most important factor in density changes by far. Next is temperature changes with humidity and general pressure changes being of lesser importance. As an example, a hot, humid day at 10,000 feet (3048 m) may have an air density 45% less than that on a cool, dry day at sea level. This will result in a 22% increase in all flying speeds at the 10,000 ft altitude compared to the sea level condition.

The aviation world has developed the concept of *density altitude* to account for the various density changes. To do this we accept a standard atmosphere as described in Chapter II with a sea level pressure of 29.92 inches (1013.25 mb), temperature of 59° F (15° C) and a lapse rate of 3.57 °F/1,000 ft (6.5 °C/km). The following table presents the standard atmosphere values for selected altitudes.

The Standard Atmosphere

ALTITUDE Feet (m)	*TEMPERATURE* °F (°C)	*PRESSURE* inches hg (millibars)	*DENSITY* Slugs/ft³ (gm/m³)
Sea level	59.0 (15.0)	29.92 (1013.3)	.002378 (1225)
1000 (305)	55.4 (13.0)	28.86 (977.4)	.002309 (1189.5)
2000 (610)	51.9 (11.0)	27.82 (942.1)	.002242 (1154.9)
3000 (914)	48.3 (9.1)	26.82 (908.3)	.002176 (1120.9)
4000 (1219)	44.7 (7.1)	25.84 (875.1)	.002112 (1088.0)
5000 (1524)	41.2 (5.1)	24.89 (842.9)	.002049 (1055.5)
6000 (1829)	37.6 (3.1)	23.98 (812.1)	.001988 (1024.1)
7500 (2286)	32.3 (0.1)	22.65 (767.0)	.001898 (977.7)
10000 (3048)	23.3 (−4.8)	20.57 (696.6)	.001756 (904.6)
12500 (3810)	14.4 (−9.8)	18.65 (631.6)	.001622 (835.6)
15000 (4572)	5.5 (−14.7)	16.88 (571.6)	.001496 (770.6)
18000 (5486)	−5.2 (−20.7)	14.94 (505.9)	.001355 (698.0)
20000 (6096)	−12.3 (−24.6)	13.74 (465.3)	.001267 (652.7)

Using the standard atmosphere table we can find our pressure altitude if we know the local pressure (from a barometer or an altimeter set to 29.92 at sea level). The pressure altitude is the standard altitude we would be at given the measured pressure in our location. Altitude and pressure system changes are included in pressure altitude.

The next step is to find our density altitude by adding the temperature factor to the pressure altitude. The following table gives the density altitude given pressure altitude and temperature.

Density Altitude

TEMPERATURE °F (°C)	PRESSURE ALTITUDE (Ft) 8500	8000	7500	7000	6500	6000	5500	5000	4500	4000
100 (37.8)	13000	12300	11700	11000	10500	9800	9300	8700	8000	7500
90 (32.2)	12500	11700	11200	10500	9900	9200	8700	8100	7500	6900
80 (26.7)	11800	11100	10500	9900	9200	8600	8100	7500	6900	6300
70 (21.1)	11200	10500	10000	9300	8800	8000	7500	6900	6300	5700
60 (15.6)	10600	9800	9300	8800	8000	7400	6900	6300	5700	5100
50 (10.0)	10000	9200	8700	8100	7500	6900	6100	5700	5100	4400
40 (4.4)	9200	8600	8100	7500	6800	6100	5600	5000	4500	3800
30 (−1.1)	8500	8000	7500	6900	6200	5500	5000	4400	3900	3200

From this chart we see the density altitude increases 500 to 800 feet for every 10 °F increase in temperature (135 to 220 m per 5 °C change). An example will illustrate how to find density altitude.

Assume our local pressure is 23.98 inches and the temperature is 80 °F. Looking at the standard atmosphere chart we see that 23.98 relates to a pressure altitude of 6,000 feet. Now our *real* altitude may be different from this (higher or lower) depending on what pressure system is in our area. Next we turn to the density altitude table and go down the 6,000 feet pressure altitude column to the 80° row and find our density altitude to be 8,600 feet.

Most of sport aviation is not concerned with blind navigation, so we are not too worried about in-flight pressure changes that can occur with time as systems move or when we fly across isobars. However, we should understand the effect of changing pressure on our altimeters. As higher pressure moves into our area our altimeter reads lower as if it were moving down to a lower altitude. Therefore an adjustment must be made in order to read true altitude. We can do this adjustment if we know our true local altitude only. Once we have this true altitude our altimeter will read true until the local pressure changes.

There is an international system called the Q code which relates to altitude and pressure. The code meaning is as follows:

QFE—This is the barometric pressure at a given point on the ground. We set to QFE when we zero our altimeters at a site and we then read true height above that site until the local pressure changes.

QNH—This is our altitude given the correct current local sea level pressure and assuming a standard lapse rate. An altimeter set with the QNH reading will read true altitude if a standard atmosphere exists.

QNE—This is the altitude reading assuming sea level pressure is 1013.25 mb. QNE will only give true altitude if a standard atmospheric exists. QNE is identical to the pressure altitude. QNH equals QNE when the sea level pressure is 1013.25.

In truth, our altimeters never read true altitudes unless a standard atmosphere exists for they are calibrated to show the altitude change according to the standard pressure change. However, this is a universal problem and all altimeters suffer it to the same degree so they all read the same within the limits of their accuracy.

APPENDIX II

CORIOLIS EFFECT

By vector analysis, the deviating force on a particle is found to be: $D = 2mV\omega \sin\phi$.

m = mass of the particle
V = velocity of the particle.
ω = angular velocity of the earth $= 2\pi/24$ radians/hour
ϕ = angle of latitude.

From this we can see that at higher latitudes (greater ϕ) the coriolis effect will be greater. The force is maximum at the poles and zero at the equator. Also, it is obvious that the force is proportional to the velocity of the particle. When the particle is at rest, there is no deviating force.

In the northern hemisphere, this force is directed to the right no matter which way the particle is moving. If a particle is moving away from a high pressure center it will curve right and move around the center. This circular motion initiates a centrifugal force which tends to lessen the pressure gradient. The opposite situation occurs for a low pressure cell. Thus, winds around a low will tend to be much higher than around a high. Waterspouts, tornadoes, dust devils and hurricanes revolve around low pressure areas.

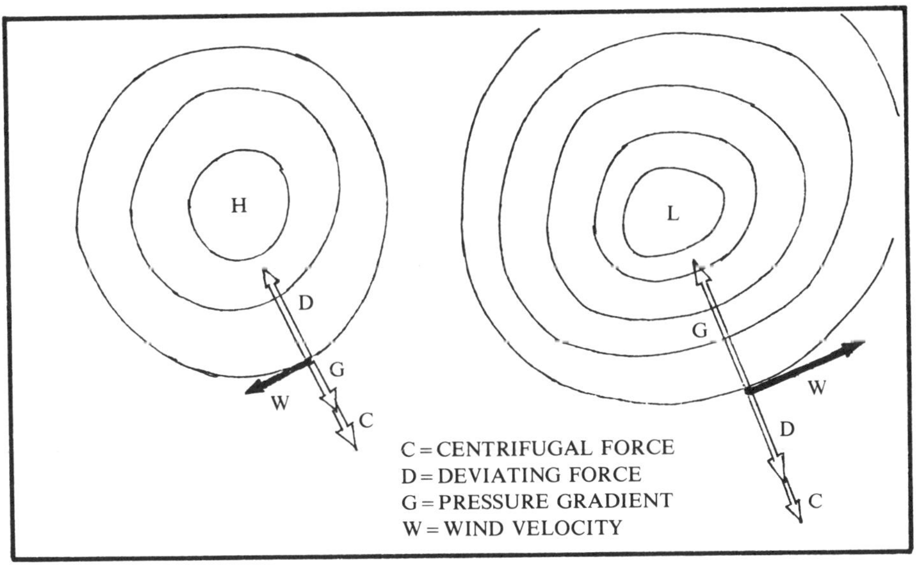

Forces Around Pressure Systems

APPENDIX III

DRYING WINDS

The warm, dry winds (chinooks, foehns, etc.) that occur on the lee side of mountain ranges can be understood by looking at the change in temperature of the air as it moves over the range. In the following figure, the temperature of the air is given on the left side of the slash, while the dew point is on the right.

The air rises adiabatically (without a change in heat, content, but cooling 5.5 °F /1,000 feet because of expansion) until it reaches the dew point (4,000 feet in this case) and condensation begins. Since the change of water vapor to rain is adding heat to the air, the cooling is less—only 4.5 °F /1,000 feet. This continues to the top of the mountain. On the lee side, the air is quickly heated by compression above the dew point and condensation stops. In addition, this heating is 5.5 °F/1,000 feet through the entire descent, arriving at the bottom warmer than the air at the same level on the other side of the range. The air on the lee side, having lost much of its moisture content during the precipitation process, is very dry.

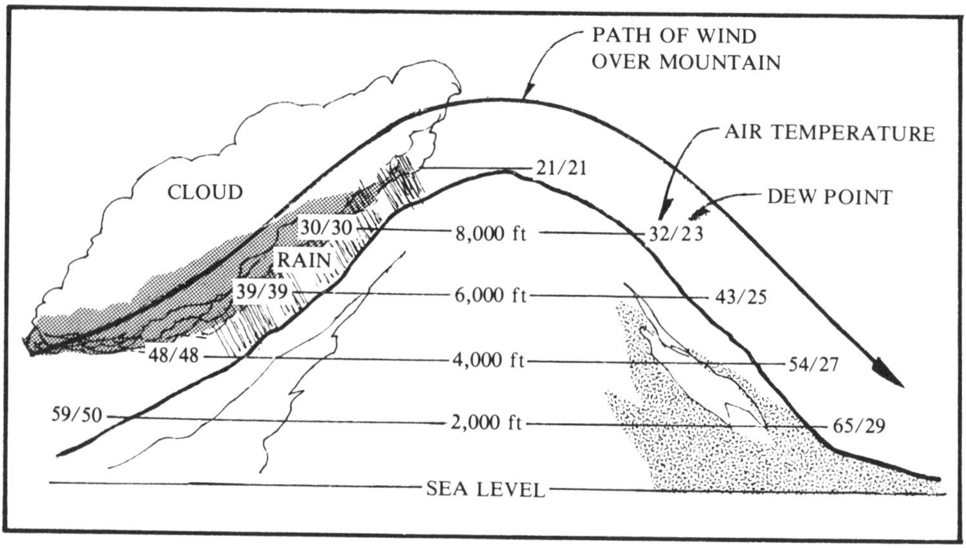

Drying and Heating of the Air Over a Mountain

APPENDIX IV

THERMAL BOUYANCY

The bouyancy of a thermal greatly increases after a cloud forms due to the release of latent heat when condensation takes place. Before the cloud forms the upward velocity can be found by balancing the bouyant forces with the drag forces. The bouyancy based on Archimedes principle is:

$$\text{Bouyancy} = \text{Mass} \times \text{gravity force} \times \frac{\text{temperature excess of thermal}}{\text{temperature of air}}$$

Since mass equals volume times density, we have:

$$B = \frac{\pi d^3}{6} \rho g \frac{\Delta t}{t}$$

Where:

d = Thermal diameter ($\frac{\pi d^3}{6}$ is the thermal volume)

ρ = air density

g = gravitational constant

Δt = difference in thermal temperature and air temperature.

t = air temperature

The drag forces can be established by assuming the thermal is a sphere. The equation for drag in the air is:

$$\text{Drag } (D) = 1/2\, C_D\, \rho\, SV^2$$

Where:

C_D = the drag coefficient. This is 0.5 for a sphere.

S = the cross-sectional drag area ($\frac{\pi d^2}{4}$ here).

V = the thermal velocity

The steady state condition will occur when the drag forces equal the bouyancy forces.

$$\text{SO } D = B \text{ or } 1/2\,(0.5)\,\rho \frac{\pi d^2}{4} V^2 = \frac{\pi d^3}{6} \rho g \frac{\Delta t}{t}$$

Solving for V, we have: $\sqrt{\frac{8}{3} g \frac{\Delta t}{t} d}$

The significance of this equation is that the upward velocity of a thermal depends on two factors—the temperature difference and the diameter. The temperature difference depends on how hot the thermal was when it started and the lapse rate. As the diameter gets larger so does V. Thus we can conclude that *larger thermals climb faster* for a given lapse rate.

Generally a thermal accelerates until it reaches the point when drag and bouyancy balance. Later it may slow with height as the lapse rate decreases and the thermal dilutes by mixing with the surrounding air. We can conclude that a thermal slowing with height is moving through more stable conditions while one that accelerates and stays strong is rising in an unstable lapse rate as shown previously in figure 180.

APPENDIX V

TRIGGER TEMPERATURE

An important bit of information for soaring pilots is trigger temperature and trigger time which determine when usable thermals will appear. To see how this works look at the figure below. Here we see a lapse rate with a nighttime inversion (solid line). In order for thermals to rise very high they must be heated at the surface to the temperature at point A since their temperature follows the dashed line as they rise. Cooler thermals will be stopped by the inversion.

The shaded area in the triangle is proportional to the heat required to change the lapse rate. We can calulate this area as follows: multiply the height of the triangle (2000 feet in this example) by ½ the temperature difference (60°–35°=25° here). For this example we have 2,000 x 12.5° = 25,000 degree-feet.

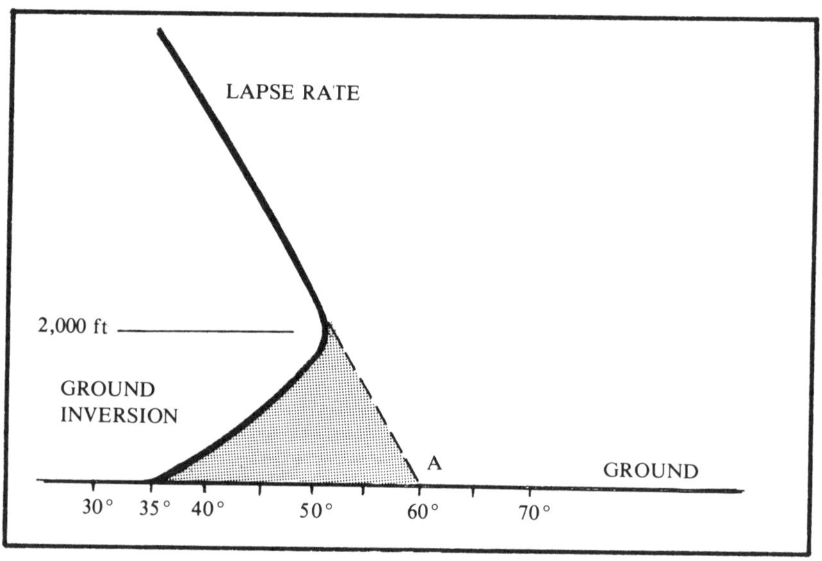

Heat Needed to Overcome Ground Inversion

The next step is to find out how much heat is available to make this change. The chart below shows the expected heating on a clear morning for a latitude of 45°. If your latitude is less or greater, move the outside edges of the curve down or up ½ hour for every 5° of latitude change. The solid lines are temperature × 1,000 ft lines. For example, the 30° line means there is enough heating to raise a layer 1,000 feet thick 30° or a layer 2,000 feet thick 15°, etc.

In our examples we needed to acquire 25,000 degree-feet of heating. From the chart we can see that this amount is reached in mid-June by about 11:00 am. We conclude that thermals slowly rise up to the 2,000 foot level by 11:00 am, then quickly increase in height and strength thereafter. The 11:00 am time is trigger time and 60° is the trigger

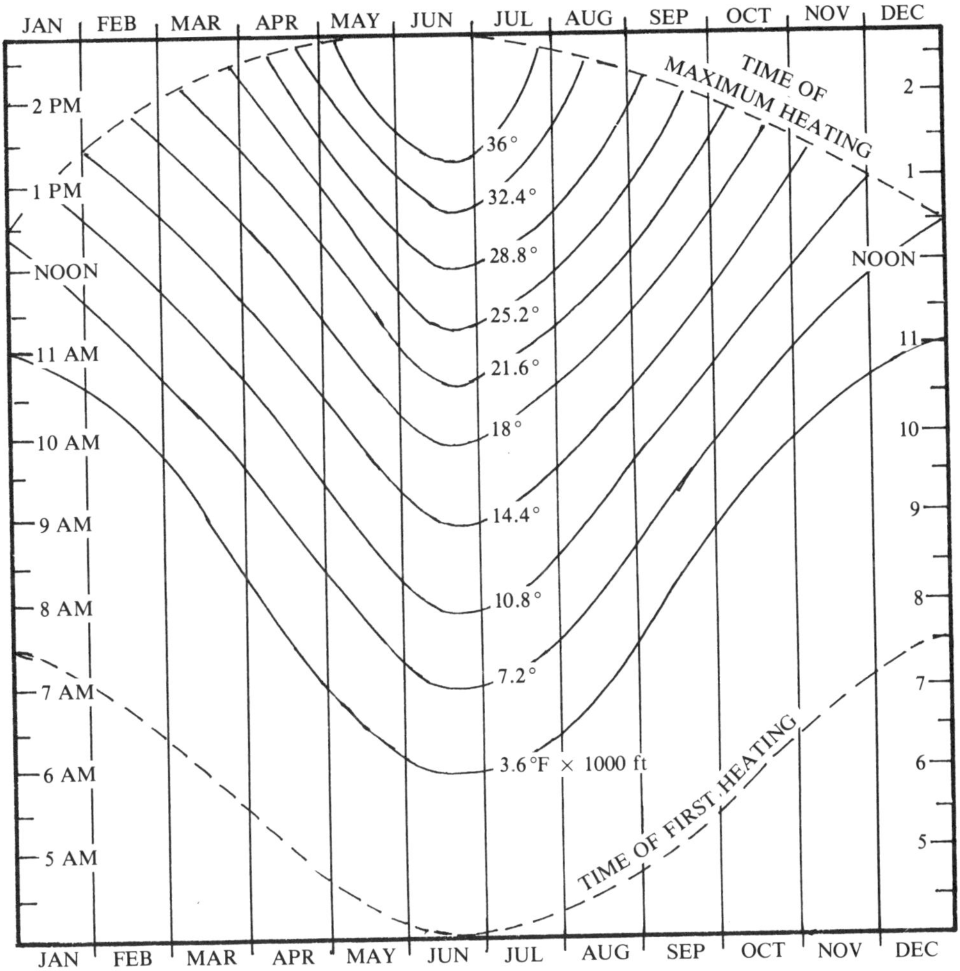

Available Heating Chart

temperature.

We can calculate in this manner the height of thermals at any time during the day if we know the actual lapse rate. We simply determine how much heat is available at any time of the day from the chart, then apply this heat to the area between the lapse rate and the dry adiabat (lapse rate line) that a thermal will follow.

Note that clouds, fog, smog and pollution will cut down the solar heating considerably. Take these effects into consideration when making such calculations. With practice you can achieve very reliable results.

GLOSSARY

ABSOLUTE HUMIDITY - A measure of the actual amount of water vapor in the air given as a weight per volume.

ADIABATIC - A process of expansion or contraction of a gas (the air) without adding or subtracting outside heat.

AIR MASS - A large volume of air with a fairly uniform humidity, temperature and stability throughout the volume.

ANABATIC FLOW - See upslope winds.

BACKING WIND - A wind changing directions in a counterclockwise manner.

BLUE HOLE - An open area of blue sky in an otherwise cloudy sky. An area displaying little or no lift.

BLUE STREET - A thermal street on a cloudless day.

CATABATIC FLOW - See downslope breeze.

CLOUD BASE - The bottom of a cloud identical with the dew-point level.

CLOUD STREET - Long lines of cumulus clouds usually formed in parallel rows.

COLD FRONT - The leading edge of an advancing cold air mass.

CONDENSATION - The change of water vapor (gas) to water droplets (liquid).

CONDENSATION NUCLEI - Small airborne particles that promote the condensation of water vapor.

CONVECTION - Currents of rising warm air.

CORIOLIS EFFECT - The turning of any free moving particle or body due to the earth's rotation (rightward in the northern hemisphere, leftward in the southern hemisphere).

CUMULUS - One of two major cloud types (see stratus). Tumbled or cauliflower type clouds.

DALR - Dry Adiabatic Lapse Rate. The change in temperature a parcel of air undergoes due to expansion as it is lifted.

DENSITY ALTITUDE - The apparent altitude you are at given the local deviations of temperature and pressure from the standard atmosphere.

DEW POINT - The altitude where the air's temperature is cool enough to cause condensation (cloud formation) in air of a given humidity.

DIVERGENCE - A separating of the airflow resulting in sink.

DOWNDRAFT - The sometimes violent falling air below a collapsing thunderstorm.

DOWNSLOPE BREEZE - The local winds that blow down a mountain slope in the evening as cooling sets in.

DUST DEVIL - A small core of dust-laden swirling winds created by a thermal lifting off in superadiabatic conditions.

EDDY - A swirl in a gas or liquid.

FALLOUT - See downdraft.

FOEHN - A drying wind that flows over mountain chains, especially in the Alps.

GEOSTROPHIC WIND - The wind produced by the pressure gradients out of the influence of surface friction. The geostrophic wind follows the isobars.

GRAVITY WIND - A cold, dense wind caused by strong surface cooling such as above a glacier.

GUST FRONT - The spreading cold, dense air that plummets to the ground beneath a thunderstorm.

HIGH PRESSURE SYSTEM - A large area of elevated pressure resulting in divergence and lightly sinking air.

HEAT FRONT - The interface of cool and warm air caused by local differences in heating of the air.

HOUSE THERMAL - A semi-reliable thermal source near a flying site.

IMBEDDED THUNDERSTORMS - Storms concealed in a stratus cloud or haze layer.

ITCZ - Intercontinental convergence zone. The belt around the globe near the equator where surface flow comes together.

INVERSION - A warming (or lesser cooling) of the air with increased altitude resulting in a layer of great stability.

ISOBAR - A line connecting points of equal pressure on a weather map.

JET STREAM - A river of high speed winds circling the globe in an undulating fashion. See polar and sub-tropical jet.
K INDEX - A measure of the thunderstorm potential given the humidity and instability.
KNOT - A unit of wind speed measurement. One knot equal 1.15 mph and 1.84 km/h.
LAND BREEZE - An evening wind blowing from the land to the sea.
LAPSE RATE - The change in temperature of the air with changing altitude.
LATENT HEAT - Heat required to change water vapor to liquid (heat of condensation) or vice versa (heat of vaporization).
LEE SIDE - The downwind side of an object.
LIFT - Rising air
LIFTED INDEX - A measure of the difference in the actual temperature at 18,000 ft and the temperature a parcel of air from the surface would acquire if lifted to 18,000 ft.
LOCAL WIND - Wind due to small-scale effects such as sea breezes and slope circulations.
LOW PRESSURE SYSTEM - A large area of depressed pressure resulting in an inflow (convergence) and slow rise of air.
MALR - Moist adiabatic lapse rate. The change in temperature a saturated air parcel undergoes as it is lifted.
MECHANICAL TURBULENCE - Swirls or eddies in the air caused by obstructions altering the air's flow.
METEOROLOGY - The technical term for weather.
MICROMETEOROLOGY - Weather on the small scale. Local effects.
OCCLUDED FRONT - The leading edge of a cold front that catches up to and passes a warm front.
OROGRAPHIC LIFT - See ridge lift.
POLAR FRONT JET - A jet stream located above the temperate zone cold fronts.
PRESSURE - A measurement (in the atmosphere) of the weight of the air at any point on the surface or altitude.
PRESSURE ALTITUDE - The altitude you would be at in relation to the standard atmosphere given your actual local pressure.
PRESSURE GRADIENT - A change in pressure over a given distance.
PRESSURE GRADENT FORCE - The force on the air due to a pressure gradient.
PREVAILING WIND - The most frequent wind direction in an area.
RELATIVE HUMIDITY - A measure of the moisture content of air at a given temperature as a percentage of its maximum capacity.
RIDGE - A long mountain or hill. Also a line of high pressure.
RIDGE LIFT - Lift created by a ridge blocking the wind flow.
ROTOR - An organized eddy of air on the lee side of a mountain or below a wave that remains in place.
SATURATION - The state where the air reaches 100% humidity.
SEA BREEZE - A wind blowing from the sea to the land during the day.
SEA BREEZE FRONT - The leading edge of the cool sea breeze air as it pushes inland.
SHEAR TURBULENCE - Turbulence caused by adjacent layers of air moving with different velocities.
SINK - Downward moving air.
SLR - Standard lapse rate. The lapse rate in conditions accepted as the Standard Atmosphere.
SQUALL LINE - A line of thunderstorms that often preceeds a vigorous cold front.
STABLE AIR - An atmosphere condition that suppresses vertical motion because of a lapse rate of less than 5.5 °F per 1,000 ft.
STANDARD ATMOSPHERE - The accepted values of pressure and temperature at each altitude allowing uniform altimeter calibration.
STATIONARY FRONT - A warm or cold front that has stopped moving. Such a front may die or begin moving later.
STRATOSPHERE - The layer of atmosphere just above the troposophere where weather takes place.
STRATUS - One of the two main cloud types (see cumulus). Flat, layered clouds.
SUBLIMATION - The changing of water vapor directly to ice, bypassing the liquid stage.
SUBSIDENCE - A slow sinking of the air (usually under a high).

SUBTROPICAL JET - The short itinerant jet stream above the tropics.
THERMAL - A rising current or bubble of warm air.
THERMAL INDEX - A measure of atmosphere instability taken as the difference in the measured temperature at a given altitude and the temperature a parcel of air would acquire if lifted to that altitude.
THERMAL PAUSE - A period of little or no thermal activity in the late morning and early evening.
THERMAL SOURCE - An area of ground or terrain heating to produce thermals.
THERMAL STREAM - A line of thermals emanating from a single source.
THERMAL STREETS - Parallel rows of thermals produced by a particular atmospheric condition (see text). Also called cloud streets or blue streets depending on the presence of clouds.
THERMAL TRIGGER - A point on the terrain that induces the lifting off of thermals.
THERMAL TURBULENCE - Swirls or gusts in the air caused by thermals pushing upward.
THUNDERSTORM - A very large growing cumulus cloud that develops in strong conditions to produce lightning, thunder, high winds, hail and rain.
TROPOPAUSE - The layer of air between the troposphere and the stratosphere.
TROPOSPHERE - The lower layer of the atmosphere where the changes take place that we identify with weather.
TROUGH - A line of low pressure in the atmosphere.
TURBULENCE - Swirls, eddies or vorticies in the air that are more or less random except when organized as rotors.
UNSTABLE AIR - An atmosphere condition that promotes vertical currents due to a lapse rate greater than 5.5° F per 1,000 ft.
UPSLOPE BREEZE - A wind blowing up hillsides during the day due to heating effects.
VALLEY WINDS - Winds blowing up or down valleys due to heating or cooling effects.
VEERING WINDS - Winds that change direction in a clockwise manner.
VIRGA - Sheets of falling rain that dries before reaching the ground.
VORTEX - A swirl or eddy
VORTEX RING - An organized circular eddy that turns inside out like a smoke ring.
WARM FRONT - The leading edge of a mass of warm air.
WAVE - An undulation in the air often formed by hills, ridges or mountains deflecting the flow.
WEATHER CHART (MAP) - A chart depicting large-scale weather conditions over a given area. Such charts are produced periodically for the surface and selected altitudes.
WIND GRADIENT - A slowing of the wind near the earth's surface due to friction.
WIND SHADOW - A severe wind gradient or complete elimination of the wind behind a large obstruction.

INDEX

Absolute Humidity: 10
Absolutely Stable: 20
Adiabatic: 5
Adiabatic Drying: 272
Air Masses: 56
Air Structure: 2
Anabatic Winds: 154
Anemometer: 88
Anticyclone: 75
Arctic Front: 65
Atmosphere: 1
Atmospheric Circulation: 53
Atmospheric Pressure: 3
Backing Wind: 99
Banner Cloud: 42
Barometer: 4, 254
Billow Cloud: 42
Blue Norther: 110
Bora: 107, 155
Cap Cloud: 41
Catabatic Wind: 154
Circulation: 54, 73, 77, 78, 187, 152
Circumpolar Westerlies: 70
Cirrus Clouds: 37
Chinook: 109
Clouds: 25
  Altitudes: 36
  Bases: 29, 216
  Causes: 25
  Height: 26
  Life: 30
  Flying In: 50
  Formation: 28
  Signs: 47
  Tops: 29
  Types: 34, 40
  Wave: 42, 173
Cloud Barrier: 232
Cloud Streams: 223
Cloud Streets: 224
  Behavior: 226
  Flying In: 226
Cloud Suck: 230
Cloud Thermal Clues: 235
Cold Front: 59, 60, 62
Complex Hill Shapes: 162
Conergence: 71, 141, 178
  Clouds: 184
  Evening: 182
  Flying In: 183
  Near Water: 141, 180
Cooling: 7
Coriolis Effect: 22, 54, 158, 271
Cumulus Cloud: 30, 34, 47
Cyclone: 75
DALR: 14, 205
Density: 3, 269
Density Altitude: 20, 269
Depression: 75
Dew Point: 10, 26
Diurnal Variation: 103
Divergence: 71, 142, 181
Downdraft: 247
Downslope Winds: 154
Down Valley Winds: 157
Dust Devils: 205
  Flying Near: 208
Eddies: 113
Evening Thermals: 197, 220
Foehn: 107
Foehn Gap: 109, 111, 174
Foehn Wall: 108, 111
Fracto-Clouds: 46
Freestream Velocity: 100
Fronts: 59
  Action: 62
  Variations: 63
Frontal Lift: 185
Geostrophic Wind: 100
Gradient, Wind: 99
Ground Inversion: 13, 200
Gust Front: 247
Heat Fronts: 150
Heating: 3
Heating, Cycles: 8
  Seasonal: 8
  Solar: 6
High Pressure Systems: 68, 82, 95
House Thermal: 193
Humidity: 10
Instability: 14
Inversions: 13, 202
  Eliminated: 203
Island Weather: 85, 148
Isobars: 75
  Spacing and Wind: 76
ITCZ: 56, 83
Jet Stream: 81, 102
K-Index: 260
Land Breeze: 140, 149
Lapse Rate: 13, 174, 201, 203
Latent Heat: 12
Layering of Air: 105
Lee Side Cloud: 43
Lenticular Cloud: 42
Lift: 49, 162
  Convergence: 178
  Indicators: 186
  Ridge: 162
  Wave: 169
Lifted Index: 26
Lifting: 26
Lightning: 245
Local Lows: 71
Local Winds: 109
Long Waves: 79
Low Level Jets: 106
Low Pressure Systems: 70, 82, 96
Magic Air: 155
MALR: 19
Mammata Clouds: 44
Mechanical Turbulence: 115
Mistral: 110
Mono Wind: 109
Mountain Wind: 154
Nimbo-: 37
Nitrogen: 2
Occluded Front: 65
Old Clouds: 31
Orbital Changes: 9
Overdevelopment: 30
Oxygen: 2
Pileus Cloud: 45
Polar Front: 65, 81
Polar Front Jet: 102
Polar Waves: 55
Pollutants: 2
Pressure: 3, 81
Q Code: 270
Radiation: 6
Rain: 31, 50
Relative Humidity: 10
Ridges (pressure): 75, 95
Ridge Lift: 162
  And Thermals: 168
Roll Cloud: 173
Rotor Cloud: 42
Rotors: 115, 123, 129
Santa Ana: 109
Sea Breeze: 139
  Daily Change: 141
  Effects: 145
  Front: 146
  General Wind: 140
  Prediction: 142
  Strength: 144
Seasonal Changes: 67
Seasonal Heating: 8
Secondary Cold Front: 64
Shear Lines: 90
Shear Turbulence: 119
Smog Front: 147, 150
Soaring: 161
Soarable Envelope: 162
Soaring Forecasts: 260
Solar Heating: 6
Squall Lines: 61, 242
Stability: 14
  Indications: 16
  Layer: 17
Standard Atmosphere: 20
Standard Lapse Rate: 13
Stationary Fronts: 59
Steering Winds: 81
Stratosphere: 2
Stratus: 34
Subcloud Layer: 230
Subtropical Jet: 102
Superadiabatic: 16, 204
Surface Chart: 258
Surface Heating: 195
Temperature: 5

Thermal: 30, 189
Acceleration: 204
Barrier: 232
Bouyancy: 273
Cloud Height: 216
Cloud Shapes: 228
Cycles: 195, 218
Daily Variation: 215
Demise: 215
Duration: 215
Evening: 197, 220
Heights: 214
House: 193
Ideal Conditions: 209
In Wind: 217
In Stable Air: 204
In Turbulence: 199
Lee Side: 191
Lift In: 209
Locating: 233
Origins: 189
Patterns: 222
Pauses: 196
Rising: 198
Shapes: 220
Sink Around: 209, 210
Sizes: 213
Sources: 192
Spacing: 220
Streams: 223
Streets: 223
Strengths: 213
Tracks: 218
Trigger: 190, 231
Types: 219
Thermal Index: 203, 260
Thermal Turbulence: 118
Thermal Waves: 178
Thunderstorms: 237
Bench: 243
Dangers: 244
Escaping: 250
Frequency: 237
High Level: 242
Imbedded: 241
Isolated: 241
Judgement: 248
Life Cycles: 238
Squalls: 242
Tornados: 251
Tramontane: 110
Trigger Time: 260, 262
Trigger Temperature: 200, 262, 274
Tropical, Cyclone: 84
Depression: 83
Highs: 55
Storm: 84
Weather: 83
Tropopause: 2
Troposphere: 1
Troughs: 75, 79
Turbulence: 49, 106, 113
Causes: 115
Canyons: 130
Cycles: 125
Definition: 113
Flying In: 135
Gaps: 130
Inland: 126
Mechanical: 115
Nature of: 114
Near Cliffs: 121
Rear Ridges: 128
Seaside: 125
Shear: 119
Signs: 124
Thermal: 118
Trees: 132
Vortex: 123
Upper Level Circulation: 70
Upper Level Front: 64
Upslope Winds: 152
Up Valley Wind: 157
Valley Winds: 154
Veering Winds: 99
Virga: 32
Visibility: 33
Warm Front: 59
Warm Occlusion: 66
Water Vapor: 10
Properties: 11
Waves, Dangers: 174
Finding: 177
Flying In: 175
Properties: 172
Wave Cloud: 42, 173
Weather Chart Heights: 78
Weather, Cycles: 253
Information: 257
Maps: 257
Signs: 40, 48, 256
Symbols: 259
Wind: 21
Aloft: 100
Backing: 99
Changes with Altitude: 101
Diurnal Variation: 103
Gradient: 99
Layers: 92
Local: 109
Near Fronts: 93
Near Low: 96
Near Pressure Ridge: 95
Prevailing: 90
Profile: 99, 106
Shadow: 134
Surface: 91
Table: 90
Veering: 99
Velocity: 47, 87, 89
Wonder Winds: 155